Digital Twins and Extended Reality: Opportunities and Challenges of Integrated Applications

Digital Twins and Extended Reality: Opportunities and Challenges of Integrated Applications

Guest Editors

Daniela Oreni
Fabrizio Banfi
Davide Mezzino

MDPI

Basel • Beijing • Wuhan • Barcelona • Belgrade • Novi Sad • Cluj • Manchester

Guest Editors
Daniela Oreni
Politecnico di Milano
Milano
Italy

Fabrizio Banfi
Politecnico di Milano
Milano
Italy

Davide Mezzino
IULM University
Milano
Italy

Editorial Office
MDPI AG
Grosspeteranlage 5
4052 Basel, Switzerland

This is a reprint of the Special Issue, published open access by the journal *Drones* (ISSN 2504-446X), freely accessible at: https://www.mdpi.com/journal/drones/special_issues/4M9A03M24O.

For citation purposes, cite each article independently as indicated on the article page online and using the guide below:

Lastname, A.A.; Lastname, B.B. Article Title. *Journal Name* **Year**, *Volume Number*, Page Range.

ISBN 978-3-7258-2763-3 (Hbk)
ISBN 978-3-7258-2764-0 (PDF)
https://doi.org/10.3390/books978-3-7258-2764-0

Contents

About the Editors

Daniela Oreni

She is an Associate Professor of Architectural Representation, Drawing, Survey, and Modeling at Politecnico of Milano, Department of Architecture, Built environment, and Construction Engineering. Prior to becoming an architect, she obtained a Ph.D. in the Conservation of Architectural Heritage and a post-graduate degree at the School of Conservation of Architectural and Landscape Heritage of Politecnico of Milano. Since 2017, she has been a Visiting Professor at the University of Applied Sciences and Arts of Southern Switzerland, Bachelor of Conservation and Restoration, course of Documentation and drawing of cultural heritage. From 2011 to 2018, she was part of the Erasmus teaching staff of the Workshop Architectural Heritage Documentation for Conservation, at Raymond Lemaire International Centre for conservation, RLICC—Unesco Chair, Katholike the Universiteit Leuven. She is part of the teaching staff of the Erasmus Mundus Joint Master's degrees, "BIM A+6: BIM based rehabilitation and sustainability analysis". She has been actively participating in architectonic and archaeological international campaigns of survey since 2003, and she has been involved in a series of international projects, collaborating with several European universities. She has also participated in national and international research projects as a scientific leader. She is a member of the ICOMOS and CIPA international authorities, as well as an expert in heritage survey and documentation, cataloguing, and digital technologies applied to cultural heritage, including historic building information modeling for conservation activities. She is the author of about 120 scientific publications in national and international journals in surveying, documentation, representation, and dissemination of cultural and built heritage (2004–2024).

Fabrizio Banfi

Dr. Fabrizio Banfi, Ph.D. is a Senior Researcher in Architecture at Politecnico di Milano, specializing in virtual heritage, BIM, extended reality (XR), and Digital Twin technologies for cultural heritage conservation. With over a decade of experience, he is a leading expert in applying digital tools to preserve and represent architectural and archaeological heritage. In 2016, Dr. Banfi received funding from the Social Sciences and Humanities Research Council (SSHRC) for the New Paradigm/New Tools for Architectural Heritage program. His research, conducted at Carleton University's Carleton Immersive Media Studio (CIMS) and Autodesk Research in Toronto, led to a Ph.D. with honors and the SIFET 2018 Award for the best national thesis. Dr. Banfi has worked on high-profile heritage projects, including the Parliament Building of Canada, Basilica di Collemaggio, Villa dei Quintili, Appian Way, Mausoleum of Cecilia Metella, Arquata del Tronto, and Lamalunga Cave (site of Neanderthal remains). His expertise in 3D surveying and modeling has been crucial in preserving these sites. As co-director of the Gicarus ABClab at Politecnico di Milano, Dr. Banfi leads research in digital heritage, with nearly 100 publications and innovative methods integrating XR, BIM, and Digital Twin technologies into heritage management. He is also an educator at Politecnico di Milano, KU Leuven, and the AUIC School. A sought-after speaker, Dr. Banfi has led workshops and seminars globally, influencing the future of digital heritage. Dr. Banfi continues to drive research in digital preservation, shaping the future of cultural heritage conservation through the integration of cutting-edge technologies.

Davide Mezzino

He is currently an Associate Professor in the Department of Humanities, IULM University. He graduated with a Ph.D. in Architecture from Carleton University, as well as a Doctorate in Cultural Heritage from the Politecnico di Torino. Since 2023, he has overseen one of the three research units of the Project of Significant National Interest (PRIN) "Know.it" (MUR P2022WRPY5—CUP J53D23016530001). In 2022, he had an expert contract at the EUROPEAN COMMISSION—Research Executive Agency (REA) for the evaluation of projects submitted to the HORIZON calls. From 2018 to 2021, he worked as a Digital Expert at the Fondazione Museo delle Antichità Egizie di Torino for the development of new paradigms for the management and dissemination of cultural heritage through digital techniques and tools. From 2014 to 2017, at the Carleton lmmersive Media Studio, Ottawa, he developed advanced skills and abilities in the field of three-dimensional surveying and modeling, as well as skills in the use of cataloguing and management systems for built heritage at the architectural and urban scales. Within CIMS, he has also conducted field research projects in Bahrain, Canada, Egypt, Italy, Morocco, Mexico, Myanmar, and Santa Lucia. Since 2013, he has been qualified to practice as an Architect in Italy and in the European Union. Since 2010, he has carried out research and training activities at Universities, Foundations, Companies, and Institutions in Italy and abroad (including Carleton University, Università di Genova, UniNettuno University, Universidad de Guadalajara, Politecnico di Torino, Politecnico di Milano, Yangon Technolgical University, Fondazione CRT, Fondazione Scuola dei Beni e delle Attività Culturali, Istituto Superiore sui Sistemi Territoriali per l'Innovazione-SiTI, and United Nations System Staff College). His research activities focus on digital representation and visual communication, with particular reference to the possibilities of eXtended Reality (XR) and the knowledge and dissemination of cultural heritage.

Preface

The architecture, engineering, and construction (AEC) industry, along with the digital heritage sector, is increasingly harnessing the transformative potential of augmented reality (AR) and virtual reality (VR). These technologies, paired with advanced digital visualization tools, greatly enhance the value of digital models, providing dynamic and interactive representations of complex built environments.

Architects and researchers are now able to produce highly detailed digital models that span both historical and modern architecture, as well as infrastructure, archaeological sites, and museums. Leveraging innovative techniques such as scan-to-BIM, HBIM, BIM-to-FEA, and CoSim, they convert raw 3D survey data (point clouds) into comprehensive, information-rich digital models.

Interoperability and seamless data sharing are crucial for developing digital models that can be used across a wide range of applications, including restoration, construction design, energy analysis, cost estimation, and virtual representation. However, despite significant advances, current digital models often lack the interactivity and immersion required for meaningful engagement. A context-sensitive approach is essential to accurately capture geometries, levels of detail (LOD–LOA), and construction features, ensuring that each element aligns with the intended purpose of the digital twin (DT).

Integrating VR and AR with digital twin technologies can bridge these gaps, enhancing the practical application of 3D models across various fields, from building lifecycle management to virtual tourism. Drones—both indoor and outdoor—play a pivotal role in data acquisition, particularly for hard-to-reach or damaged areas. By combining aerial survey data with ground measurements, drones create accurate 2D and 3D models, essential for conservation, restoration, and reuse projects.

These models can be continuously updated throughout a building's or site's lifecycle, becoming indispensable tools for managing, monitoring, and simulating conservation activities. They serve as vital resources for professionals, including surveyors, archaeologists, architects, conservators, engineers, and facility managers. Additionally, digital models captured through photogrammetry or LiDAR offer public engagement opportunities, with immersive VR and AR experiences that allow for the remote exploration of heritage sites, raising awareness of their historical and architectural significance.

This collection highlights the innovative approaches and practical applications aimed at professionals in documentation, surveying, digitization, visualization, and simulation, with a focus on addressing complex scenarios in architecture, engineering, and heritage preservation.

Daniela Oreni, Fabrizio Banfi, and Davide Mezzino
Guest Editors

drones

Article

Digital Strategies to Enhance Cultural Heritage Routes: From Integrated Survey to Digital Twins of Different European Architectural Scenarios

Sandro Parrinello [1] and Francesca Picchio [2,*]

1 Department of Architecture, University of Florence, 50121 Florence, Italy; sandro.parrinello@unifi.it
2 Department of Civil Engineering and Architecture, University of Pavia, 27100 Pavia, Italy
* Correspondence: francesca.picchio@unipv.it

Abstract: This paper focuses on a research project for the acquisition and post-production of digital data to create informative virtual representations and digital twins of different European Cultural Heritage sites. The goal was to establish a reliable database for a multi-scalar web platform, also accessible through extended reality (XR) tools. This initiative aims to support the promotion and management of cultural and historical monuments within the context of European Cultural Routes supported by the Council of Europe. The project involves different case studies spanning European geographic regions, such as the Upper Kama in Russia, the Valencian Routes of Jaime I in Spain, and the Gdańsk fortresses in Poland. The methodology employed in this effort primarily relies on integrated rapid survey techniques. Unmanned aerial vehicles (UAVs) and simultaneous localization and mapping (SLAM) technologies were used for data collection. These methods contribute to the creation of accurate 3D databases and models that transform the cultural routes into a digital format accessible via an informative platform. The actions presented in this paper are part of the European project "PROMETHEUS", which is funded by the Horizon 2020 program of the European Union. The project involves collaboration between universities and enterprises, fostering inter-sectoral cooperation. Various techniques such as photographic archives, census analysis, and scan-to-BIM (building information modeling) processes are employed to develop this method further. In fact, the ultimate goal of the project is to establish a framework that can be replicated in other cultural contexts, enhancing the digital documentation and valorization of heritage sites.

Keywords: integrated survey; digital documentation; UAVs; scan-to-BIM; digital twins; extended reality; cultural routes; interactive informative platform; H2020 Prometheus

Citation: Parrinello, S.; Picchio, F. Digital Strategies to Enhance Cultural Heritage Routes: From Integrated Survey to Digital Twins of Different European Architectural Scenarios. *Drones* **2023**, *7*, 576. https://doi.org/10.3390/drones7090576

Academic Editors: Daniela Oreni, Fabrizio Banfi and Davide Mezzino

Received: 26 July 2023
Revised: 6 September 2023
Accepted: 7 September 2023
Published: 12 September 2023

1. Introduction

The recognition and protection of cultural heritage routes in Europe have stimulated the development of new approaches and guidelines focused on preserving the intricate relationships between the sites involved [1,2]. This entails a deeper understanding of the tangible and intangible connections that exist within a specific territory. By comprehending the historical, social, and cultural context of these routes, it becomes possible to safeguard their unique characteristics and cultural identity [3].

Therefore, it is essential to document and preserve the historical and architectural values associated with each element of the route. By highlighting these values, it becomes possible to raise cultural and tourist awareness about the entire route on a regional scale, enhancing its significance and appeal to visitors. Furthermore, as Cultural Heritage routes continue to evolve, it is essential to adopt dynamic intervention protocols that move away from static approaches and embrace more flexible and adaptive methodologies. This transformation in investigation methods includes the documentation of heritage, the development of dynamic digital archives, and the establishment of standardized languages to communicate the material and immaterial values of the route.

Qualifying a cultural heritage route corresponds in a way to defining a landscape. It is a matter of setting limits to something indefinite, to include textiles of a mosaic that at first does not appear clear, but whose beauty can only be perceived. Each portion of this mosaic must be framed, analyzed, and defined, and then be positioned so as to compose a new design of the original landscape phenomena. It certainly does not express the same image as the real one, but thanks to new technologies and digital applications certain aspects of its complexity can be enhanced. Cataloging the acquired information and interconnecting metadata and databases enable the creation of complex systems of models and digital products that elucidate the elements of the route. To manage this multitude of products, systems that support the interpretation and understanding of the technical and historical–cultural features become necessary. By analyzing these features, intervention and implementation procedures can be devised to establish connections between the past, the present, and the future of a cultural route or its associated architectural elements.

This holistic approach involves comprehensive documentation, the preservation of cultural identity, and dynamic intervention protocols to describe each route. Integrated surveying technologies and methodologies, coupled with digital databases and standardized languages, play a vital role in achieving these objectives. They are employed to create digital databases and digital twins (DTs) that encompass detailed and extensive representations of the heritage [4]. By leveraging these data, DTs can provide virtual replicas of architectural environments, complete with historical and cultural information, technological and constructive aspects, material and immaterial values are all datasets that are useful to connect a virtual model (digital) with its physical entity (real) [5].

The research presented in this paper addresses the issue of insufficient documentation and subsequent undervaluation of the inherent cultural values associated with historical routes. This deficiency leads to a gradual detachment from the tangible heritage, underscoring the necessity for a fresh approach to represent these sites. This approach should serve as both a vessel and a source of information encompassing architectural assets and the surrounding landscape. In this sense, digital twins serve as a solution, representing highly accurate and morphometrically precise 3D representations of these monuments. By enriching these models with specific content, they establish a connection with the heritage they portray, unveiling aspects that might not always be readily apparent, e.g., past configurations or possible future developments on the artefact.

The data gathered and employed in constructing a DT of an architectural artifact encompasses not only its architectural intricacies, but also its historical and cultural significance within the wider context of its surroundings. This extends to its interplay with nearby monuments, regardless of their inclusion within the designated route. Consequently, the amassed information not only ensures a parallel and verifiable alignment with the physical object being replicated through the digital twin, but also establishes correlations with analogous artifacts that share proximity in both geographic and thematic terms (date of build, functions, constructive materials, style, etc.).

The strategy employed to generate these digital twins embraces an all-encompassing and multidisciplinary documentation methodology. The resilience of this experimental methodology is rooted in its pragmatic procedural frameworks, which have been derived from a series of field tests conducted over diverse pilot cases. This lends the methodology a heightened efficacy and a capacity for seamless adaptation and repeatability to diverse contexts for future implementation.

First of all, adopting expeditious survey methodologies and integrating heterogeneous data was used to create reliable databases for 3D models for all analyzed contexts. Consequently, different modeling strategies (parametric modeling, reverse-modeling, and mathematical–geometric modeling) were adopted according to the different descriptive objectives of each monument and each of the analyzed routes. Lastly, the project culminated in the establishment of a digital user–space interaction system, achieved by developing 3D interpretative web-platforms dedicated to cultural routes.

The experimentation of this methodology inside PROMETHEUS H2020 has been conducted by 35 international researchers, affiliated with different universities (University of Pavia, University of Florence, Polytechnic University of Valencia, Gdańsk University of Technology, and Perm National Research Polytechnic University) and several enterprises (EBIME, CTA, SISMA, BLESARQ, and Metaheritage) (Figure 1). The objective of this extensive collaboration was to formulate collective strategies for constructing comprehensive architectural digital 3D models. These models aimed to encapsulate the architectural heritage's evolution across its specific route, encompassing its present, historical, and potential future configurations.

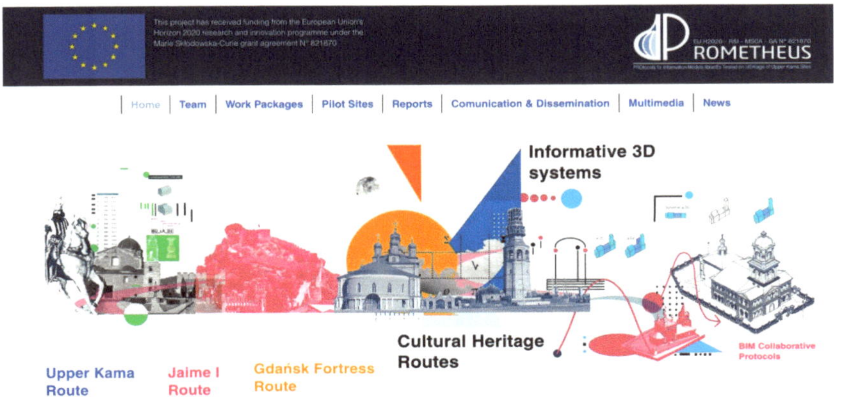

Figure 1. H2020 PROMETHEUS website home page (Graphic Elaboration by A. Dell'Amico).

Pilot cases have been selected always considering their scale and extension, aiming to develop specific knowledge strategies for each route and define useful designs for the digital storytelling of this tangible and intangible heritage. The results generated, comprehensive 3D multi-scaled models, connected information, and virtual platforms in which this connection happens, enable immersive and interactive experiences, allowing visitors to explore and engage with the heritage sites virtually. New ways of interactive communication in the domain of multimedia Cultural Heritage applications and the development of advanced technologies are facilitating access and increasing the value and public awareness of Cultural Heritage [6]. Augmented reality and virtual reality can be integrated to offer enhanced storytelling and educational experiences [7], bringing the past or future of currently unknown and endangered routes to life in a dynamic and accessible way, also ensuring their longevity and accessibility for future generations [8].

2. Materials and Methods

2.1. Site Description

Analyzing cultural landscape contexts differing in history, culture and architectural typology becomes fundamental to structuring a methodological protocol for the acquisition and processing of heritage information platforms. This analysis helps in understanding the different requirements and challenges associated with acquiring and processing heritage information.

The application was undertaken on some European contexts—identified in the project partner universities—which are heterogeneous in terms of the type of Cultural Heritage routes and investigated at different scales. These contexts included the widespread heritage of Upper Kama (Russia) at territorial scale, the Jaime I sites (Spain) at the provincial scale, and the fortification system of the city of Gdańsk (Poland) at the urban scale [9] (Figure 2).

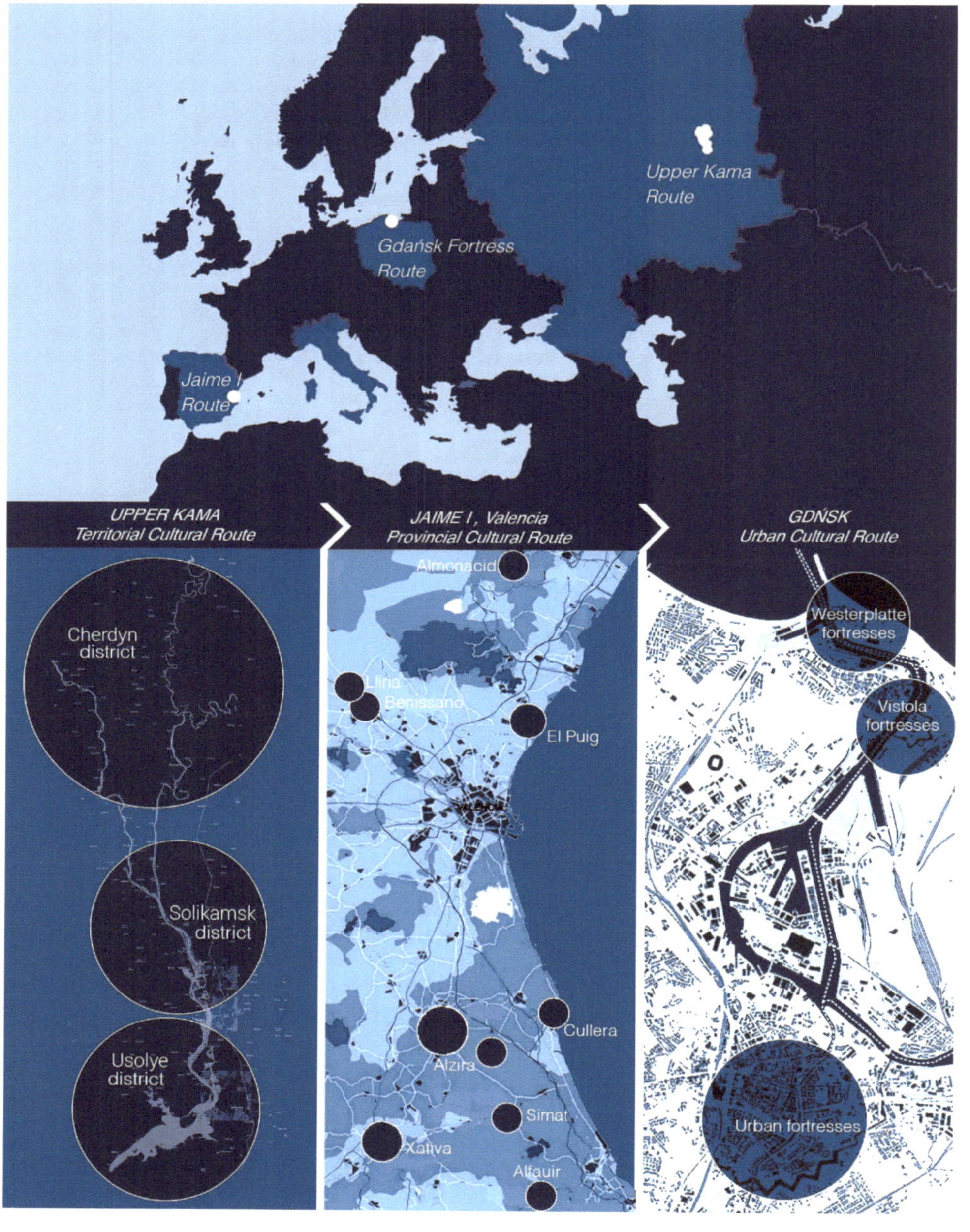

Figure 2. European Cultural Routes: territorial level (Upper Kama), provincial level (Valencia), and urban level (Gdańsk).

These are three case studies with different extensions, and which bring together complex architectural phenomena in which the value of the cultural route is fully manifested. First and foremost are the cultural–historical influences. In these case studies, the tradition, the antecedent—be it the Tartar, Moorish, or Teutonic Knights' pre-existences—is modified as a result of a series of historical events that give rise to cultural contaminations. These contaminations modify language, decoration, and stylistic models, producing new forms and patterns or, more simply, promoting the development of signs that renew and characterize the landscape. Signs becomes historicized and becomes an unforgettable part of a cultural identity, definitively modifying the place and conditioning the events that mark, from that moment on, the future of that particular socio-political and economic context. In addition to influences, secondly, time must be considered, both the time in which these architectural evidences are built, and the time of propagation of the cultural phenomenon, the stability or invariance of the same influences in the different time frames. Time becomes an essential reference factor for the classification of the route, because it allows us to determine limits and order the events that have characterized the history of these landscapes.

- The Upper Kama region is located north of Perm Krai, limited to the west by the chain of Ural Mountains, and to the north by the Komi region. Since the 15 century, the salt trade, from Europe beyond the Urals to China [10], has increased trade flows and intermediate settlements along this route, centered around the main districts of Solikamsk (1430), Cherdyn (1535), and Usolye (1606) [11,12]. The Upper Kama region is still composed of more than 50 monumental Orthodox complexes, spread over the territory but interconnected by a single cultural language, but where management planning aimed at protecting these assets is absent.
- The route of Jaime I in Spain is a cultural itinerary that runs through those territories affected by the Catholic Reconquista of 1238 and represented by the current Provinces of Valencia, Castellón, and Alicante [13]. Although there is no real official source on the itinerary associated with Aragonese King Jaime I, the tourist and cultural promotion of this route is aimed at ideally connecting the 112 sites (religious buildings, defensive and strategic buildings, and some minor buildings) located between Teruel, where the itinerary begins, and Peñíscola, where it ends. The selection rationale of the sites investigated was dictated, on the one hand, by their consistency from a historical, geographical, and architectural point of view with the various strands belonging to the route; on the other hand, by the numerous transformations many of them have undergone over time, which have led to their progressive deterioration, abandonment and, in some cases, consequent ruin.
- The fortifications in Gdańsk define an organic system, closely integrated with the city, but defining a clearly visible spatial boundary. This system consists of fortifications dating from different historical periods and therefore classifiable typologically and temporally into macro-groups (from the 10th century to the 20th century): line fortifications from the medieval period, early modern outer fortifications, and 20th century fortifications, erected and expanded to defend the New Port and the Westerplatte peninsula [14–16].

Documentation of the image of a place and its monuments, framed in their current condition, becomes an objective of territorial monitoring of the evolutionary processes that involve culture and local identities [17], and it defines a path of initiative and evolution for the development of adequate maintenance and management plans [18].

The primary objective of the project was to represent the complex relationships that define the identity of these places, including the monuments and their connections to different cultural contexts, and thus, the 'phenomenon' of the specific cultural route, through a specific and new digital configuration. The digitization process served as a methodological approach accompanied by a thorough analysis involving the decomposition and critical reconstruction of the architectural units. This was achieved through the semantic classification of its sub-elements—from landscapes to buildings, and from the buildings to their details—in all their relationships with the contexts [19] (Figure 3).

Figure 3. Images of some of the investigated sites along the routes.

In the case of the Upper Kama, a large part of the investigation involved verifying the existence of monuments and finding them, as sometimes this built heritage is hidden in the forests, within an extremely large area. In the other two cases, the work was very different. For the Spanish route, the interaction with colleagues from the Polytechnic University of Valencia and the presence of numerous studies on the route facilitated the recognition of the sites and the organization of the maps that identified the monuments to be considered.

For the Gdańsk fortresses, on the contrary, there was no unified study that intended to analyze the defensive heritage according to cohesion logic with respect to the evolutionary grid of the city. Interaction with archives and colleagues in the history of architecture at the Polytechnic University of Gdańsk was therefore crucial in order to be able to systematize each monument, and then to standardize the language of communication with that of other case studies [20].

2.2. Integrated Survey and 3D Databases

Acquisition and survey campaigns were conducted to create digital duplicates of the monuments along the different European context routes. These duplicates served as the base for a detailed analysis of the historical and evolutionary phases of each church, castle, or fortress.

Several tools were employed to produce a digital reference archive of the city and its historical–architectural image. Aiming to define repeatable acquisition protocols for the development of 3D models of the built heritage, all the case studies included the testing of expeditious documentation methodologies. These actions included the joint application of image-based and range-based instruments whose data, suitably integrated, contributed to the production of three-dimensional multi-scalar databases [21], from which it was possible to understand and interpret the digitized cultural route: from the building to the territory and from the territory to the building. An integrated survey methodology was adopted, utilizing terrestrial laser scanners (TLS), mobile laser scanners (MLS), and terrestrial and aerial photogrammetry (UAV). This strategy ensured the acquisition of accurate spatial data and photographic information, which was crucial for structuring and validating the geometric database [22] (Figure 4).

Figure 4. Different digital tools used for data acquisition.

Furthermore, the use of rapid-survey instrumentation was necessary to understand the extent of the three routes and to ensure the replicability of the method [23], both in terms of timing and the type of product obtained. The archives obtained with these tools and other data (3D panoramas, high-resolution landscape images, etc.) provided the necessary information to ensure the reliability, in terms of geometric and colorimetric accuracy, of the architectural models.

The first experience gained from the Upper Kama territory provided valuable insights and knowledge that guided the acquisition standards and methods for the other European contexts. The documentation methodology was adapted to specific typological features, enabling efficient and expedient data acquisition. The photogrammetric tool, both from close-range and UAVs, was tested to generate comprehensive 3D models capable of effectively describing each monument [24,25] (Figure 5).

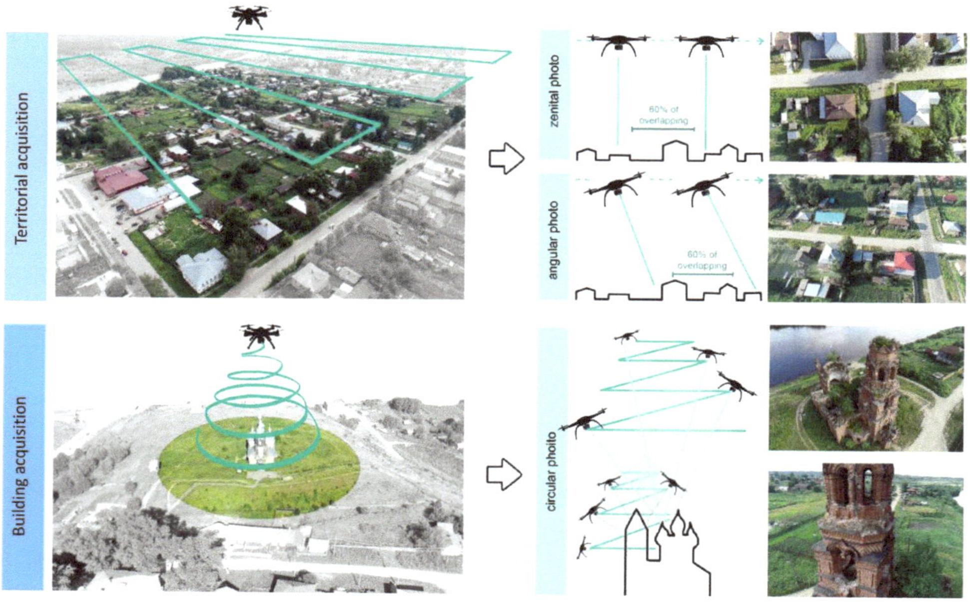

Figure 5. Scheme for close-range data acquisition with UAVs designed for the Upper Kama context.

The use of drones was crucial to acquire data from inaccessible areas or architectural surfaces (high heights of some vertical elements, interdicted areas that are not visible from the driveways or pedestrian streets roads, the presence of construction sites or debris in the proximity of some monuments, etc.) and ensure uniformity in describing each analyzed complex. The georeferenced database of the route obtained by drone in some cases enabled the georeferencing of individual architectural databases obtained with LIDAR instruments. For this reason, the drone was used at an altitude 30 to 60 m maximum above the ground that ensured a high level of resolution of the acquired images (designed GSD of 1–2 cm/pixel maximum) (Figure 6).

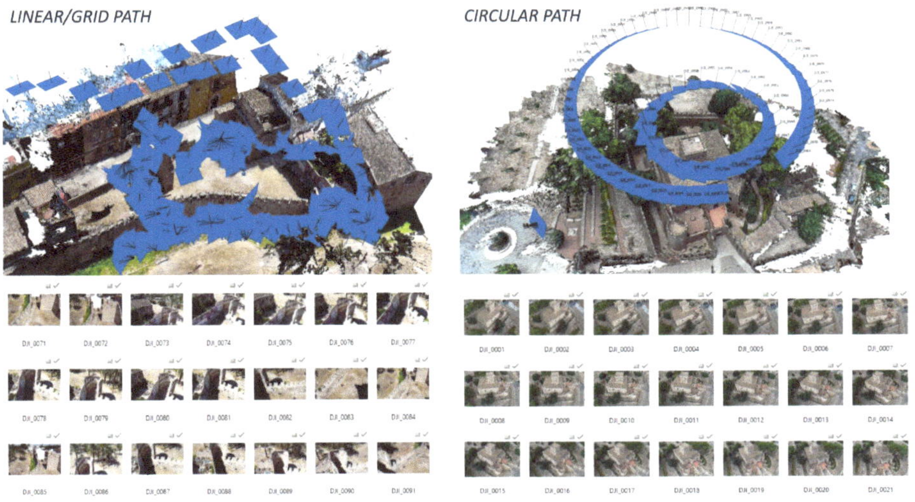

Figure 6. UAV acquisition with DJI mavic mini drone in Valencia province.

In cases where certain areas were not accessible by drone due to several factors, such as tree cover or flight restrictions, mobile systems were utilized for the rapid interior surveys of buildings and those areas. Close-range photogrammetry complemented the documentation process (Figure 7), focusing on surfaces requiring a more detailed description, especially for masonry textures or areas with specific decay conditions.

Figure 7. Integrated data acquisition of Lowland Gate (Gdańsk). On the left, Laser Scanner RTC360 point cloud; on the right, point cloud from UAV DJI Phantom RTK.

To achieve the project's objectives, it became necessary to assess and compare the effectiveness of various digital survey methodologies. The aim of creating 3D archives and informative models of heritage required the definition of a specific methodology that balanced the principles of metric reliability and survey precision with procedural efficiency.

Being "swift" within the available time on site proved crucial both for the specific projec and in attempting to envision replicable procedures in other cultural contexts.

Specifically, the methodology aimed for data integration, creating a comparativ matrix among different types of measurements, information, and metadata, describing an constructing a heterogeneous database for each monument at various levels. The diversit of data, in relation to the diversity of analyzed architectures, becomes functional both fo evaluating process replicability and for the proper functioning of the interactive platform the model utilization system.

Digital surveying, despite striving to provide a wealth of information about locations is always directed toward a specific purpose. The interactive platform, the digital twin o the cultural route, must consider numerous aspects in which the acquired information's specificities take on different connotations—designs, photographs, 3D models, texts, etc Surveying and modeling design must always compare the real dimensions of the cultura asset, the one that is surveyed, with its digital translation, envisaging a distinct dialogue between these two images.

The cultural landscape expressed by the routes is thus narrated through dynamic databases that contribute to updating the management and valorization systems of the built environment. It is precisely in the interaction with digital data that the specificity of its informational potential must be characterized. For instance, the management aspect is quite distinct from the enhancement and the various ways in which heritage can be valorized, both in terms of increasing awareness and in explicating the historical events that have shaped its evolution.

In this regard, 3D models must be capable of gathering all the necessary information to enable freedom of utilization coupled with understanding.

In order to structure a modeling strategy that could respond to the multiple descriptive needs mentioned above (knowledge for conservation, informatization, and alternative fruition of the route), different scan-to-modeling processes [26,27], also characterized by different levels of development (LODs), were tested at the urban scale for the methodology of data elaboration.

First of all, four levels of interpretation and LODs were defined:

1. Territorial level: At this level, different centers within the districts are identified using a geomatic hyper-simplification model of the landscape, territory, infrastructure, and connectivity systems, as well as the potential service systems that can be linked to this scale. Specifically, during this phase, the development involved creating a simplified digital terrain model (DTM) with the main contour lines based on existing cartographic data and metric information obtained from integrated surveys (utilizing tools such as mobile laser scanners and UAVs). This model delineated the connectivity and accessibility attributes of individual monuments, outlining potential connecting pathways between religious and fortified complexes. This corresponds to LOD 0 (digital terrain model) with an accuracy level of approximately 10 m [28].
2. Urban level: In this model, all the different activities and services present within the single center are identified. This level corresponds to LOD 1 (one-block model), with an approximate accuracy of 5 m.
3. Monumental area level: Within the single urban center, the model of the monumental area will be deepened, in order to identify the different areas for safeguarding the historical character of the single territorial centers [29]. To this end, the landscape features of the area identified in relation to the aesthetic monumental elements within which it will be possible to identify the typological features of the monumental structures present in the area will be described in a single model. This level corresponds to LOD 2 (architectural scaled model), with an approximate accuracy of 3 m.
4. Monumental building level: The various buildings present in the monumental areas, identified by the previous level of analysis, will be further investigated in this level of investigation where they will be individually described through the construction of a specific HBIM information model in which the decorative and technological

features of the entire building will be represented. This level corresponds to LOD 3 (full architectural scaled model), with an approximate accuracy of 0.2 m.

Subsequently, other types of 3D modeling were developed, including a reverse-modeling approach (surface mesh and surface NURBS), aiming to obtain heterogeneous outputs for experimenting with different cognitive and communicative phases of the route. In reality, although these modeling methodologies were tested on various case studies to create different informative models for the three routes, some of these modeling techniques have been integrated and interchanged, often producing outputs and results that bridge between HBIM (historic building information modeling) and GIS (geographic information systems) informative models, virtual reconstructive models of historical phases, and navigable point clouds, among others.

3. Results

3.1. Post-Processing Data and Digital Twins

The amount of information acquired through the use of digital survey tools on the three cultural routes was massive. Detailed point clouds on the buildings were flanked by more sparse—but still highly descriptive—point clouds of the landscape in which the monuments were located (Figure 8).

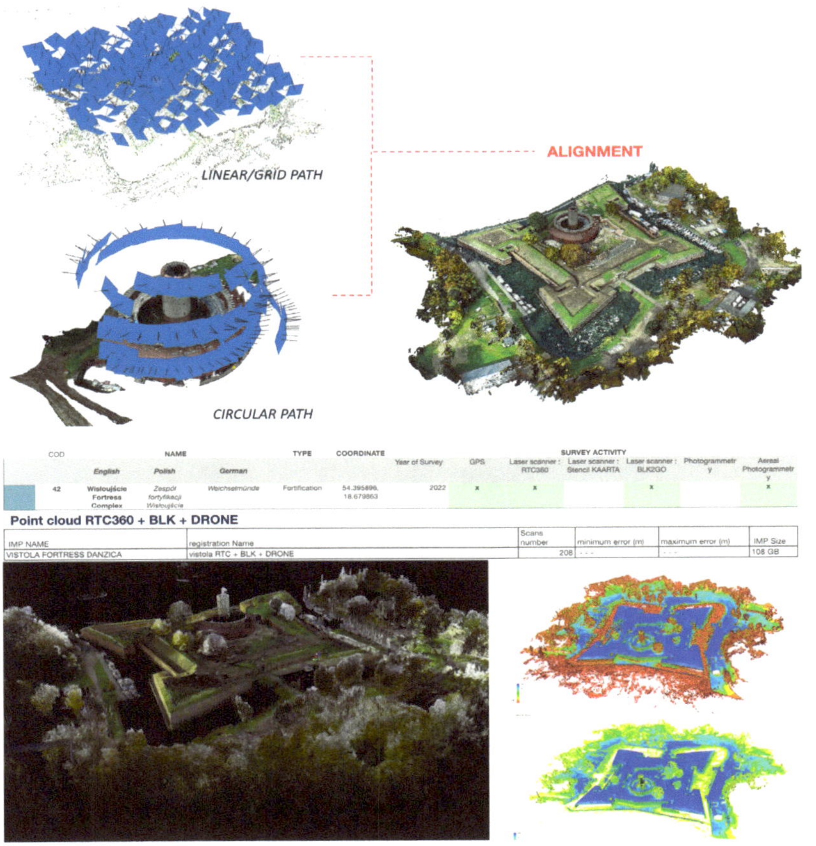

Figure 8. Data acquisition of Vistola Fortress (Gdańsk). Above, UAV data integration of linear and circular path acquisition; below, data integration between MLS, TLS, and UAVs.

3.1.1. Upper Kama 3D Modeling

Different modeling methodologies were attempted to achieve such levels of detail and test the fruition of the obtained 3D models [30]. The scan-to-BIM process is a commonly used procedure to convert point cloud data into informative 3D models [31]. It involves capturing the geometric and spatial information of a physical structure using 3D laser scanning or similar technologies and then using those data to create a building information model (BIM). The BIM represents, in this sense, a very accurate picture of the geometry, materials, and other relevant information of a building, enabling analysis, visualization, and management of the structure.

Regarding the Upper Kama route, the scan-to-BIM process was extensively tested. The architectural richness of the decorative element of the buildings along this route presented a complex case for experimentation with the simplification of 3D parametric models. The experimentation was tested to reduce the complexity of the models (decimation), while retaining the necessary information by preserving the overall shapes and structures. This simplification helps to optimize the performance and usability of the models, making them more manageable for analysis, visualization, and other applications [32] (Figure 9).

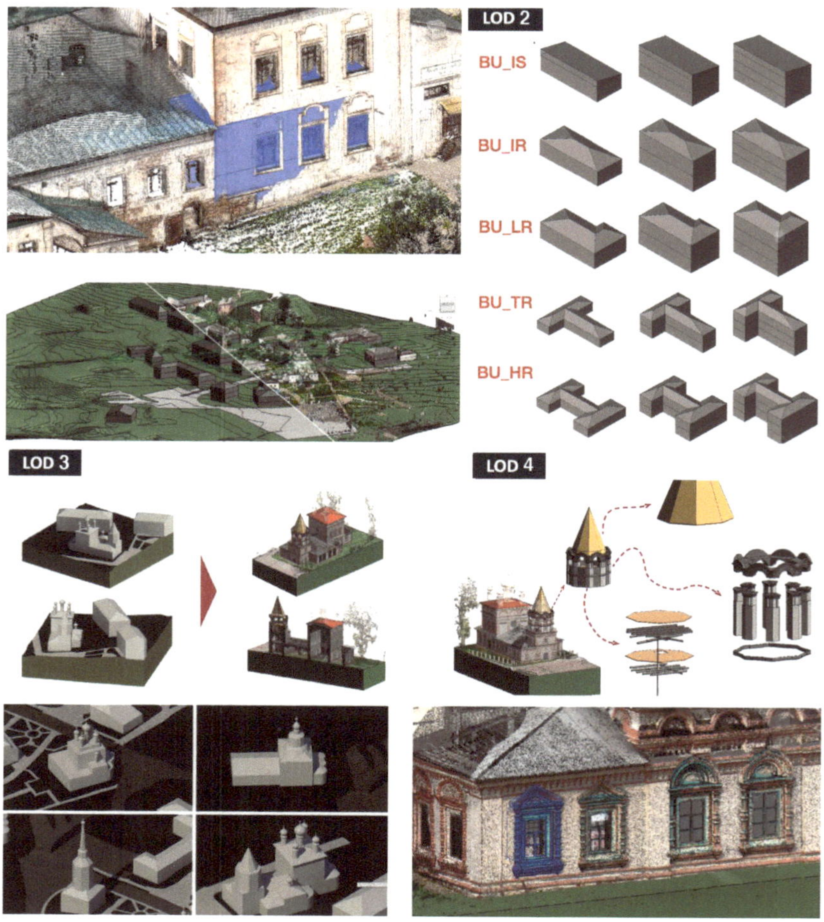

Figure 9. Scan-to-BIM process for Upper Kama informative models (data elaboration performed by H. Fu and A. Dell'Amico).

For the description of the spatial, landscape, and urban character levels, the volumes and encumbrances of the buildings were modeled in simple volumes, useful for defining the dimensional features of the encumbrances and orientation [33]. These were then placed on a simplified DTM, modeled from a photogrammetric cloud obtained by drone. The technological and decorative features of some monumental buildings were then explored in a second level of modeling. This branched system allows querying by levels, from the general to the particular, with a logic of the simplified reading and gradual deepening of information related to route knowledge. Each modeled element, both of the landscape and the monument, was imputed into a shared abacus with the aim of building an initial library of digitized elements of the architectural features identifying the monuments of Upper Kama. The modeling phase of the monuments was set in worksharing mode on a scan-to-BIM basis. This process passed from a preliminary phase of recognition and decomposition of the different hierarchical components that constitute the building and that defined the parametric digital twin [34,35].

The information components that qualified the models as digital twins of specific Upper Kama districts routes were a result of the analysis and instrumental acquisition of the survey activities and the filling of the thematic sheets. The sheets were organized according to a structure that included both general information (for recognizing the architectural unit) and morphological–spatial characterization (specific to the volumetric and overall distribution system, facades, architectural details, down to the specifics of ornamentation), as well as historical considerations (linked to archival information), the state of conservation, and the technological characteristics of the building (encompassing both construction aspects related to masonry and the pathologies of present deterioration).

These data engage in a continuous dialogue between what is presented in the model, which explicitly conveys the history that can be discerned from the analysis of architectural elements, and the information provided in the sheets. The digital twin speaks of itself and how the building has come down to the present day, while also envisioning the potential for further development in the future. The digital twin should exist in parallel to the real one, with the capability to transform and update itself as the real structure evolves. In the case of the Upper Kama project, out of the 80 sites surveyed, over 50 have been digitally recorded, with 7 parametrically modeled in BIM following the aforementioned procedures.

3.1.2. Jaime I 3D Modeling

Similarly, reverse-modeling procedures were tested from integrated photogrammetric–lidar databases [36]. This procedure was mainly tested for the Jaime I route, where the aim was to produce, in a short timeframe, databases that were exhaustive but at the same time not overloaded with information. The models obtained consisted of geometric and colorimetric information that guaranteed the descriptiveness of the mesh models obtained. These, decimated in the number of polygons to be more easily manageable, nevertheless maintained a high level of realism that can be shown within visualization platforms (Sketchfab) [37]. However, to ensure that these models were equally informative representations of the route, additional information sheets or descriptive elements needed to be incorporated into the systems. These supplementary components can be appropriately linked to the entire model or specific portions of it, enhancing the understanding of the artefact and its connections with the surrounding environment.

For the Jaime I route, over 20 sites have been identified, including castles, city walls, religious buildings, and minor architecture. Integrated and efficient digital surveys were conducted for these, with the aim of subsequent reverse-modeling procedures. Notably, the sites along the Valencian route are larger in comparison to the Russian route or even the Spanish route, which does not exhibit a constellation of isolated episodes across an extensive territory. Instead, it comprises sites of memory that blend into a more intricate landscape system. While most Russian sites consist of scattered episodes, in the Spanish case, the castles, ruins, and fragments of James I's history are found within large architectural complexes that have grown in size and territorial significance over centuries. This is why the three-dimensional databases and models aimed to more accurately represent the surrounding territory of each individual monument, often embedded within historic centers.

Throughout the stages of data acquisition, development of the digital archive, and representation of the route through various illustrative systems, this specific provincial dimension of Valencia influenced the different methodological choices. This research project was designed to foster dialogue among researchers; therefore, alongside the scientific goals suggested by the proposal, the Spanish experience led to intense debates where two cultural dimensions (Italian and Spanish), historically accustomed to dealing with architectural heritage, sought compromises to construct a language that transcended national ideological boundaries in representation and graphic expression. For this reason, the organization of historical–constructive, technological, and conservation-related information was structured to constitute the informative framework of the realized reverse-modeling models. These models, appropriately semanticized first in their historical–constructive phase, and then in their technological components, are queryable (by volumes or surfaces) and linked to respective infographic information within a simplified interactive platform (a web-GIS system for interaction and Sketchfab for individual model visualization).

By including information sheets or descriptive apparatuses, viewers or users of the models can access more specific knowledge about the route, such as historical context, architectural details, cultural significance, or any other relevant information. These additional resources contribute to a deeper understanding and appreciation of the artefact being represented, enhancing the overall experience and educational value of the models [38].

Apart from visualization, another practical use of these models is the ability to physically reproduce these monuments through solid prototyping. The mesh models derived from point clouds were appropriately optimized, including decimation where necessary, and filling in any portions that may not have been fully captured by digital instrumentation. This optimization process prepares the models for further processing through 3D printing, enabling the creation of physical replicas of the monuments for an alternative and more inclusive fruition of CH routes (Figure 10).

3.1.3. Gdańsk Fortresses 3D Modeling

The last modeling strategy from databases is the mathematical tool that uses non-uniform rational basis spline (NURBS); it develops 3D geometric surfaces from 2D vector data obtained by detailed drawings. This process was mainly tested for the last route, that of the Gdańsk fortresses. The decision to operate in this mode in this context was dictated by several factors: on the one hand, for a logistical and temporal question of model development; on the other hand, for methodological experimentation to be carried out on the research output (Figure 11).

As far as the first point is concerned, it should be emphasized that the logistical events connected with the project (the initial blockade for COVID-19 and the subsequent war between Russia and Ukraine) forced us to reshape both the partnership and the subject of the research. In place of the Perm Polytechnic University (Russia) from 2023, less than a year after the project's closure, a new academic partner (the Polytechnic University of Gdańsk) and a local company (CTA) were involved. The timeframe for the development of information models on this route was drastically reduced, forcing a more flexible and less time-consuming methodology to be structured. The idea was to create easily usable models from a navigable information platform (embedded in a GIS environment).

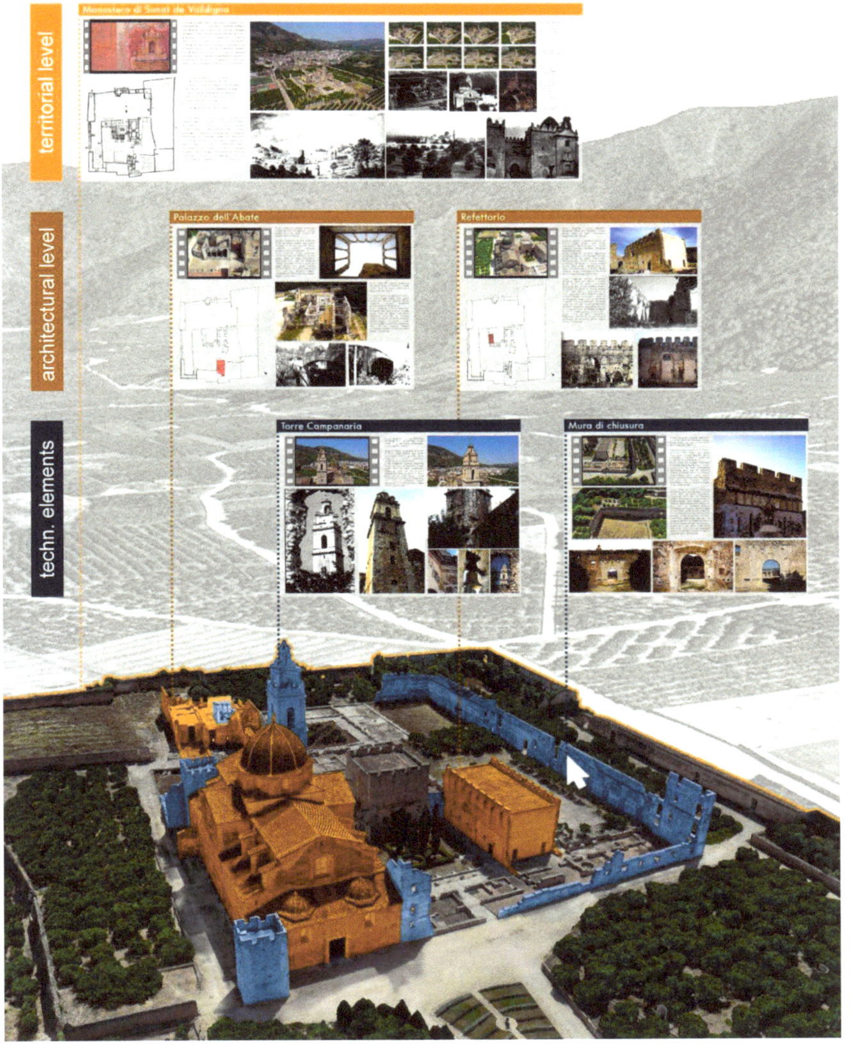

Figure 10. Data segmentation for informative models of Simat religious complex on the Jaime I route (data elaboration by G. Porcheddu).

Out of the 50 surveyed and recorded sites, 3D models have been developed for 12 case studies, aimed at creating a comprehensive database of the architectural, constructional, and geometrical components of the current state of various monuments. Additionally, a phase currently underway within the project involves modeling historical reconstructions of some of the previous configurations of the fortified structures. This investigation is being conducted by analyzing past drawings and photographs present in historical archives, as well as through an examination of the current object's configuration. This includes analyzing data produced by 3D surveys and studying existing marks on historical walls and planimetric developments.

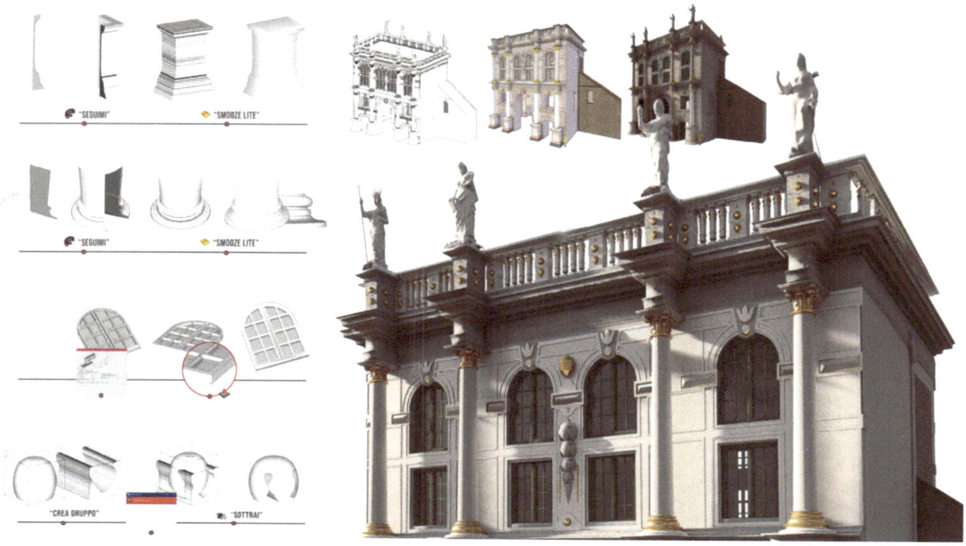

Figure 11. Three-dimensional modeling in Rhinoceros, realized from 2D drawings, for the Golden Gate (Gdańsk) (data elaboration by F. Picchio, A. Dell'Amico, S. Bonanomi, C. Caroli, G. Venanti).

In particular, the team is focusing on reconstructing the image of the city's access system, represented by monumental gates, as well as the defensive walls (boundary walls and Renaissance bastions), the transformation of which has significantly influenced the overall cityscape. Once again, the NURBS modeling phase interacted with the BIM and GIS approaches, aiming to develop models of various historical phases that can converge within the interactive platform of the route. This aims to provide an understanding of the transformations undergone over the centuries.

3.2. *Interactive Informative Platform*

With the aim of complying with the open accessibility features of the results required by Europe regarding H2020 projects, each digital twin realized for the monuments of the three cultural routes will be inserted into an interactive and searchable platform to make them visible and usable. The georeferenced 3D database platform provides the main basis for data integration and analysis, allowing comparisons between real objects and their digital representations. Specifically, this involves the realization of spatial information systems (SISs) for cultural heritage applications of GIS technology [39]. This product seems to have several key features that contribute to a range of benefits:

1. Integration of different data: The product connects different types of information content, such as cards, images, parametric models, technical drawings, and analyses. This integration likely allows for a more comprehensive understanding of the architectural sites and their spatial context.
2. Significance of the spatial component: By combining spatial data with other architectural information, the product can help researchers and users better comprehend the importance of the spatial component in their analyses. This could lead to insights into the relationships between each architectural element and its surrounding.
3. Accessibility for different users: The product will be designed to be user-friendly and accessible to various types of users, not just experts in the field. This inclusivity could lead to a broader range of users benefiting from the technology and contributing to architectural heritage research.

4. SIS technology: Integrating SIS technology into the product can enhance the accuracy and efficiency of spatial analyses, that take into account each building with its context.
5. Increase in knowledge: The product's capabilities likely lead to a deeper understanding of the value and challenges associated with cultural heritage monuments. This increased knowledge could be instrumental in making informed decisions regarding land planning strategies and preservation efforts.

Such a georeferenced 3D database platform seems to offer a powerful tool to explore, analyze, and preserve cultural heritage sites holistically (Figure 12). Its integration of different types of data and spatial information can lead to valuable insights and foster better appreciation of the significance of these monuments in their respective contexts.

Figure 12. Insertion of image-based point cloud of Gdańsk southern bastions into ArcGIS platform (data elaboration by F. Picchio, J. Badach and G. Porcheddu).

This platform is an ambitious project that was initiated during the last year of the project and is still in the development phase. The experimentation conducted thus far has allowed for the design of a multi-level and multi-informational information system, where all currently processed data must find their placement and hierarchical organization in terms of reading and digital system enhancement strategy.

As of now, each of the three platforms (Upper Kama, Valencia, and Gdańsk) has been conceived as an independent and separate web-GIS system.

- For the Upper Kama project, the three districts that comprise the various analyzed sites will be represented by a platform integrating GIS and BIM systems. The inserted models of monuments and their context (LOD 0 and LOD 1) will be associated with different information (texts related to the toponymy of historical centers and historical and current images), accessible through specific choices in a menu within the developed platform. From here, it will be possible to access a higher level of detail (LOD 2), which will feature interactive elements describing the monumental system of various buildings and their relationships with the context. Clicking on each of these buildings will delve into the detailed BIM of the individual church or monument (LOD 3), from which one can access specific information about the building.
- For the Valencia route of Jaime I, the platform will not utilize simplified terrain models, but rather a web space built upon a cartographic map of the Province of Valencia. On this map, the locations of various investigated sites will be marked, represented by clickable volumes, in accordance with the geographical layout of the route. The

clickable points within this virtual space will allow interaction with individual reality-based models, optimized in terms of polygon count to ensure ease of management and navigation. Furthermore, these models will be semantically decomposed according to a technological–constructive organization of the historical phases that have impacted the sites. In this manner, by selecting a portion or element of a specific phase, it will be possible to access cards containing textual, graphical, and photographic data that enhance the reading and comprehension of the navigable 3D models.

- For the route of the fortifications in the center of Gdańsk, an idealized platform of the city has been designed, where distances and dimensions of various elements have been rethought in relation to their perception within the urban complex. Houses, streets, and minor urban features have been simplified and placed with variable dimensions based on their perceived significance. Similarly, fortified elements have been modeled based on conducted surveys and proportionally inserted with larger dimensions compared with other elements on the platform. Structured on Unreal Engine 4, the platform aims to emphasize the defensive system of the Polish route. It becomes clickable, enabling access to detailed models of each fortified element. These models have been obtained through mathematical NURBS modeling processes, as well as mesh or parametric modeling for some of the most representative buildings (Figure 13).

Figure 13. An informative platform of an "imaginary" Gdańsk city center (elaboration by S. Parrinello, H. Fu). From this web-GIS platform, it will be possible to access the different 3D models and information elaborated for the cultural route.

The three platforms and all related specific content will be accessible online through the website of the European project (https://www.prometheush2020.eu/, accessed on 26 July 2023). As the system is embedded within the Internet, this must allow accessibility and simplicity for all users, and above all, be independent of any software [40]. A web-GIS platform thus constitutes a new configuration of the route. In particular, on the basis of drone-based reality models, virtual storytelling is realized narrating cultural, architectural, or conservation status aspects of the cultural route, useful reading for non-experts as well as for technicians and experts in the field of conservation and restoration. Each of the route platforms therefore allows different levels of interaction with the monuments on the route: from the more general, where one visualizes and interacts with maps and points; to the more detailed, where one learns about the history of the artifact by reading its current spatial components and reconstructions of its historical phases.

Another aspect that the project is considering and trying to develop as a further and more in-depth knowledge phase of each site is the possibility of using extended reality, VR, and AR applications [41]. By integrating VR and AR technologies into the project, it becomes possible to extend the reach of the cultural route beyond physical boundaries, enabling people from different parts of the world to explore and learn about the heritage virtually. The use of VR and AR applications in combination with 3D digital models can significantly enhance the development of immersive storytelling experiences along the cultural route, as well as in-depth learning of the object (Figure 14).

Figure 14. A 3D printing for tactile fruition of the Prison Tower monument in Gdańsk. In the middle, AR applications for visualization in 1:5 scale of the monument; on the right, 3D printing and QR codes to access to the textured model on the Sketchfab app.

In particular, some of the 3D models produced (such as the Prison Tower building model from the gate access system in Gdańsk) have been made accessible via mobile devices in augmented reality (AR) mode. This accessibility enables viewing in relation to the real object (the monument) as well as in alternative contexts. The AR visualization application was developed using Unreal Engine 4 software and prepared in apk format for Android devices. On-site access can be initiated through a physical QR code or by displaying the mobile device. This automatically loads and displays the model on the device.

For off-site access, the application "Outdoor guides and explorers" can be utilized. Using the Sketchfab platform, a scale model (ranging from 1:1 to smaller) can be accessed through a link and in a markerless mode. Upon activating the AR function, a short period (around several seconds) is dedicated to recognizing the physical space and establishing a support plane for the virtual placement of the model. Once the model is viewed, users can move around the 3D building model or approach it freely with their smartphones to explore the detailed model from different angles.

Additionally, another mode of access has been developed through a VR application. This mode has been predominantly experimented with for the monumental buildings in the Upper Kama region, given the physical inaccessibility of some monuments and the limited overall understanding of such systems.

In particular, the embedded Sketchfab models were made viewable via oculus. This makes it possible to fully visit the building, both inside and outside, access to which is often forbidden to the public due to structural instability or private use [42]. The future goal is to also include the entire platform designed for the three routes within applications that exploit VR technology, making building–building and building–context relationships of the route visible (Figure 15).

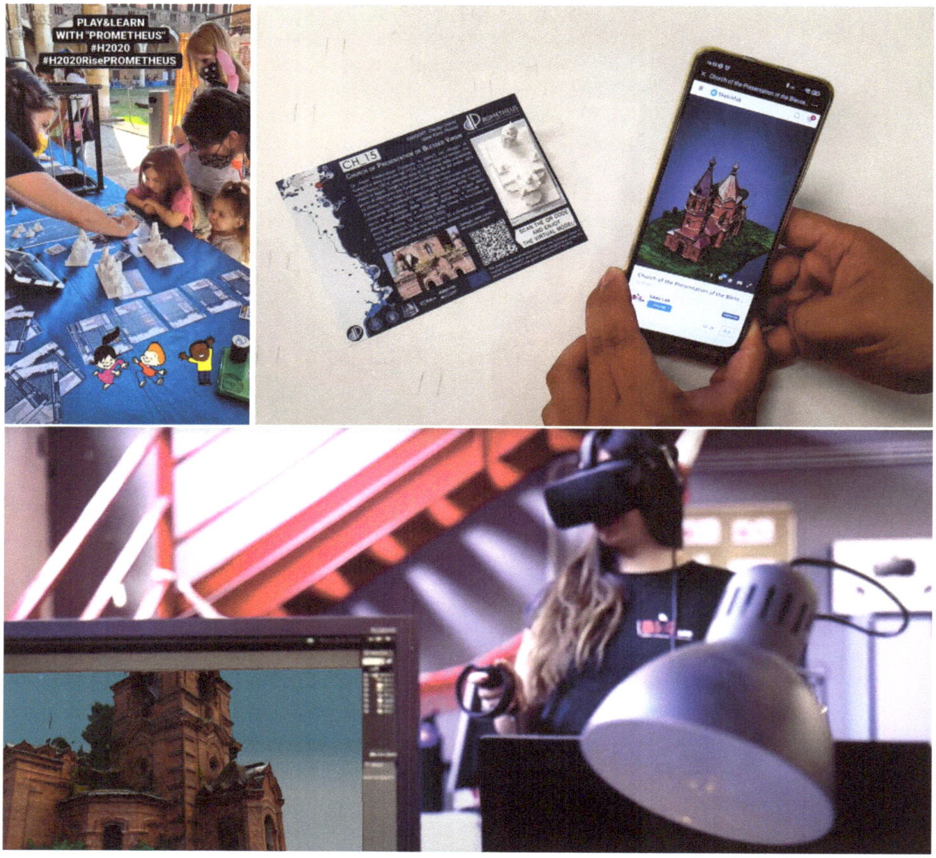

Figure 15. Informative cards and 3D models with a QR code linked to Sketchfab, designed for Research Night in Pavia, 2021; below, the design and the test of VR immersive experience with oculus of digital cultural routes of the Upper Kama.

4. Discussion

Although the project is in the development phase for these final stages of realization and fruition of the web-GIS platform, it is possible to present some reflections on what has been achieved so far, both from a methodological and a product point of view.

The semantized models serve as visual support for a wealth of available information (historical, geographical, technological) related to each case study of the route. Organizing the elements of the cultural route through segmentation procedures of the 3D database (point cloud or mesh model) becomes a crucial structure for archiving and interpreting the data [43].

This process directly stems from the mental efforts of designing, integrating, and converting a physical element into its digital counterpart, where each digitized element can be part of, or function as, an autonomous, complete, and searchable database. The project's requirement to make research findings accessible through open-access platforms promotes the sharing of these databases, fostering deeper understanding of the different European territories. To ensure easy and immediate access to information, the proposed information system will be developed as Web Apps, allowing the connection of three-dimensional models (optimized for the web and developed on BIM, NURBS, and mesh modeling software) to related multimedia and iconographic resources [44].

The architectural elements of the cultural routes, differentiated by typology, location, and historical evolution processes, constitute independent databases, each queryable on its own. However, when integrated within the same platform, they can reveal aspects and values that are not always easy to understand [45]. A semantized model, enriched with informative content in this manner, could be configured as a "wunderkammer", a chamber of wonders, containing several configurations of the object itself: historical reconstructions, typological–constructive information, and material and immaterial aspects referring to each part of the semantic model.

With this foundation, the "transition", starting from the survey method and extending to the heritage values, revolves around the interaction between users and the cultural route. Digitizing the itinerary and providing access to "enhanced" content compared with what is available along the route shifts the experience towards the virtual realm, making the itinerary more inclusive [46]. Offering virtual accessibility could attract more informed users, raising awareness about the value of these routes. Consequently, these sites may be considered "protected" areas, recognized as architectural and landscape complexes within specific cultural itineraries.

However, the actions undertaken reveal some critical points in the results obtained and being developed, which also point to some weaknesses in the survey method adopted. The limitations mainly concern the heterogeneity and lack of completeness of some route site information, which makes the platform 'incomplete'. This aspect is mainly attributable to a problem of tight deadlines both in the information acquisition phase and, above all, in the phase of processing this information, which is still not completed. The project, initially dedicated only to the Upper Kama route, was only reformulated in its final year, forcing the adaptability of the method developed in the Russian route to be verified in the other case studies as well. In both the Valencia and Gdańsk cases, the acquisition and processing protocol was readjusted to the different case studies of the routes' sites, inviting a rethinking of more rapidly realizable outputs in order to fit within the project's timeframe. To Upper Kama's parametric information model structure, which required lengthy timeframes and architectural–technological comparisons between similar elements of the same route, two more have been added, envisioning the structuring of information models on much more heterogeneous sites. Defensive monuments and religious sites often presented elements that were not comparable to each other, understandable only when framed with related typological elements of the same geographical contexts, even those not part of the route. As a result, the planning of an investigation protocol designed for a multi-year project but executed over the course of a single year required the simplification and selective acquisition of information, as well as the implementation of pilot tests on some routes rather than the entire cultural itinerary.

On the other hand, we can consider that this forced restructuring of actions and objectives has brought added value to the project. The reimagining of obtaining digital twins more quickly (through rapid survey technologies) and modeling them more efficiently (via reverse-modeling processes) has compelled the team to engage with diverse data and assess multiple solutions, aiming to consistently achieve the custom result that is most functional in terms of timing and available resources, activities perfectly aligned with the project's goal of cultural and experiential exchange.

Author Contributions: Conceptualization, S.P.; Methodology, S.P. and F.P.; Investigation, S.P. and F.P.; Data curation, S.P. and F.P.; Writing—original draft, S.P. and F.P.; Writing—review & editing, F.P.; Supervision, S.P.; Project administration, S.P. All authors have read and agreed to the published version of the manuscript.

Funding: This project has received funding from the European Union's Horizon 2020 research and innovation program under the Marie Skłodowska-Curie; grant agreement N° 821870.

Data Availability Statement: All updated information about the ongoing project is visible on the official website: https://www.prometheush2020.eu/, accessed on 26 July 2023. Part of the outpu is visible at the following online platform: https://sketchfab.com/DAda-LAB, accessed on 26 July 2023.

Acknowledgments: The study is strictly linked to and supported by the research project entitled Prometheus: "PROtocols for information Models librariEs Tested on HEritage of Upper Kama Sites"—a project for the documentation of the architectural heritage of the monuments within Cultural Heritage Routes. Part of the work has been developed inside International Summer schools coordinated by University of Pavia, in Russia (2019), Spain (2022), Poland (2023). They have seen professors, researchers, experts, and students from different universities involved participate in the development of digital strategies to build digital twins of cultural heritage.

Conflicts of Interest: The authors declare no conflict of interest. The funders had no role in the design of the study; in the collection, analyses, or interpretation of data; in the writing of the manuscript; or in the decision to publish the results.

References

1. Parrinello, S.; Picchio, F.; De Marco, R.; Dell'Amico, A. Documenting The Cultural Heritage Routes. The Creation of Informative Models of Historical Russian Churches on Upper Kama Region. *Int. Arch. Photogramm. Remote Sens. Spat. Inf. Sci.* **2019** *XLII-2/W15*, 887–894. [CrossRef]
2. Logan, W. Cultural diversity, cultural heritage, and human rights: Towards heritage management as human rights-based cultural practice. *Int. J. Herit. Stud.* **2012**, *18*, 231–244. [CrossRef]
3. Zhou, L. Cultural routes and the network construction of the World Cultural Heritage Conservation. In Proceedings of the ICOMOS 15th General Assembly and Scientific Symposium Monuments and Sites in their Setting-Conserving Cultural Heritage in Changing Townscapes and Landscapes, Section IV: Cultural Routes: The Challenges of Linear Settings for Monuments and Sites, Xi'an, China, 17–21 October 2005; pp. 17–21.
4. Oikonomopoulou, E.T.; Delegou, E.; Sayas, J.; Moropoulou, A. An innovative approach to the protection of cultural heritage: The case of cultural routes in Chios Island, Greece. *J. Archaeol. Sci. Rep.* **2017**, *14*, 742–757. [CrossRef]
5. Ćosović, M.; Maksimović, M. Application of the Digital Twin concept in cultural heritage. In Proceedings of the 1st International Virtual Conference on Visual Pattern Extraction and Recognition for Cultural Heritage Understanding, Pescara, Italy, 12–13 September 2022.
6. Okanovic, V.; Ivkovic-Kihic, I.; Boskovic, D.; Mijatovic, B.; Prazina, I.; Skaljo, E.; Rizvic, S. Interaction in eXtended Reality Applications for Cultural Heritage. *Appl. Sci.* **2022**, *12*, 1241. [CrossRef]
7. Liestøl, G. *Augmented Reality Storytelling–Narrative Design and Reconstruction of a Historical Event in Situ*; International Association of Online Engineering: Vienna, Austria, 2019; pp. 196–209. Available online: https://www.learntechlib.org/p/216412/ (accessed on 26 July 2023).
8. Sullivan, A.M. Cultural heritage & new media: A future for the past. *J. Marshall Rev. Intell. Prop. L.* **2015**, *15*, 604.
9. Available online: https://www.prometheush2020.eu (accessed on 26 July 2023).
10. Brumfield, W.C. The Architectural Heritage of Solikamsk and the Northern Districts of Perm Province. *Cah. Slaves* **2008**, *10*, 317–355. [CrossRef]
11. Brumfield, W.C. Usolye: Stroganov outpost in the Urals. Russia and India Report. Open Source. Available online: https://www.rbth.com/articles/2011/08/11/usolye_stroganov_outpost_in_the_urals_12850 (accessed on 1 July 2019).
12. Parrinello, S.; Cioli, F. Un progetto di recupero per il complesso monumentale di Usolye nella regione della Kama Superiore. *Restauro Archeol.* **2018**, *26*, 92–111.
13. Swift, F.D. *The Life and Times of James the First the Conqueror*; Clarendon Press: Oxford, UK, 1894.
14. Biskup, K. Rejony ograniczeń budowlanych twierdzy gdańskiej w wiekach XIX i XX [Areas of construction limitations of the Gdańsk fortress in the 19th and 20th centuries]. In *Gdańsk Pomnik Historii*; Kostarczyk, A., Ed.; PL_ Regionalny Ośrodek Studiów i Ochrony Środowiska Kulturowego: Gdańs, Poland, 1998.
15. Borucka, J.; Gatermann, H. *Architekturführer Danzig: Gdańsk Sopot Gdynia. [Gdańsk Architectural Guide Danzig: Gdańsk Sopot Gdynia]*; DOM Publishers: Berlin, Germany, 2016.
16. Szczepański, J.; Dymnicka, M. Dilemmas of Identity in Contemporary Cities. The City of Gdańsk as an Example. *Procedia Eng.* **2016**, *161*, 1225–1229. [CrossRef]
17. Cullen, G. *Concise Townscape*; Routledge: London, UK, 2012.
18. Bandarin, F.; Van Oers, R. *The Historic Urban Landscape: Managing Heritage in An Urban Century*; John Wiley & Sons: Hoboken, NJ, USA, 2012.
19. Santagati, C.; Lo Turco, M. From structure from motion to historical building information modeling: Populating a semantic-aware library of architectural elements. *J. Electron. Imaging* **2017**, *26*, 011007. [CrossRef]
20. Garau, C. Emerging technologies and cultural tourism: Opportunities for a cultural urban tourism research agenda. In *Tourism in the City: Towards An Integrative Agenda on Urban Tourism*; Springer: Cham, Switzerland, 2017; pp. 67–80.

21. Pierrot-Deseilligny, M.; De Luca, L.; Remondino, R. Automated image-based procedures for accurate artifacts 3D modeling and orthoimage generation. *Geoinformatics Fce Ctu* **2011**, *6*, 291–299. [CrossRef]
22. Daneshmand, M.; Helmi, A.; Avots, E.; Noroozi, F.; Alisinanoglu, F.; Arslan, H.S.; Anbarjafari, G. 3d scanning: A comprehensive survey. *arXiv* **2018**, arXiv:1801.08863.
23. La Placa, S.; Doria, E. Reliability of Dtms Obtained with Mobile Fast Surveys Techniques. *Int. Arch. Photogramm. Remote Sens. Spat. Inf. Sci.* **2022**, *46*, 299–306. [CrossRef]
24. Bevilacqua, M.G.; Russo, M.; Giordano, A.; Spallone, R. 3D reconstruction, digital twinning, and virtual reality: Architectural heritage applications. In Proceedings of the 2022 IEEE Conference on Virtual Reality and 3D User Interfaces Abstracts and Workshops (VRW), Christchurch, New Zealand, 12–16 March 2022; IEEE: Piscataway, NJ, USA, 2022; pp. 92–96.
25. Russo, M. AR in the Architecture Domain: State of the Art. *Appl. Sci.* **2021**, *11*, 6800. [CrossRef]
26. Spanò, A.; Patrucco, G.; Sammartano, G.; Perri, S.; Avena, M.; Fillia, E.; Milan, S. Digital Twinning for 20th Century Concrete Heritage: HBIM Cognitive Model for Torino Esposizioni Halls. *Sensors* **2023**, *23*, 4791. [CrossRef] [PubMed]
27. Badenko, V.; Fedotov, A.; Zotov, D.; Lytkin, S.; Volgin, D.; Garg, R.D.; Liu, M. Scan-to-BIM methodology adapted for different application. *Int. Arch. Photogramm. Remote Sens. Spat. Inf. Sci.* **2019**, *42*, 1–7. [CrossRef]
28. Over, M.; Schilling, A.; Neubauer, S.; Zipf, A. Generating web-based 3D City Models from OpenStreetMap: The current situation in Germany. *Comput. Environ. Urban Syst.* **2010**, *34*, 496–507. [CrossRef]
29. Donato, V.; Biagini, C.; Bertini, G.; Marsugli, F. Challenges and opportunities for the implementation of H-BIM with regards to historical infrastructures: A case study of the Ponte Giorgini in Castiglione della Pescaia (Grosseto-Italy). *Int. Arch. Photogramm. Remote Sens. Spatial Inf. Sci.* **2017**, *XLII-5/W1*, 253–260. [CrossRef]
30. Müller, P.; Wonka, P.; Haegler, S.; Ulmer, A.; Van Gool, L. Procedural modeling of buildings. In *ACM SIGGRAPH 2006 Papers*; ACM: New York, NY, USA, 2006; pp. 614–623.
31. Laing, R.; Leon, M.; Isaacs, J.; Georgiev, D. Scan to BIM: The development of a clear workflow for the incorporation of point clouds within a BIM environment. *WIT Trans. Built Environ.* **2015**, *149*, 279–289.
32. Chiabrando, F.; Donato, V.; Lo Turco, M.; Santagati, C. Cultural heritage documentation, analysis and management using building information modelling: State of the art and perspectives. *Mechatron. Cult. Herit. Civ. Eng.* **2018**, *92*, 181–202.
33. Sanseverino, A.; Messina, B.; Limongiello, M.; Guida, C.G. An HBIM Methodology for the Accurate and Georeferenced Reconstruction of Urban Contexts Surveyed by UAV: The Case of the Castle of Charles V. *Remote Sens.* **2022**, *14*, 3688. [CrossRef]
34. Dell'Amico, A. HBIM and digital documentation for development reliable parametric models on complex systems. In *AIP Conference Proceedings*; AIP Publishing: Long Island, NY, USA, 2021; Volume 2428, p. 1.
35. Dell'Amico, A. Memoria e modello digitale. La costruzione di un sistema informativo per la salvaguardia del patrimonio architettonico diffuso dell'Upper Kama. *Restauro Archeol.* **2022**, *30*, 32–53. [CrossRef]
36. Canevese, E.P.; De Gottardo, T. Beyond point clouds and virtual reality. Innovative methods and technologies for the protection and promotion of cultural heritage. *Int. Arch. Photogramm. Remote Sens. Spat. Inf. Sci.* **2017**, *42*, 685–691. [CrossRef]
37. Statham, N. Scientific rigour of online platforms for 3D visualization of heritage. *Virtual Archaeol. Rev.* **2019**, *10*, 1–16. [CrossRef]
38. Rahaman, H.; Champion, E.; Bekele, M. From photo to 3D to mixed reality: A complete workflow for cultural heritage visualisation and experience. *Digit. Appl. Archaeol. Cult. Herit.* **2019**, *13*, e00102. [CrossRef]
39. Colucci, E.; De Ruvo, V.; Lingua, A.; Matrone, F.; Rizzo, G. HBIM-GIS integration: From IFC to cityGML standard for damaged cultural heritage in a multiscale 3D GIS. *Appl. Sci.* **2020**, *10*, 1356. [CrossRef]
40. Myers, B.; Hudson, S.E.; Pausch, R. Past, present, and future of user interface software tools. *ACM Trans. Comput.-Hum. Interact. (TOCHI)* **2000**, *7*, 3–28. [CrossRef]
41. Tzima, S.; Styliaras, G.; Bassounas, A. Augmented reality applications in education: Teachers point of view. *Educ. Sci.* **2019**, *9*, 99. [CrossRef]
42. Skarlatos, D.; Agrafiotis, P.; Balogh, T.; Bruno, F.; Castro, F.; Petriaggi, B.D.; Rizvic, S. Project iMARECULTURE: Advanced VR, iMmersive serious games and augmented REality as tools to raise awareness and access to European underwater CULTURal heritagE. In Proceedings of the Digital Heritage, Progress in Cultural Heritage: Documentation, Preservation, and Protection: 6th International Conference, EuroMed 2016, Nicosia, Cyprus, 31 October–5 November 2016; Springer International Publishing: Cham, Switzerland, 2016; pp. 805–813.
43. De Marco, R.; Pettineo, A. The recognition of Heritage qualities from feature-based digital procedures in the analysis of historical urban contexts. *Int. Arch. Photogramm. Remote Sens. Spatial Inf. Sci.* **2022**, *XLVI-2/W1-2022*, 175–182. [CrossRef]
44. Rechichi, F.; Mandelli, A.; Achille, C.; Fassi, F. Sharing high-resolution models and information on web: The web module of BIM3DSG system. *Int. Arch. Photogramm. Remote Sens. Spat. Inf. Sci.* **2016**, *41*, 703–710. [CrossRef]
45. Saccucci, M.; Pelliccio, A. Integrated BIM-GIS System for the enhancement of urban heritage. In Proceedings of the 2018 Metrology for Archaeology and Cultural Heritage (MetroArchaeo), Cassino, Italy, 22–24 October 2018; pp. 222–226.
46. Lombardo, V.; Damiano, R. Storytelling on mobile devices for cultural heritage. *New Rev. Hypermedia Multimed.* **2012**, *18*, 11–35. [CrossRef]

drones

MDPI

Article

Scan-to-HBIM Reliability

Stefano Brusaporci *, Pamela Maiezza, Adriana Marra, Alessandra Tata and Luca Vespasiano

Department of Civil, Construction-Architectural and Environmental Engineering, University of L'Aquila, 67100 L'Aquila, Italy; pamela.maiezza@univaq.it (P.M.); adriana.marra@univaq.it (A.M.); alessandra.tata@univaq.it (A.T.); luca.vespasiano@graduate.univaq.it (L.V.)

* Correspondence: stefano.brusaporci@univaq.it

Abstract: The modeling of the historical architecture can be characterized by different levels of geometric development, more or less advanced, which correspond to different values of deviation between the real object and its three-dimensional representation. The aim of the paper is a critical study of the different levels of geometry (LoG) that the architectural elements of a model can have in relation to the efforts required to achieve them and the modeling objectives. More specifically, the contribution of this study proposes—on the basis of a survey campaign of a case study conducted by drone—the evaluation of the deviation between the point cloud and HBIM models with the different levels of development in order to assess their reliability according to specific objectives, such as the documentation of the architectural asset, feasibility study, restoration project, etc.

Keywords: HBIM modeling; level of geometry (LoG); accuracy; deviation; reliability

Citation: Brusaporci, S.; Maiezza, P.; Marra, A.; Tata, A.; Vespasiano, L. Scan-to-HBIM Reliability. *Drones* **2023**, *7*, 426. https://doi.org/10.3390/drones7070426

Academic Editors: Daniela Oreni, Fabrizio Banfi and Davide Mezzino

Received: 25 May 2023
Revised: 20 June 2023
Accepted: 23 June 2023
Published: 27 June 2023

1. Introduction

The well-known advantages of building information modeling (BIM) in the architecture, engineering and construction (AEC) sectors have made this process a reality in the existing as well as the architectural heritage.

However, HBIM (historic building information modeling) poses the problem of how to represent the complex forms of historic architecture within a standardized parametric environment. In addition to the lack of BIM libraries for the architectural components of historic buildings, the question arises as to the means with which to reconcile the irregularity and uniqueness of the architectural heritage with a modeling logic based on the typification of the elements [1]. In particular, it is necessary to choose whether to return the modeling to parametric and reusable objects at the expense of model accuracy, or to prefer a greater adherence to the survey data by modeling the architectural component with the specific features that distinguish it, thus setting aside, or reducing, the benefits of standardization.

The modeling of the historical building can then be characterized by different levels of geometric development, more or less advanced, which correspond to different values of deviation between the real object and its three-dimensional representation. The acceptability of these deviations, i.e., the reliability of the model, will have to be assessed on the basis of the purposes of the representation, as well as in consideration of the characteristics of the study objective [2,3].

The paper presents a critical study of different levels of geometry (LoG) of the model's parametric architectural element, highlighting the differences between the real geometry and the modeling, and taking into account the modeling efforts and the modeling goals. More specifically, the contribution proposes the evaluation of the deviation between point clouds by drone photogrammetry—a solution often necessary to survey architectural elements of building façades whose height above the ground does not allow for direct surveys or the use of laser scanning—and HBIM models with different levels of development, in order to assess their reliability according to specific objectives, such as the documentation of the architectural heritage, feasibility study, restoration project, etc.

The experimentation to support the study was conducted on a window on the second floor of Palazzo Camponeschi (16th century), the current headquarters of the Rector's Office of the University of L'Aquila (Figure 1).

Figure 1. Window of Palazzo Camponeschi, L'Aquila (16th century): textured mesh model.

2. State of the Art

The three-dimensional modeling of an architectural heritage—especially the HBIM—is onerous in terms of time, effort, and thus also of cost due to its often complex and irregular shapes. Several lines of research have sought an optimization of the scan-to-BIM process, from the point of view of both the quality of three-dimensional representation and the reduction in time consumed.

With regard to the first line of research, we recall the studies focused on the attempt to make up for the shortcomings of BIM in the modeling of irregular shapes through the integration with non-uniform rational basis spline (NURBS) surfaces [4]. In 2017, Banfi [5] proposed protocols called the GOG (grade of generation) to develop and support the generation of HBIM models, ranging from simplified types (GOGs 1–8), which use simplified functionalities integrated into the BIM applications (i.e., based on extrusion, subtraction, and other modeling functionalities) to the NURBS-based types (GOGs 9, and other mod10). In particular, the multislice-based wireframe model (GOG9) and the cloud-based model (GOG10) aim to maintain the morphologic richness and precision acquired by the survey for all those architectural elements whose ideal modeling would lead to an excessively high deviation between the model and the real object. Brumana et al., in 2018, proposed protocols based on GOG to define new LOGs for complex architectural element [6].

Banfi [7], again in the context of the grades of generation (GOG), focuses on modeling various architectural vaults with the goal of demonstrating how the concept of te parameterization of a NURBS surface can be associated with a BIM object and BIM database.

The second line of research, aimed at improving productivity and containing costs for the three-dimensional representation of historical architecture, includes experiments relating to parametric modeling, also based on the visual programming language (VPL) [8]. Such research focuses on identifying mathematical rules and parameters to create flexible and adaptable models from which the desired model can be obtained as a variation of the generic kind [9]. In particular, Lo Turco et al. [10] apply the semantic decomposition method to model the decorative elements of architectural heritage windows, opening up the possibility of encoding geometric parameters in the VPL environment, which are useful for defining freeform or complex geometries within the BIM environment. On the other hand, Lanzara and Capone [11] define through VPL an algorithmic generative tool based on the definition of geometric rules and construction techniques for modeling the complex decorative apparatus of domes.

The irregularity and complexity of the forms of historical architecture, from which these studies are based, pose the question of assessing the accuracy of 3D representation with respect to reality [12]. Moreover, the issue of the geometric reliability of the model becomes even more essential for historic/heritage building information modeling (HBIM), because of the specific nature of BIM based on the use of standardized and typed 3D objects, which are difficult to reconcile with the uniqueness of historical buildings [13].

The accuracy of the 3D model is measured in terms of deviation, that is, the distance between the model and the point cloud, a three-dimensional cast of the real object [3]. Bonduel et al. [14] propose a standardized method for assessing scan-to-BIM accuracy based on a quality evaluation on a macro-scale followed by an analysis of the individual BIM elements (micro-scale). On the other hand, Wang et al. [15] apply the issue of reliability assessment to the BIM model of a subway station.

Finally, in defining the level of accuracy (LOA), it is crucial to cite the interesting distinction made by the U.S. Institute of Building Documentation [16] between measured accuracy—"the range of standard deviation that must be obtained from the final measurements taken regardless of the method used to acquire those measurements"—and represented accuracy—"the range of standard deviation that must be obtained once the measured data are processed in some other form, such as a linework or model" (p. 12). Measured accuracy can also be unspecified and left to the building documentation professional who will determine the measured accuracy LOA needed to achieve the specified represented accuracy.

With regard to the level of accuracy (LoA), Maiezza [17] and Maiezza and Tata [18] propose three possible levels of geometric accuracy level (LoA). These are defined as a function of the deviation measured on most of the elements considered, i.e., the standard deviation resulting from the comparison between the point of cloud and the surface of the model or the one resulting from the generatrix and directrix of the same surface and the slices extracted from the point of cloud.

As reported by these studies, in the presence of models with a low level of model development, the level of accuracy (LoA) can be: low, when the deviation between the point of cloud and model is greater than 70 mm; medium, when the value ranges between 50 mm and 70 mm; high, when the value of the standard deviation is less than 50 mm.

In the case of models with a medium level of model development, the LoA is evaluated by comparing the deviation values between the surface's generatrix and directrix with respect to the slices extracted from the point of cloud. In this case, the level of accuracy is low for values above 50 mm, medium for values between 20 mm and 50 mm, and high for values below 20 mm.

For models with a high level of model development, the LoA is estimated in relation to the deviation values of both the generatrix and directrix, both to the entire surface. When the deviation related to the generatrix and directrix is greater than 20 mm, the deviation measurement of the point of cloud and the surface is omitted, and the LoA is low. Otherwise, the LoA is low when the measured deviation between the surface and point of the cloud is greater than 50 mm, medium when the deviation is between 20 mm and 50 mm, and high when this deviation is less than 20 mm.

As highlighted by these studies [17,18], the acceptability of these deviations, and therefore the reliability of the model, has to be assessed on the basis of the purposes of the representation, as well as in consideration of the characteristics of the object of study [2].

3. Methodology

As previously mentioned, in recent years, there has been a widespread use of BIM methodologies for new and existing constructions, confirmed by the spread of international and national standards and the obligatory use of BIM in public procurement (in addition to the Italian standards D.M. n.560 of 2017 and D.M. n.312 of 2021, we note the Road Map document toward the mandatory BIM of the German Bundestag Federal Ministry of Transport and Digital Infrastructure). As a consequence, several lines of research aimed at optimizing scan-to-BIM processes and assessing the accuracy of the 3D model obtained from the critical reprocessing of acquired data have been developed.

The research presented here is part of this line of investigation and, drawing on the authors' previous experience [1,2,17,18], defines a workflow for assessing the accuracy of HBIM models in relation to the specific project purposes for which they were developed.

The survey of the artifact represents the first phase of the entire process as, through the use and integration of different procedures (laser scanning or terrestrial and/or aerial photogrammetry), the geometric and dimensional data necessary for the successive modeling and reliability assessment of the model are acquired. The reliability of the integrated survey procedures is recognized by numerous experiments [19–21]. However, in this specific case, it was considered appropriate to proceed with drone photogrammetry as this solution is able to survey architectural elements of building façades located at a considerable height, obtaining results, such as point clouds, with high levels of precision.

The crucial moment of the workflow is that of HBIM modeling. This is based on the critical analysis of the data deriving from the survey phase but is also influenced by the goals and objectives to be pursued. From the earliest design phases, it is important to define the level of detail and geometric development that the models must respect according to the different purposes and uses.

Finally, to assess the accuracy of the model (LoA) in relation to the modeling purposes, it is necessary to compare the HBIM model with the data acquired from the survey carried out using UAV digital photogrammetry. The comparison was carried out by means of a cloud-to-mesh (C2M) analysis, available within the open-source software Cloud Compare, which permits the assessment of the formal similarity degree between the two models by highlighting, with a false-color map, the less corresponding areas. The deviation in the values of the standard deviations and the mean distance between the mesh model and the point cloud is due to the different LoGs used in the modeling and can therefore be attributed to the approximation of the geometric detail of the model with a low LoG compared to the others.

Similarly to the level of geometric development (LoG), the accuracy (low, medium, or high LoA) can also be defined a priori according to the use of the model. Therefore, if at the end of the process, the geometric accuracy assessment (LoA) does not satisfy the deviation values identified in the initial phases, i.e., the LoA is inadequate for the intended purpose, it is necessary to adjust the model according to an iterative process that leads to meeting the minimum requirements for the different levels of development.

4. Results

This section focuses on some features concerning the workflow illustrated in the previous one through the application of the procedure to one of the windows of Palazzo Camponeschi, located in the historical center of the city of L'Aquila. The experimentation is part of the University projects started after reconstruction and restoration activities and aimed at digitizing the building heritage. The choice to analyze the window on the second floor of Palazzo Camponeschi is also associated with its type and characteristics. Indeed being a Baroque window, it presents unique peculiarities that require considerable effort in the modeling phase. However, it is the complexities of the architectural element that allow the proposed procedure to be stressed and validated.

4.1. Drone Survey

The studied window is on the second floor, at a height of about 10 m from the ground The street on which it faces is quite busy and has a width that does not exceed 8 m Considering these circumstances, the height at which the architectural element is located, and the narrowness of the road, it was considered that the most effective technique to employ in the survey was that of the SfM (structure from motion), upon realizing the pictures by a drone [21,22]. The use of the drone permits to obtain horizontal photos along with a reduction in the areas of shadows that would otherwise be present with a TLS (terrestrial laser scanner), but also with photos taken from the ground, which would have also presented a very disadvantageous angle.

The used drone (DJI Mini 2 drone with sensor: 1/2,3″ CMOS; lens: FOV: 83, format 35 mm equivalent: 24 mm, aperture: f/2.8), because of its small size and its handling, is well suited to the urban context characterized by relatively narrow spaces. In addition, the very low weight allowed it to be operateed without having to stop vehicular traffic. The greatest difficulties were found in the constant presence of several aerial cables along the elevation, and the instability of the wind at varying heights, because the different heights of the surrounding buildings caused irregular and relatively changeable local currents.

In the photographic campaign, 42 shots were taken at a distance between 1.5 and 6 m from the subject. The photographs were generally taken with a horizontal visual axis, but also with a visual axis facing downward, in an attempt to minimize shadows. The obtained dataset was processed with the AgiSoft Metashape 2.0.2 software, according to the workflow provided by the program (Figure 2). From the first step of the photo alignment, a sparse cloud of about 40k pt was obtained, which, after some filtering operations, was reduced to 20k pt. On the basis of the latter, the dense cloud was elaborated on a more circumscribed area of the model, consisting of 740k pt. A new filtering operation reduced the cloud to 670k pt. The mesh, characterized by 175k faces, was then processed on this basis.

Mesh processing was not performed at the highest level of detail, but at the level defined as 'medium' in the software. This is because the result obtained at the highest level of definition was excessively affected by the noise present in the dense cloud, partly due to the non-optimal lighting conditions of the portion of the façade where the window is located, and partly due to the presence of reflective surfaces [23]. As for the lighting, it was realized during the processing that the presence of the gutter directly above the window greatly reduced the brightness of the photos in the highest part of the window frame. On the other hand, the lower portion of the window and especially the adjacent portions of the wall were strongly illuminated by direct sunlight. Direct sunlight also gave rise to reflection phenomena, both on the windows and especially on the glass parapet. The juxtaposition of the glass parapet and the window glass panes generated further reflections that created problems for the software algorithm, partially compromising the results obtained not only on the reflective surfaces, but also in the surrounding areas, like the window's inner frame.

Upon analyzing retrospectively the photos that make up the dataset, we realized that in the time taken to make the photographs, the light conditions had undergone some slight variations that, although limited, were still able to produce repercussions on the results that were obtained with this technique. One of the most difficult aspects to manage is the variability of lighting, because often when the variations in brightness are progressive and not rapid, the operator, who is focused on drone piloting, does not realize the change in lighting, thus sometimes compromising the outcome. The other problem encountered was in relation to the presence of reflecting surfaces, which is also common to other survey techniques. However, the results offered by the SfM technique, in this case, do not seem to be particularly poor or compromising with respect to the final result.

The model obtained was finally exported in '.e57' format to be used in the subsequent steps of the study.

Figure 2. Processing of photogrammetric drone survey: sparse cloud, dense cloud, mesh model, textured mesh model.

4.2. The Modeling of the Historical Building: Critical Analysis and Identification of Levels of Development

The HBIM procedure is based on the digital translation of the historical building into its virtual reproduction through the integration of different survey techniques and the use of the resulting data (i.e., point clouds) as a starting point for the creation of the digital replica. The modeling process, in this case, constitutes the critical act of restitution, based on the understanding of what has been surveyed, the semantic decomposition of the point cloud, and finally, the realization of the HBIM model. Although the phase of acquisition of the current status of the asset, especially from a geometric point of view—thanks to the use of instruments and technologies capable of generating a real clone of reality in a digital environment—is characterized by high speed and accuracy, there is a gap between the acquired data and the three-dimensional HBIM models' restitution.

This gap is above all amplified, usually, by parametric modeling, which leads to a more difficult modeling process and to resulting in models that deviate more from reality, especially if compared to other kinds of models (mesh, NURBS, etc.).

As established first for LoDs by the UNI 11337-4:2017 standard, and then for LOINs by the UNI EN ISO 19650:2019 and UNI EN 17412-1:2021 standards, Italian Levels of Development or Level of Information Need of the three-dimensional objects are mainly defined on the basis of the purposes and objectives to be met with the models (Appendix A).

Both levels depend on the project or process phase and a measurement of the deepening and the stability, or quality, of the information content linked to each of the three-dimensional elements.

In the case of historic buildings, in particular, the choice of the level of development to be achieved must include considerations relating to the type of element, its characteristics, its possible present or absent uniqueness, the availability or unavailability of the information necessary for the knowledge of the asset (e.g., information relating to the construction equipment), and their reliability.

Finally, as also highlighted in the LOINs by the UNI EN 17412-1:2021 standard, the whole modeling process also depends on the degree of parameterization, and consequently of the reproducibility that is intended to confer to the modeled elements.

All these aspects contribute to define which is the best modeling method and procedure to follow, and concur, together with the level of geometric development, to determine the effort, both in terms of difficulty and timing, which is necessary for the realization of the parametric model of the architectural component.

Therefore, The HBIM parametric modeling is influenced by numerous factors, more or less objective, and derives from a critical phase of analysis carried out by the modeler—and, in general, by the client, which, in the case of a public tender, is the one to define the characteristics of the model elements—who determines the choices on the basis of their previous knowledge and experience.

Based on the circumstances, the modeling of the same element can therefore be different and the three-dimensional model could be considered acceptable with a more or less high level of development, and consequently a more or less high deviation.

Although the UNI EN ISO 19650 standard has already been adopted in Italy, and thus LoINs are already applicable at a regulatory level, it was decided to base the present study on LoDs, which are still in force today, despite the continuing absence of a reference scale required from the new Public Contracts Code to facilitate its applicability in common practice and, specifically, in public procurement. This choice finds further justification in the flexibility of the proposed procedure and in its applicability to other scales of detail or development degrees, including LOINs.

In fact, precisely because the choice of a level of development of a three-dimensional object derives from multiple subjective and objective factors, three different levels of development of the object model, deriving from grouping the LoDs provided by the UNI standard in three main levels, were identified:

- Low LoG, corresponding to a simplified model from the point of view of geometric information content (Italian LoD A and LoD B);
- Medium LoG, corresponding to a model defined from the point of view of geometric complexity (LoD C, LoD D);
- High LoG, corresponding to a detailed and realistic model (LoD E, F and G).

Therefore, assuming, for example, the creation of a model for one of the first design stages recognized at a regulatory level, such as the Technical and Economic Feasibility Project, a simplified model with a low LoG, corresponding to a LoG A or B, and consequently a higher deviation, might be acceptable. At the same time, depending on the purpose, aims, and design stage, it may not be worthwhile to waste excessive time on the preparatory phase for modeling (i.e., the analysis, recognition, and breakdown of the components that make up the architectural element) or in the modeling phase.

On the contrary, assuming the creation of the HBIM model for a conservation or restoration project, a model with a high LoG (LoG E, F, or G) and a minimum deviation from the real object will be required. The latter is fundamental especially in the hypothesis of using the model to evaluate out of plumb or swellings. Therefore, in this case, it will be appropriate to devote time both to the first phase of analysis—identifying, for example, all the components that make up an architectural element such as a window—and all the moldings which constitute them (Figure 3). This effort, as previously mentioned, could be accentuated in the case of a non-unique element, which could therefore be modeled with a high degree of parameterization according to the BIM logic (e.g., parameterization on the various axes and creation of nested families) and make it possible to take advantage of the benefits deriving from the possibilities of its future reuse in spite of the major effort required for the realization of the 3D parametric element.

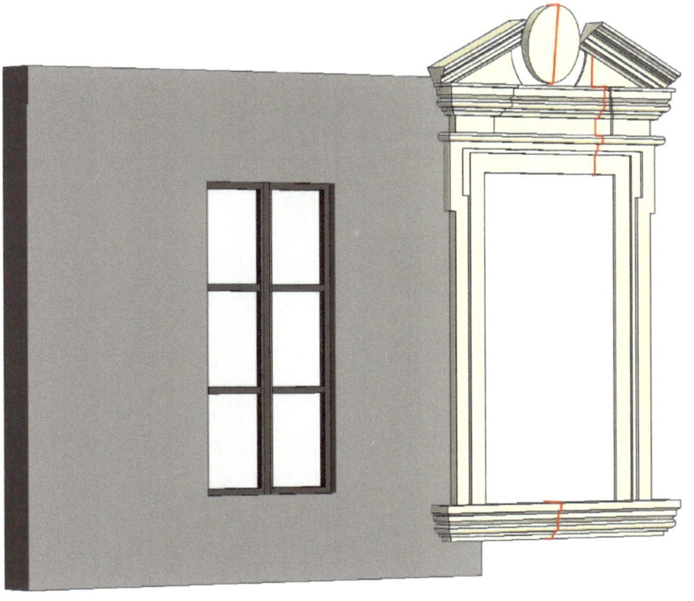

Figure 3. Three-dimensional modeling of the window of Palazzo Camponeschi with the multiple forms of directrix of the highlighted geometric elements.

In the case of the studied window for the modeling with a low LoG, it was decided to considerably simplify the decorative apparatus present on the external façade. All the elements that compose it have therefore been modeled as simple parallelepipeds extruded along different directions. On the contrary, in the case of the high LoG, the profiles of the moldings have been identified and the latter have been modeled in detail. Finally, in the case of the medium LoG, an intermediate approach between the two was adopted, that is, the elements have all been modeled but with simplified moldings compared to the high LoG.

4.3. Evaluation of the Model Accuracy

The level of geometry reliability of the HBIM model (LoA) is closely related to the objectives of the project for which it was created. As previously highlighted, the LoA of the three-dimensional representation with respect to the real object (i.e., the point cloud derived from the survey performed using different acquisition techniques) is evaluated according to the standard deviation obtained by comparing the two entities [24]. If the distance between the three-dimensional model and the point cloud is greater, the model is closer to ideal restitution; on the other hand, if the distance between the two is smaller, then the model is closer to the real object: the as-built model [25].

In the case of models developed for the first design phases recognized at the regulatory level, since a simplified geometric development level is required, it is acceptable that a low-to-medium level of accuracy is achieved [17,18]. In the passage to a detailed geometric development level, the LoA can take on values ranging from low to high. For example, in the case of a model realized in view of a restoration or maintenance project, it is desirable to reach a high level of geometric accuracy [17,18] as the model must be as close as possible to the construction. The case differs when the model is used for communication or enhancement, for which, even if the LoG is high, the LoA may be low or medium in view of the interrelation with other software or digital applications.

Therefore, to assess the reliability of the model's level of geometric detail according to the representation purposes as defined in the previous section, the geometric accuracy (LoA) of the HBIM model developed for the Palazzo Camponeschi window was calculated through a C2M analysis.

In particular, the test showed that:

- For the low LoG, the analysis returns a standard deviation value of more or less 30 mm (Figure 4). The level of accuracy (LoA), evaluated according to the deviation between the whole surface and the model of the architectural element, fits a high level since the deviation between the points and the surface is <50 mm [17,18].

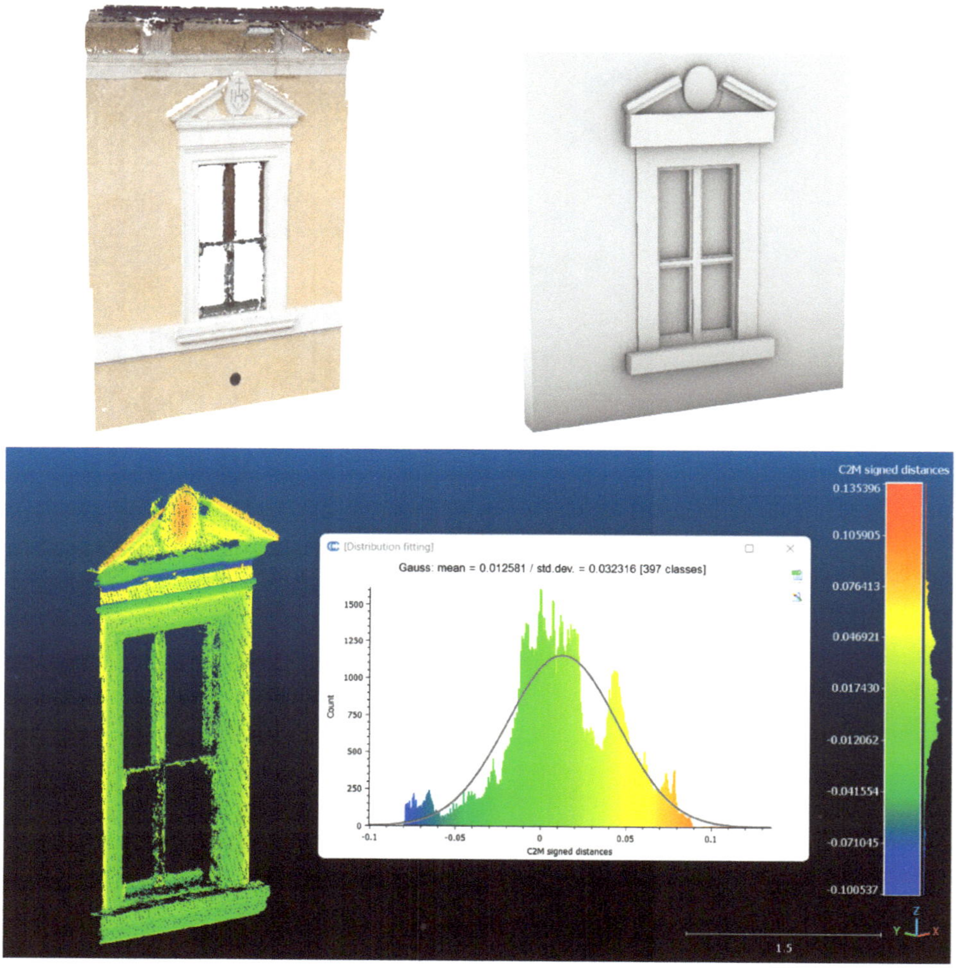

Figure 4. Geometric accuracy of the HBIM model of the Palazzo Camponeschi's upper–window: deviation between the point cloud and the model with low LoG.

- For the medium LoG, the comparison between the generatrix and directrix of the surface and the slices extracted from the point of cloud returns a standard deviation value of about 17 mm (Figure 5); therefore, the LoA achieved corresponds to a high level [17,18].

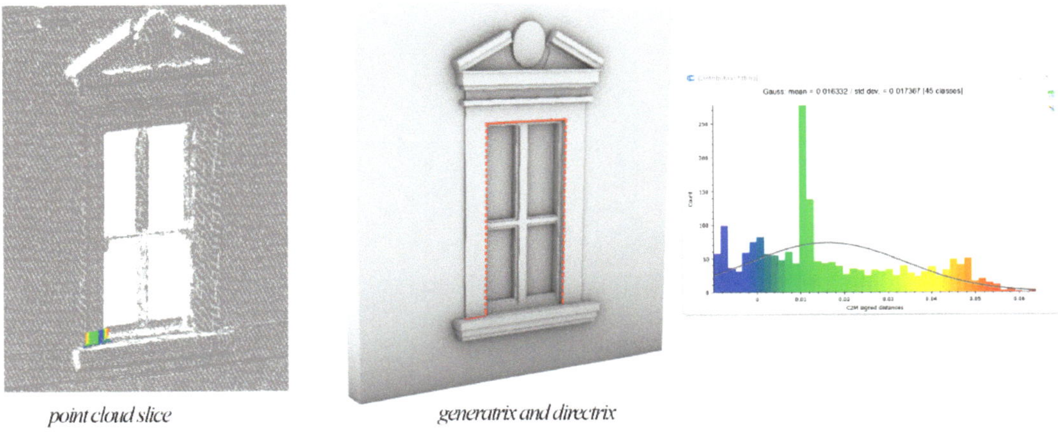

Figure 5. Geometric accuracy of the HBIM model of the Palazzo Camponeschi's upper–window: deviation between the point cloud and the model with medium LoG, evaluated with respect to the generatrix and directrix (highlighted in red in the figure).

- For the high LoG, the standard deviation value was measured of both the generatrix and directrix of the surface with respect to the slices extracted from the point of cloud (SD = 10 mm), and then the value obtained between the entire surface and the model of architectural element (st. dev < 15 mm) (Figure 6). The LoA corresponds to the high level since the deviation between the points with respect to both the generatrix and directrix and the surface is <20 mm [17,18].

For all three models, rather small deviation values were obtained (high LoA for all three LoGs).

In the case of the low LoG, the test showed that the achievement of deviation values were comparable to those considered acceptable for a high LoG and a deviation evaluated on the surface corresponding to the average LoA. Such accuracy, depending on the specific purposes and characteristics of the architectural object, may not be required. For example, in the case of a model for one of the first design stages (e.g., Technical and Economic Feasibility Project) for which a low LoG has been decided, even higher deviation values would have been acceptable, corresponding to a medium or low LoA. At the same time, the LoA achieved for the high LoG could be deemed suitable for the modeling needs required for the conservation and restoration project.

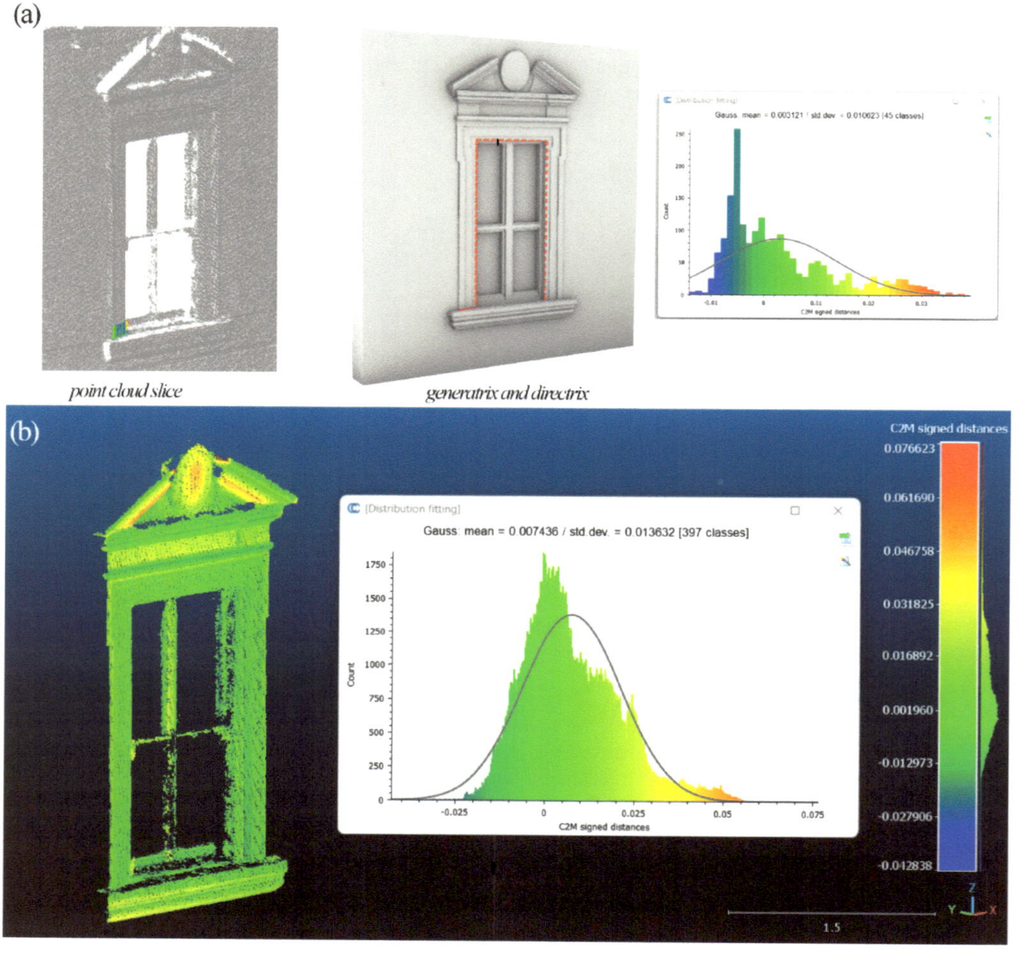

Figure 6. Geometric accuracy of the HBIM model of the Palazzo Camponeschi's upper—-window: deviation between the point cloud and the model with high LoG, evaluated with respect to the generatrix and directrix (**a**) and with respect to the entire surface (**b**).

5. Conclusions

Nowadays, building information modeling is an essential reality in the AEC (architecture engineering construction) industry, sanctioned even at the regulatory level. The advantages and potential of semantic modeling based on objects enriched by heterogeneous information have by now been incorporated by both international and national standards, which are increasingly moving toward the introduction of mandatory BIM, at least for public tenders.

The increasingly imminent mandatory use of BIM processes for the performance of public contracts makes the definition of a reference standard necessary and undelayable.

If the standardization of the BIM process for new construction does not present particularly critical issues, the same does not apply to the existing one, especially in the case of buildings belonging to architectural heritage. As is well known, in this field, the lack of information availability is accompanied by difficulties in modeling the complex forms of historical architecture. The issue of the reliability of the HBIM model concerning the architectural asset then becomes crucial.

With regard to geometric modeling, in particular, standardizing the HBIM process means defining, on the one hand, the geometric detail that the model must possess in order to achieve the set goals and, on the other hand, the acceptable reliability of the three-dimensional representation with respect to the real object, evaluated in terms of deviation. Specifically, according to the objectives of the modeling (e.g., feasibility study, restoration project, etc.) and the characteristics of the architectural object (more or less the complexity of the forms, the possibility of parameterization or the uniqueness of the elements, etc.), the level of geometric development (LoG) of the components that will constitute the model must be established. In parallel, again based on these factors, the level of accuracy (LoA) to be achieved, that is, the maximum acceptable deviation, must also be defined. At the end of the modeling, the geometric reliability should be evaluated through a comparison between the model and the point cloud and, if it is not satisfactory, adjusting the model according to an iterative process thus becomes necessary.

An interesting development in this line of research concerns the systematic extension of such reasoning, in the contribution applied only to two examples (feasibility study and restoration project), to the wide range of possible purposes of an HBIM model, simultaneously also taking into consideration what may be the different characteristics of the buildings belonging to the architectural heritage.

The objective of such a study would be to correlate to these aspects (architectural purposes and architectural features) the level of geometric development to be achieved with the related level of reliability so as to define an operational standard regarding HBIM modeling, which would be useful especially in the case of public tenders.

Author Contributions: Introduction, S.B.; state of the art, P.M.; drone survey, L.V.; the modeling of the historical building: critical analysis and identification of levels of development, A.T.; methodology and evaluation of the model accuracy, A.M.; conclusions, S.B. All authors have read and agreed to the published version of the manuscript.

Funding: This research received no external funding.

Data Availability Statement: Not applicable.

Acknowledgments: The present study has been carried out in the framework of the Project "SICURA—CASA INTELLIGENTE DELLE TECNOLOGIE PER LA SICUREZZA CUP C19C20000520004—Piano di investimenti per la diffusione della banda ultra larga FSC 2014-2020", whose financial support is acknowledged.

Conflicts of Interest: The authors declare no conflict of interest.

Appendix A

The levels of development (LoD) have been defined in Italy by the UNI 11337:2017, part four. This standard defines the LoD as a measure of the "nature, quantity, quality and stability of the data and information" associated with each three-dimensional element that makes up the model. The transition from one LoD to the next involves a higher degree of development and a more detailed object, based on more reliable information, both from a geometric (LOG—geometric development level) and an informative point of view (LoI—development level of the non-graphical attributes). Therefore, with an increase in the LOD, the UNI standard provides for an increase both in the quantity of attributes held by the BIM object and in their quality and stability, understood in the sense of granularity, reliability, and data consolidation.

These levels of development are now being updated and will be permanently replaced in Italy by the Level of Information Requirements (LOINs), or Level of Information Need, as defined by the English original standard, with the aim of ensuring a greater flexibility and effectiveness in BIM processes.

LOINs are one of the most significant introductions made by one of the latest BIM regulations, which is the international UNI EN ISO 19650:2019 standard, of which UNI 11337 stands out as complementary, which follows the transposition by the CEN of the standards EN ISO 19650-1, EN 19650-2, EN ISO 19650-3, and EN ISO:19650-5.

LOINs must be respected by each element of the model, and the standard underlines that the required level of information is strictly dependent on the type of use and needs and should be determined by the minimum amount of information necessary according to its purpose, with the aim of "meeting each relevant requirement, including the information requested by the other persons in charge and no more". The purpose of defining this level of information requirement is therefore to avoid an overabundance of unnecessary information, which negatively affects the development of the project and the fluidity of information exchange flows.

This makes the definition of the contents to be included in the models flexible and adaptable, in particular for historical building; therefore, rather than attempting to reach and satisfy specific LoDs classified and defined for new buildings, it is possible to establish a priori, based on the purposes of the model, the kinds of information that must be inserted within the elements and exclude those that are not necessary. Finally, of those that are necessary, in the case of historic buildings, it is necessary to distinguish which ones can be investigated and are thus available in case of need (for example, information regarding the construction equipment of the architectural components), and which ones, instead, are simply available or absent (for example, those associated with history deriving from archival and bibliographic research) and therefore cannot be searched for and inserted into the model.

References

1. Brusaporci, S.; Maiezza, P.; Tata, A. A Framework for Architectural Heritage HBIM Semantization and Development. *Int. Arch. Photogramm. Remote Sens. Spat. Inf. Sci.* **2018**, *XLII–2*, 179–184. [CrossRef]
2. Brusaporci, S.; Maiezza, P.; Tata, A. Trasparenza e affidabilità dei modelli HBIM. In *BIM Views: Esperienze e Scenari*; Papa, M.L., D'Agostino, P., Eds.; CUA: Fisciano, Italy, 2019.
3. Quattrini, R.; Clini, P.; Nespeca, R.; Ruggeri, L. Measurement and Historical Information Building: Challenges and opportunities in the representation of semantically structured 3D content. *Disegnarecon* **2016**, *9*, 14.1–14.11.
4. Oreni, D.; Brumana, R.; Della Torre, S.; Banfi, F.; Barazzetti, L.; Previtali, M. Survey turned into HBIM: The restoration and the work involved concerning the Basilica di Collemaggio after the earthquake (L'Aquila). *ISPRS Ann. Photogramm. Remote Sens. Spat. Inf. Sci.* **2014**, *2*, 267–273. [CrossRef]
5. Banfi, F. BIM orientation: Grades of generation and information for different type of analysis and management process. *Int. Arch. Photogramm. Remote Sens. Spat. Inf. Sci.* **2017**, *42*, 57–64. [CrossRef]
6. Brumana, R.; Della Torre, S.; Previtali, M.; Barazzetti, L.; Cantini, L.; Oreni, D.; Banfi, F. Generative HBIM modelling to embody complexity (LOD, LOG, LOA, LOI): Surveying, preservation, site intervention—The Basilica di Collemaggio (L'Aquila). *Appl. Geomat.* **2018**, *10*, 545–567. [CrossRef]
7. Banfi, F. HBIM Generation: Extending Geometric Primitives and BIM Modelling Tools for Heritage Structures and Complex Vaulted Systems. *Int. Arch. Photogramm. Remote Sens. Spat. Inf. Sci.* **2019**, *42*, 139–148. [CrossRef]
8. Murphy, M.; Meegan, E.; Keenaghan, G.; Chenaux, A.; Corns, A.; Fai, S.; Chow, L.; Zheng, Y.; Dore, C.; Scandurra, S.; et al. Shape grammar libraries of European classical architectural elements for historic BIM. *Int. Arch. Photogramm. Remote Sens. Spat. Inf. Sci.* **2021**, *46*, 479–486. [CrossRef]
9. De Luca, L.; Véron, P.; Florenzano, M. A generic formalism for the semantic modeling and representation of architectural elements. *Vis. Comput.* **2007**, *23*, 181–205. [CrossRef]
10. Lo Turco, M.; Tomalini, A.; Bono, J. Proportions, Constraints and Semantics for a Parametric Model. *Nexus Netw. J.* **2023**, *25* (Suppl. 1), 333–340. [CrossRef]
11. Capone, M.; Lanzara, E. Parametric tools for Majolica Domes Modelling. *Nexus Netw. J.* **2022**, *24*, 673–699. [CrossRef]
12. Bianchini, C.; Nicastro, S. La definizione del Level of Reliability: Un contributo alla trasparenza dei processi di Historic-BIM. *Dn Build. Inf. Model. Data Semant.* **2018**, *2*, 46–59.
13. Bianchini, C.; Inglese, C.; Ippolito, A.; Maiorino, D.; Senatore, L. Building Information Modeling (BIM): Great Misunderstanding or Potential Opportunities for the Design Disciplines? In *Handbook of Research on Emerging Technologies for Digital Preservation and Information Modeling*; Ippolito, A., Cigola, M., Eds.; IGI Global: Hershey, PA, USA, 2017.
14. Bonduel, M.; Bassier, M.; Vergauwen, M.; Pauwels, P.; Klein, R. Scan-to-BIM output validation: Towards a standardized geometric quality assessment of building information models based on point clouds. *ISPRS Int. Arch. Photogramm. Remote Sens. Spat. Inf. Sci.* **2017**, *XLII-2/W8*, 45–52. [CrossRef]

15. Wang, Q.; Qian, P.; Liu, Y.; Li, T.; Yang, L.; Yang, F. Geometric Accuracy Evaluation Method for Subway Stations Based on 3D Laser Scanning. *Appl. Sci.* **2022**, *12*, 9535. [CrossRef]
16. U.S. Institute of Building Documentation. USIBD Level of Accuracy (LOA) Specification Guide. 2019. Available online: https://lidarmag.com/2019/08/20/usibds-version-3-of-the-level-of-accuracy-loa-specification-version-3-0-publishes-today/ (accessed on 2 May 2023).
17. Maiezza, P. As-Built reliability in architectural HBIM modeling. *Int. Arch. Photogramm. Remote Sens. Spat. Inf. Sci.* **2019**, *XLII-2/W9*, 461–466. [CrossRef]
18. Maiezza, P.; Tata, A. Standard for geometric and informative reliabilities in HBIM models. *Disegnarecon* **2021**, *14*, 15.1–15.10.
19. Allegra, V.; Di Paola, F.; Lo Brutto, M.; Vinci, C. Scan-to-BIM for the Management of Heritage Buildings: The Case Study of the Castle of Maredolce (Palermo, Italy). *Int. Arch. Photogramm. Remote Sens. Spatial Inf. Sci.* **2020**, *43*, 1355–1362. [CrossRef]
20. Barba, S.; Di Filippo, A.; Limongiello, M.; Messina, B. Integration of active sensors for geometric analysis of the Chapel of the Holy Shroud. *Int. Arch. Photogramm. Remote Sens. Spatial Inf. Sci.* **2019**, *XLII-2-W15-149-2019*, 149–156. [CrossRef]
21. Picchio, F.; Parrinello, S.; Barba, S. Drones and Drawings-methods of data acquisition, management, and representation. *Disegnarecon* **2022**, *15*. [CrossRef]
22. Ulvi, A. Documentation, Three-Dimensional (3D) Modelling and visualization of cultural heritage by using Unmanned Aerial Vehicle (UAV) photogrammetry and terrestrial laser scanners. *Int. J. Remote Sens.* **2021**, *42*, 1994–2021. [CrossRef]
23. Teppati Losè, L.; Sammartano, G.; Chiabrando, F.; Spanò, A. Challenging multi-sensor data models and use of 360 images. The Twelve Months Fountain of Valentino park in Turin. *IOP Conf. Ser. Mater. Sci. Eng.* **2020**, *949*, 012060. [CrossRef]
24. Apollonio, F.I.; Gaiani, M.; Sun, Z. Characterization of Uncertainty and Approximation in Digital Reconstruction of CH Artifacts. In *Heritage Architecture Landesign Focus on Conservation Regeneration Innovation Le vie dei Mercanti XI Forum Internazionale di Studi*; Gambardella, C., Ed.; La Scuola di Pitagora: Napoli, Italy, 2013.
25. Apollonio, F.I.; Gaiani, M.; Sun, Z. A reality integrated BIM for Architectural heritage conservation. In *Handbook of Research on Emerging Technologies for Architectural and Archaeological Heritage*; Ippolito, A., Cigola, M., Eds.; IGI Global: Hershey, PA, USA, 2017; pp. 31–65.

drones

Article

Analysis and Precision of Light Detection and Ranging Sensors Integrated in Mobile Phones as a Framework for Registration of Ground Control Points for Unmanned Aerial Vehicles in the Scanning Technique for Building Information Modelling in Archaeological Sites

Juan Moyano [1,*], Juan E. Nieto-Julián [1], María Fernández-Alconchel [1], Daniela Oreni [2] and Rafael Estévez-Pardal [3]

1 Department of Graphical Expression and Building Engineering, University of Seville, Ave. Reina Mercedes 4A, 41012 Seville, Spain; jenieto@us.es (J.E.N.-J.); mfalconchel@us.es (M.F.-A.)
2 Department of Architecture, Built Environment and Construction Engineering, Politecnico di Milano, Via Bonardi 9, 20133 Milano, Italy
3 Department of Graphical Engineering, University of Seville, Ave. Reina Mercedes 4A, 41012 Seville, Spain
* Correspondence: jmoyano@us.es

Abstract: The protection of heritage sites is one of the keys that our civilisation presents. That is why great efforts have been invested in order to protect and preserve movable and immovable property with a certain historical value, as is the case of archaeological sites scattered throughout the territory of southern Iberia (Spain) in the form of dolmens and negative structures dug into the ground, constituting a good sample of the megalithic culture in southern Spain. To study, manage and preserve these archaeological monuments, considered a set of cultural assets, various techniques and methodologies are required to facilitate the acquisition of three-dimensional geometric information. The Scan-to-BIM approach has become one of the most up-to-date work exponents to carry out these objectives. The appearance of LiDAR techniques, and recently their incorporation into smartphones through integrated sensors, is revolutionising the world of 3D scanning. However, the precision of these techniques is an issue that has yet to be addressed in the scientific community. That is why this research proposes a framework, through experimental measurement, comparison and knowledge of the limitations of this technology, to know the precision of the use of these smartphones, specifically the iPhone 13 Pro, as a measurement element to establish points of control with the aid of photogrammetry by unmanned aerial vehicles (UAVs) in archaeological sites. The results demonstrate a residual uncertainty of ±5 mm in the capture of GCPs from the mobile phone's LiDAR light detection and ranging sensor, and there was a deviation of the measurements in a range between 0 and 28 m of distance between the GCPs of (0.021, 0.069) m.

Keywords: hand-held mobile laser scanning; scan to BIM; aerial photogrammetry; archaeological sitesk

Citation: Moyano, J.; Nieto-Julián, J.E.; Fernández-Alconchel, M.; Oreni, D.; Estévez-Pardal, R. Analysis and Precision of Light Detection and Ranging Sensors Integrated in Mobile Phones as a Framework for Registration of Ground Control Points for Unmanned Aerial Vehicles in the Scanning Technique for Building Information Modelling in Archaeological Sites. *Drones* **2023**, *7*, 477. https://doi.org/10.3390/drones7070477

Academic Editor: Giordano Teza

Received: 9 June 2023
Revised: 5 July 2023
Accepted: 15 July 2023
Published: 20 July 2023

1. Introduction

The Industrial Revolution 4.0 is changing the way the industry and construction sectors are perceived, making the use of operational and information technology increasingly important. The paradigm of the fourth industrial revolution involves data management and the interconnection between machines, objects and processes [1]. In this context, digitisation covers both architectural and archaeological fields and is an integral process in the convergence towards industrial digitisation. In particular, the geometric characterisation of the complex forms present in archaeology and related fields implies a detailed and exhaustive description of the geometry of the objects that make up the configuration of the structures.

The idea of conservation of Cultural Heritage has to do with qualification systems where relationships of internal order (principles governing geometric shape) and relationships of shapes and volumes are described. Therefore, geometric characterisation implies the use of 3D digitisation techniques and instruments and interactive environments that bring us closer to a Historic Building Information Model (HBIM) [2] based on geomatic techniques that generate precise and detailed data on the shape, surface, texture and volume, among other geometric properties of the object.

In this sense, to acquire records in the complex framework of complex shapes that occur in archaeology, it is necessary to have certain technologies that have a high acquisition price, such as terrestrial laser scanning (TLS) technology.

However, portable devices such as mobile phones or tablets have emerged since 12 October 2020 that incorporate a sensor with a LiDAR (light detection and ranging) system in different versions of the iPhone, making them accessible to many people and professional fields. This Apple release in 2020 integrated a sensor designed for Augmented Reality (AR) without specifications on relative and absolute accuracies. Despite this, the versatility and applicability of the system challenged the geomatics and telemetry specialist community [3].

On the other hand, Structure from Motion (SfM) short-range photogrammetry has a highly developed space in the fields of architecture and archaeology [4–6] due to the enormous amount of data it provides. When this process is developed by unmanned aerial vehicles (UAVs), it acquires greater applications, especially in the registration of building roofs, where the surfaces of the zenithal planes are important since terrestrial laser scanning is not capable of covering them.

Fields such as the dynamic study of the earth's surface have been explored in various works, an example the one developed by Duró et al. [7], who tried to study the alterations in erosion on lake shores by means of UAVs, placing the ground control points (GCPs) in the alluvial plain. This type of study has been carried out in other work related to civil engineering [8] or safety at construction sites [9].

UAV-based photogrammetry typically uses ground control points to aid in measurements and establish precise scales in the model. If the points are made from a total station, georeferencing is indirect, referring to methods that assign world frame coordinates to 3D measurements collected in a relative local reference frame.

In the case of ground control points (GCPs) that have been taken with a global navigation satellite system (GNSS) receiver, the GCPs will be georeferenced in a universal navigation system. In the field of archaeology, the precision of unmanned aerial flights is an important factor, which involves flight plan indicators and georeferencing [10].

The advantages of unmanned aerial vehicles with respect to the terrestrial laser scanner are well known, due to the low cost of some flight equipment and the speed of data acquisition, although this aspect is relative if precision is currently considered for personal laser scanners [11] and the speed of their records and the taking of measurements for the GCPs.

Regarding the study of the precision of integrated mobile phones, one of the first studies was carried out in the field of geoscience [12] on small objects making comparisons with photogrammetry techniques. The choice of fourteen samples presented a singular morphology with sharp edges, and the field samples for its registration took measurements that did not exceed fifty centimetres in length.

In the case of three-dimensional models for small objects studied by some authors [13], it is stated that the limitation of the method is found at a length of 5 m. However, the case studies are varied, from the facade of a building to the sculpture and round bust of a piece of art.

One of the most interesting works on light detection and ranging (LiDAR) precision is the work of Costantino et al. [14], in which they perform a sampling of ten objects with an analysis of the quality of the cloud of points obtained in relation to the variation in the surface, loss of planarity and omnivariance.

Another interesting work is the one carried out by Chase et al. [15], who studied the accuracy of the sensor integrated in the iPhone 12/13 Pro in an interior space with a length of 25 m and with different control points that took as reference the TLS and the total station (TS).

In addition, in the framework of architectural topography, Spreafico et al. [3] evaluated the LiDAR sensor for fast 3D mapping.

In summary, there are some very current studies focused on the precision of the aforementioned sensor, which range from the analysis of small objects to interior spaces of greater dimension, reaching its maximum expression at a length of 25 m. A basic question that the researchers expose in the work of Kršák et al. [16] is the low precision of 3D models from UAV photogrammetry. It is necessary to validate the sets of points obtained through algorithms to know in advance the level of precision that one is going to achieve in the records and if these are specially intended to achieve planimetry or to obtain results in the Scan-to-BIM process.

On the other hand, the possibility of using portable devices such as mobile phones or tablets in Pro versions that are relatively cheap and easy to transport presents a new challenge in the field of geomatics, replacing large-scale TS or TLS equipment. Therefore, an operator with a simple drone and a mobile phone could perform a suitable data capture record for Scan-to-BIM in archaeological sites.

For all these reasons, this research proposes a framework where, through measurement and comparison techniques and knowledge of the limitations of the experimental methods, the precision of using the iPhone 13 Pro as a measurement element to establish control points is determined. The aim is to create a framework that assists in photogrammetry through unmanned aerial vehicles (UAVs) in works related to archaeology and heritage architecture. For this purpose, a set of archaeological sites was investigated through different work procedures that will be seen below.

2. Case Study and 3D Processing

The megalithic ensembles of Los Gabrieles (Valverde del Camino), Spain, are one of the most relevant spaces in terms of quality and number of megalithic monuments. They are located in the province of Huelva (Spain) (Figure 1a). Huelva's megalithism has developed as a cultural phenomenon exclusive to the death of the Copper Age [17]. It was after the discoveries by Obermaier in 1924 of the Dolmen del Soto that this type of construction became popular, especially due to the appearance of grave goods and ceramic artefacts.

As a result of the study of funerary architecture, researchers and archaeologists have agreed that two types of culture were established in Huelva: On the one hand, there is the group of shepherds of the Neolithic tradition who inhabited the Andévalo and the Sierra regions and, on the other hand, there are the tombs with false domes that relate to the culture of the Millares [17].

The territorial context in which these deposits are found is very similar to the rest of those found in the Sierra de Huelva and Seville regions, where mountain orography and forest spaces predominate and where the georeferenced data in a Geographic Information System have been processed using the QGIS software [18], which allowed us to define priorities in the cartographic information (Figure 1b).

Massive Data Capture Systems (MDCS) like aerial photogrammetry processes are ideal for capturing geometry from nadiral planes, which ground-based laser scans cannot reach [19]. Therefore, from a technical point of view, it improves the capture of the process that we call Scan-to-BIM, that is, the obtaining of construction information models of historical buildings or, in this case, archaeological sites.

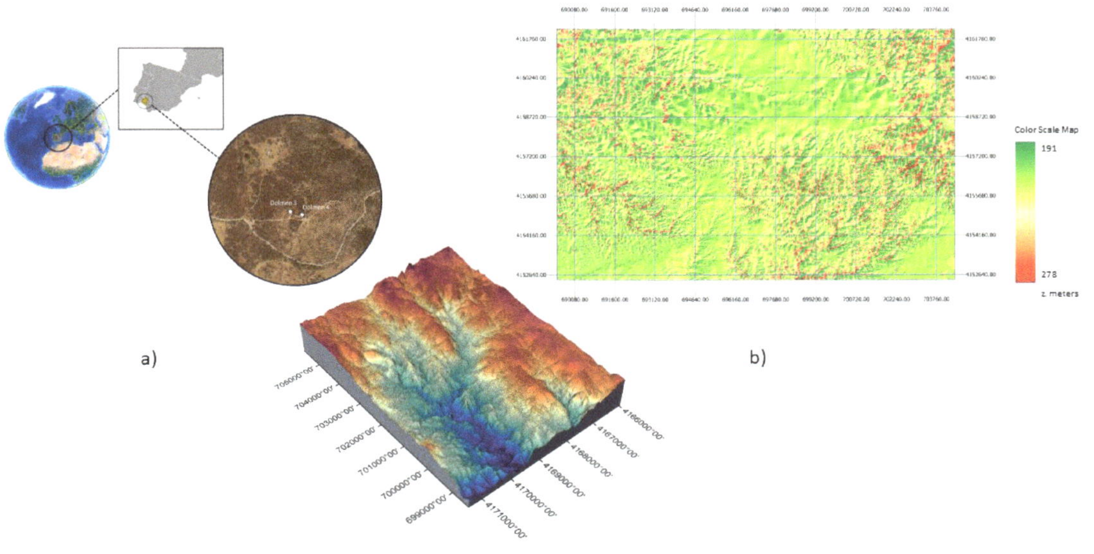

Figure 1. (**a**) Situation of the archaeological sites; (**b**) Digital Elevation Model of the cartographic context of the Gabrieles dolmens; (**c**) layout of a subregion as 3D surface models.

For the specific aerial photogrammetry work at these aforementioned sites, we used two small dolmens with an open morphological structure and small dimensions located in the Los Gabrieles complex in the municipality of Huelva. Thus, we could observe the process of photographic shots through two intelligent flights; the image in Figure 2 is a shot at a height of 12 m, and in Figure 3, a second intelligent flight mode based on zenithal and oblique exposures that determine a minimum height of 24 m can be seen. The 3D photogrammetric models that are extracted from the UAVs make it possible to collect a large amount of data in a short time, but like any system, it has its limitations, despite covering an overhead space that does not capture the terrestrial geometry. To obtain an estimate of the accuracies achieved, one of the most common methods from a scientific point of view is to make a comparison between the SfM model and the TLS.

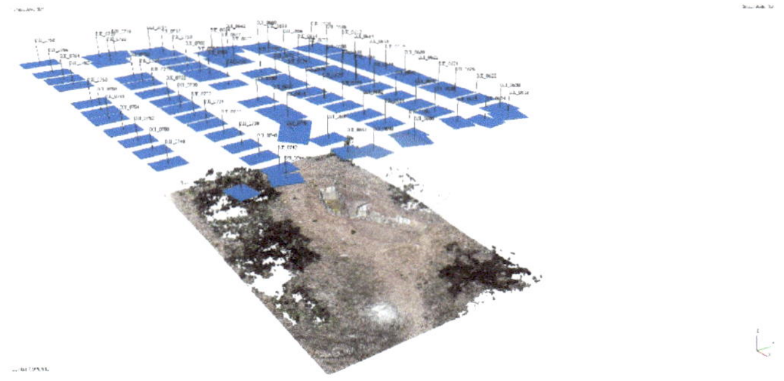

Figure 2. Photogrammetry of Dolmen no. 4 using the Linear Intelligent Flight Plan at a height of 12 m.

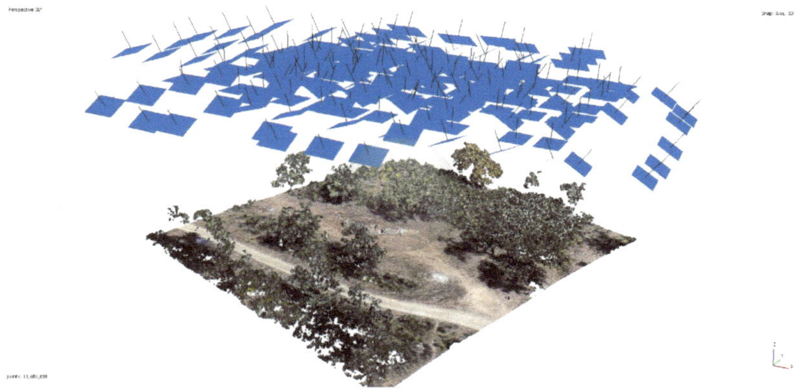

Figure 3. Photogrammetry of Dolmen no. 4 using an Intelligent Oblique Flight Plan at 24 m.

The aerial photogrammetry point cloud was processed using v. 1.8.5 of the Agisoft Metashape [20] software. In the case of the study of the dolmens, seven control points were taken using wooden targets of 37.5 × 25.3 cm strategically placed (i.e., uniformly distributed) on the ground. For the measurements of the control points, measurements were made with a LEICA Flexline TS02 total station. To cover the entire surface, the GCPs were distributed using diagonal geometry and outside the surface of interest typical of the dolmens. The coordinate system was established through axes parallel and perpendicular to some established axes. The local XYZ coordinates set were 100, 100 and 10 m, respectively. Subsequently, the coordinates of the elements in each space were recorded to achieve a uniform set of points.

Accordingly, after importing the images to a Chunk, Agisoft Metashape performed an automatic calibration process using an exchangeable image file (EXIF) [21], detecting each of the images. The processing method used is the one described by Westoby et al. [22], which is carried out from the alignment of the photographs to the postprocessing with mesh generation. The processing parameters of said software were set to "high accuracy" for dense cloud extraction. Once the dense cloud extraction results for the two processes were obtained—one dense cloud with images at a height of 12 m, obtaining a set of 7,046,235 points, and another processing at a height of 24 m with oblique images, obtaining a dense point cloud of 2,281,026 points—the processing was also repeated for Dolmen number 3, which is very close to number 4 (they are separated by approximately ninety metres).

3. Experimental Process Equipment

One of the unmanned aerial vehicles with enormous potential for field work in inspection and photogrammetric records is the DJI Mavic 2 Enterprise Dual drone, whose takeoff weight does not reach 910 g and has a GPS+GLONASS GNSS system with an operating frequency of 2.400–2.4835/5.725–5.850 GHz (Figure 4). It features two cameras contained in the same gimbal: (i) an infrared M2ED camera with an uncooled Vox Microbolometer and an HFOV lens of 57° with f/1.1 aperture and a 12-micron pixel size; (ii) an M2ED visual camera with a $^{1/2.}3''$ CMOS sensor with 12 M effective pixels and an FVO lens of approximately 85° that is equivalent to a 35 mm format with f/2.8 aperture. Both sensors have output in .jpeg format and MP4 video format.

Figure 4. DJI Mavic 2 Enterprise Dual drone.

The operator, through the Smart Controller, can see the two spectra that present different views in each of the sensors that are in the gimbal of the drone. The DJI has four intelligent flight plan modes, including the "oblique plan", but the minimum flight height in this plan is 25 m, which implies a GSD as ground sampling distance of 0.76 cm/px for images acquired with the UAV. The other configured flight plan is a flight mode in which a minimum height of 12 m can be achieved by reducing the GSD to 0.43 cm/px.

In the experimental work, the TLS was used as an auxiliary record. With the appearance of new geomatic technology equipment such as Personal Laser Scanning (PLS), which simplifies data capture and the manageability of records in architectural spaces, it was decided to carry out a first survey through BLK360, which uses digitalisation technology in waveform (WFD) with a maximum scanning speed of 360,000 points/second.

There are three HDR digital cameras with a colour sensor and fixed focal length (single image 2592 × 1944 pixels, 60° × 45° (vx hz), full-dome scanning of 30 images and automatic gap rectification, 150 Mpx, 360° × 300°), as well as an infrared thermal camera (160 × 120 pixel single image, 71° × 56° (vx hz), 10-image full-dome scan, 360° × 70°), all four included in the kit. This instrument achieves a range accuracy, according to the manufacturer, of 4 mm at 10 m and 7 mm at 20 m [11].

The result obtained is a point cloud of 1,350,000 points, confined to the space of the archaeological site. The use of two massive data acquisition techniques allows us to guarantee that the entire surface of the dolmens is covered. For this reason, another part of the data capture techniques based on Structure from Motion (SfM) images is well known by the scientific and academic community [23] and allows for the construction of 3D models widely used in the field of archaeological architecture [4]. The photogrammetry workflow must manage a data plan based on measurement control points so that they are incorporated into Metashape's Agisoft software and, arriving at the intermediate postprocessing process of the images, include the measurements that scale the final model.

Control points are widely used for formal architectural control, in which an analytical model can be made after geometric analysis. Dos Santos et al. [24] revealed the complexity of the dimensional control of the control points in the georeferencing of a record with a terrestrial laser scanner through a scenario of interior and exterior spaces. And in this complexity, several authors [25] carried out a study that examined the georeferencing and geometric quality of a SfM record through multiple combinations of ground control points (GCPs). The conclusions established the proportionality of the control points in a uniformly distributed distribution. In our case study, the control points were taken by the LEICA Flexline TS02 total station with a precision of 2 mm [26]. The full description of the photogrammetry processing methodology used is described in previous works [27,28], leaving, in this case, the workflow well defined. Regarding the process carried out for the orientation of the base station, it was fixed by means of the local XYZ coordinates established at a single point of the project, which were taken as 100, 100 and 10 m, respectively.

Regarding the iPhone 13 Pro, its technical characteristics are the same as those of its predecessor, the 12 Pro version, as described by Monsalve et al. [29], who are a reliable source since Monsalve et al. quote their predecessors Luetzenburg et al. [12]. The LiDAR light detection and ranging sensor of this device operates at 8XX nm wavelengths, using vertical-cavity surface-emitting lasers to emit the laser, and the direct time of flight of the pulses is measured with a Single-Photon Avalanche Diode (SPAD). The iPhone 13 Pro captures a mesh through a mobile app such as "Polycam", which requires a paid subscription to export the data. Another well-known and, in this case, free app is "3d Scanner App". There are other applications, such as Scaniverse [30], but the most used are those specified above that export the data into various point cloud files such as .e57, .ply or .pts, among other files.

In the case of the "3d Scanner" application, it presents six capture modes in which LiDAR Advanced stands out as a prominent mode where the resolution of the point cloud can be set in a range of 5 to 50 millimetres. When measuring with Polycam, this software captures a record through a triangular mesh on the surface of the objects. In this case study, since the objective is to measure the different distances between targets, used as GCPs, the field on view is based on following the trajectory marked by the points. It must be remembered that the iPhone 13 Pro with a LiDAR sensor is equipped with a barometer, a three-axis gyroscope and an accelerometer, which the data capture software uses to determine the three spatial coordinates in the local reference system.

4. Limitations of the Methods Used

Stringent and reproducible measurements based on SfM must proceed in a flow order (established by researchers such as James et al. [31]) that determines the parameters to be taken into account, and any evaluation of the quality of the data must include a comparison with the coordinates of the verification points, the surfaces or the measures of length. A split test to determine systematic errors is also admissible. A split test aims to produce two sets of data with different shapes in the survey design. For example, it seems interesting that measurements carried out between the difference the GCPs and those obtained by a total station are analyzed in terms of the differential between the three coordinate axes (X, Y, Z), with which a data set could also be taken into account when comparing absolute distances between point units.

In the measurements carried out, there is an uncertainty added to the method that we could call residual uncertainty. This uncertainty stems from the absence of points recorded by the UAV on the crosshairs of the target and, likewise, from the absence of points on the crosshairs of the data captured with the iPhone 13 Pro scan and the TLS. In most cases, the best estimate is the arithmetic mean $\bar{Q}$ of the n observations, according to Equation (1).

$$\bar{Q} = \frac{1}{n}\sum_{k=1}^{n} q_k \quad (1)$$

One of the parameters to take into account in the study of precision is the quality of the point cloud. This factor will be established by means of a verification procedure of the density of the cloud of points, measured in cm^2, and could be determined from the data set obtained from one of the targets, which will be the reference of the measurement model. Costantino et al. [14] determined three parameters that influence the quality of the point cloud: (i) the variation in the surface, (ii) the planarity and (iii) the omnivariance. These three parameters determine aspects that can influence the construction of a 3D model, that is, the transformation of the cloud of points to a digital mesh, but for the precision of the data sets that we are going to observe, the most important thing is to know the omnivariance, that is, the degree of inhomogeneity of the cloud of points in the three dimensions (Equation (2)), or what we could specify as the density of the cloud of points.

$$O_\lambda = (\lambda_1 \cdot \lambda_2 \cdot \lambda_3)_{13} \quad (2)$$

In our experiment, we analyse each of the results that show the density of points in a fixed element that, as already mentioned, has as its object a rectangular wooden target of 948.75 cm^2 strategically placed on the surface of the ground as GCPs. The procedure consists of taking a series of 12 samples from the points closest to the filial cross of the target on the cloud of points and doing an analysis and average of the results obtained in each of the samples involved in the analysis. The sets of points of the aerial photogrammetry series at 24 m will be taken first, followed by 12 m, whose denominations of the set of points obtained are Dsfm_24 and Dsfm_12. In the same way, we work with the set of TLS points, and this procedure will also be carried out in the cloud of points from the scan using the sensor integrated in the mobile phone. From this analysis, the random error can be naturally inferred when taking the measurements of each of the samples.

To determine the density of the 3D point cloud, CloudCompare [32] software was used, in which elements are segmented to allow one to observe the factors already exposed by Rizali et al. [33], such as roughness, reflectivity strength, RGB colours and texture.

After the analysis, the following results were obtained (illustrated in Figure 5): The Dsfm_24 obtained a value of 0.3573 pt/cm^2, the Dsfm-12 obtained a value of 1.2057 pt/cm^2, the mobile phone sensor recorded 4.0737 pt /cm^2 and, finally, the TLS obtained a value of 7.9841 pt/cm^2, obtained through surveys carried out in Dolmen no. 3. The values are reflected in red on the graph in Figure 5. Thus, from the values of the 12 samples that determine the distance between the points, the box plots of each of the selected samples were obtained.

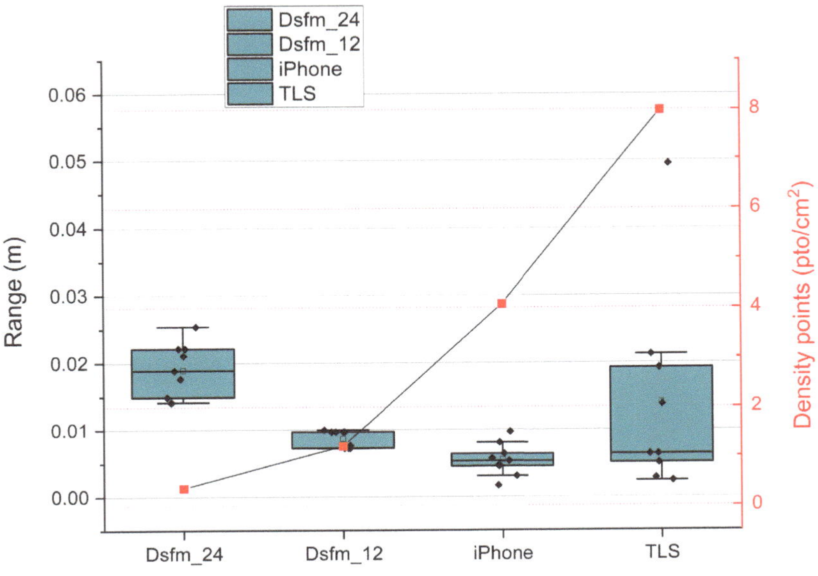

Figure 5. Analysis of the density of points and uncertainty.

Of all the surveys, the one with the greatest dispersion was the one from the TLS, and the one that performed best was the record of the iPhone 13 Pro, although its uncertainty expressed in the range of metric units was very similar to the values of the TLS. Clearly, the density values obtained from aerial photogrammetry between 24 and 12 m are very different, with the 24 m survey yielding greater uncertainty.

Point density is one of the parameters that determines the quality of the geometry of a 3D mesh. In the modelling process, the number of points and their distribution can determine the morphological parameters of the objects and their final result. The resulting triangular mesh for 3D modelling varies according to the point density parameters, the modelling algorithms [34,35] and the shape of the object surface [23]. The dot density is a parameter that is measured according to the surface area. Therefore, the dot density of the dot array surface was evaluated per square centimetre, and the pattern taken was a rectangular bull's eye. The best way to check the spatial resolution of the point cloud, that is, the 3D Euclidean distance (see Equation (3)) between the closest points, is to sample the set of points. The distance between the points and the box plots of each of the selected samples was obtained as follows:

$$d_E(P_1,P_2,P_3) = \sqrt{(x_2 - x_1)^2 + (y_2 - y_1)^2 + (z_2 - z_1)^2} \quad (3)$$

where d_E is the Euclidean distance between points in space and x, y and z are the Cartesian coordinates of those points.

5. Experimental Methodology and Results

In the first phase, an alignment is carried out with the Hand-Held Mobile Laser Scanner (HMLS), a portable mobile laser scanner that obtains a dense point cloud exported in a .ply file in the same coordinate system, and then a network of control points is measured with the total station (TS) using LEICA Flexline TS02 equipment. Therefore, throughout the work, different tests were used to obtain a set of data related to the comparison between the measurements of the total station (TS) and the set of points obtained through the iPhone 13 Scan Pro.

In the second phase of the experiment, a comparison is carried out to find out the geometric precision between the Hand-Held Mobile Laser Scanner and the SfM_{UAV} using ground control points taken with the LEICA Flexline TS02 equipment. All of this estimation processing has been carried out using the open-source software CloudCompare. This analysis will determine the behaviour of the HMLS in the morphological structures of archaeological sites that involve organic forms and complex geometries.

Finally, in a case study different from the previous ones, we proceed to evaluate the precision of the HMSL that serves as a reference for capturing the control points on the ground for photogrammetry using a UAV. In this case, the point cloud record obtained from the TLS is taken as a reference. A comparison between the two of them will determine new data on the performance of the systems in the area of the archaeological site being studied. The 3D model of the terrestrial laser scanner is used as a reference to determine the coincidence points and their dispersion in the system used.

5.1. Analysis of the Distances between TS Targets and iPhone 13 Pro

For this section, three case studies were used, the first of which was Dolmen no. 3, in which five ground control points were used (Figure 6). The data set was exported as a file .ply and TS data were exported in .txt. The data obtained from the deviation were taken to a graph to establish the existence of a direct correlation between the deviation and the measurements obtained in the images of Figure 7, 9 and 11. In this first data output, a total of ten samples were taken with a maximum distance of 9.7367 m and a minimum distance between targets of 3.0432 m. The absolute differential values obtained ranged from 20.47 cm (points 3–6) to 1.69 cm (points 1–6). This analysis is performed to be able to predict the changes and local trends observed after the field work is carried out.

Figure 6. Scanning with the LiDAR sensor of Dolmen no. 3 iOS device.

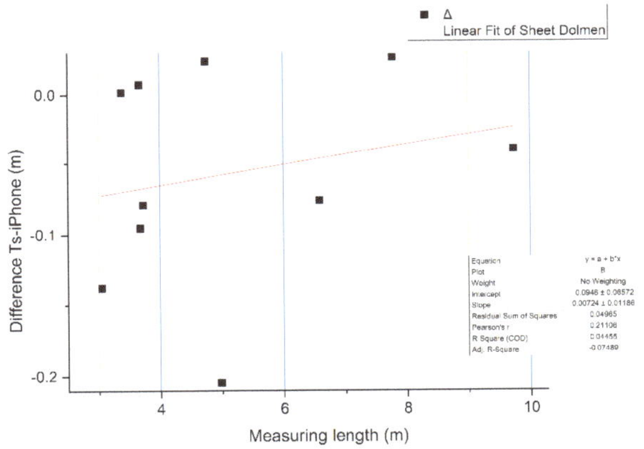

Figure 7. Linear correlation diagram of the deviation between TS and iPhone 13 Pro measurements in Dolmen no. 3.

The following work was carried out in an aerial photogrammetry of the Palma del Condado train station (Figure 8), in which seven control points were taken. In total, this analysis was carried out using twelve samples with a maximum distance of 14.1996 m and a minimum distance between targets of 2.8557 m. The absolute differential values obtained ranged from 1.98 cm (points 5–6) to 0.471 cm (points 4–5).

Figure 8. Partial aerial photogrammetry of the GCPs taken for an SfM_{UAV} at the Palma del Condado train station (Spain).

The last work (Figure 9) was carried out on an almost horizontal plane with little unevenness to check how the LiDAR sensor of the mobile phone works in a parking lot of the Higher Technical School of Building Engineering. Also in this work, twelve samples were used with a maximum distance of 23.3692 m and a minimum distance of 2.3680 m. The absolute differential values obtained ranged from 23.1087 cm (points 1–8) to 2.3560 cm (points 7–8). Therefore, the results of the following studies will determine the correlation between the accuracy of the measurements and the distance between targets.

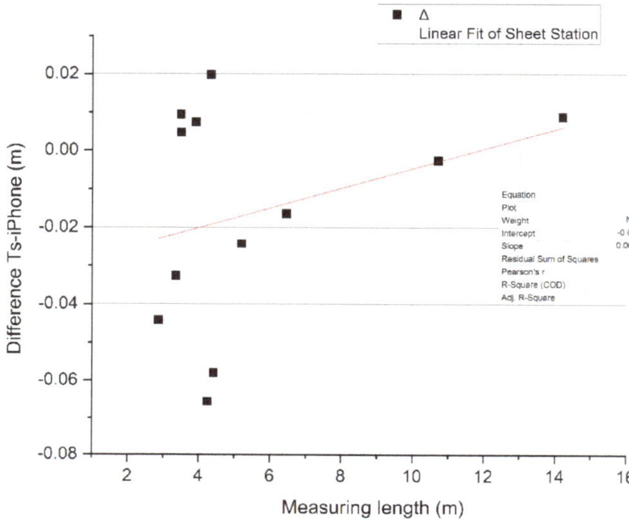

Figure 9. Linear correlation diagram of the deviation between TS and iPhone measurements at the station.

The data sets obtained from the three series, Dolmen no. 3 (Figure 6), station (Figure 8), and school (Figure 10), were mathematically analysed to see the behaviour of the mobile phone sensor Figures 7, 9 and 11.

Figure 10. Photograph of the field of experimentation at the School.

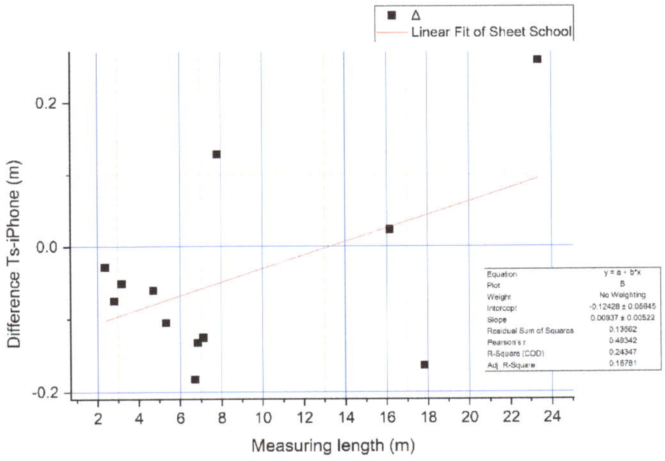

Figure 11. Linear correlation diagram of the deviation between the measurements of the TS and iPhone at the school.

From the processed data set, the range of the deviation was determined in a box plot to find out the dispersion of the values obtained Figure 12. The distance between previously selected characteristic points represented the differences between the three-point clouds in the experiment carried out at the School, the Station, and Dolmen no. 3. The standard deviation results were lower in the Station experiment since a value of 0.02125 m was obtained, compared to the two values obtained in the School and Dolmen no. 3 experiments, which were 0.06921 and 0.06343 m, respectively.

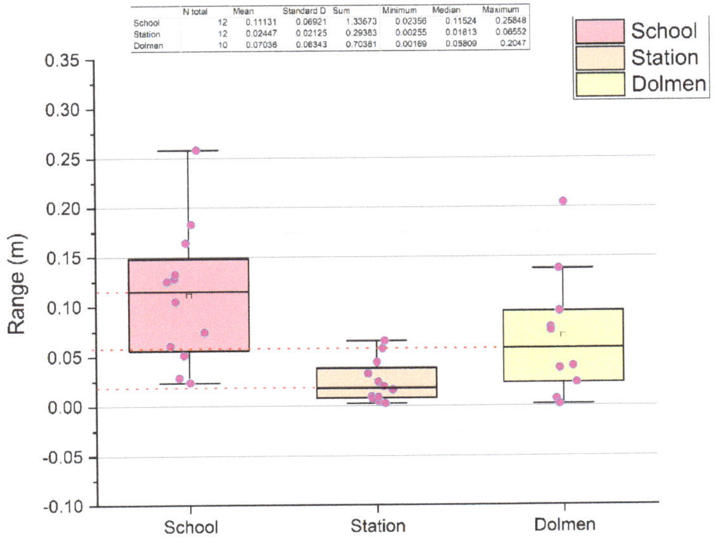

Figure 12. Box-and-whisker plot of measurement results and comparison between TS and iPhone.

The results are shown in Table 1, configured by the parameters of the total number of samples in each case study and their corresponding values.

Table 1. Table of deviations of the three case studies.

	TOTAL NO.	Mean (m)	Standard Deviation (σ) (m)	Sum (m)	Min. Distance (m)	Median (m)	Max. Distance (m)
SCHOOL	12	0.11131	0.06921	1.33573	0.02356	0.11524	0.25848
STATION	12	0.02447	0.02125	0.29363	0.00255	0.01813	0.06552
DOLMEN	10	0.07036	0.06343	0.70361	0.00169	0.05809	0.20470

Therefore, after analysing the experiment results for the three sets of points, we can observe that the school experiment was the one that obtained the worst results. However the maximum length value of 23.3692 m between targets 1 to 8 shows that the accuracy of the values decreases at greater distances in this case.

5.2. Analysis of the Deviation between Photogrammetry Point Clouds Using an UAV and the iPhone 13 Pro

From the field work carried out in Dolmen no. 3 of the Gabrieles complex, data related to aerial SfM_{UAV} photogrammetry with precision processing based on three GCPs at a distance of 12 m using an intelligent flight plan were obtained. The photogrammetry results show average error values in the scale points, a precision of 0.005 m, an error of 0.197 pix and a GSD of 0.43 cm/px, so it can be considered that the use of the UAV is reliable for subsequent modelling. The SfM_{UAV} data at a distance of 24 m yielded average error values in the scale points, a precision of 0.005 m, an error of 0.168 pix and a GSD of 0.76 cm/px.

The applications that are developed in photogrammetry for surveying purposes, land construction or Scan-to-BIM in the field of archaeology have shown that GCPs greatly increase the accuracy of a dense point cloud [36]. Three targets with a black and white photogrammetric target are enough to establish a scale model, although it is already known that there are other parameters that can influence the quality of the point cloud in UAV photogrammetry, such as the flight height, the orientation of the camera in the flight plan process and the quality of the image that the drone gimbal has [37,38]. The objectives of this part of the work are to perform a scan with the LiDAR light detection and ranging sensor of the HMLS mobile phone and, through an analysis process with the algorithms available from CloudCompare, to make a comparison.

Once the SfM point cloud data were obtained using the UAV, it was first scaled with the measurements taken in the Agisoft Metashape software and, subsequently, aligned with the registration data obtained from the light detection and ranging sensor (mobile phone LiDAR) by selecting common point pairs between these two-point clouds in CloudCompare. The three pairs of points (R0-A0, R1-A1 and R2-A2) were located at the centre of the ground control points, which, as already specified, used a 10 mm thick coloured chipboard base in black and white, yielding a final alignment RMS value of 0.00580 m.

Subsequently, the Iterative Closest Point (ICP) algorithm was applied to try to automatically optimize the alignment in order to obtain better results, but this process did not yield positive results, so the coupling between both point clouds was maintained by searching for pairs of adjacent points in the two data sets, and then the transformation parameters between them were calculated [39].

To determine the deviation between clouds of points, the C2C tool, which calculates the distance from each point, is applied. In this case, the point cloud of the aerial photogrammetry was taken as a reference to compare with the closest point in the point cloud of the iPhone 13 Pro.

In point cloud-to-point cloud comparison procedures, it is likely that there are outliers that are not visible to the naked eye but may affect the results. The best way to detect them is to conduct a first comparison analysis and then eliminate those points that can generate noise and alter the final results. These work filters in the cloud of points establish a certain adjustment that allows for the evaluation of the results in their entirety [40].

In aerial photogrammetry over forests or rocky massifs where there is abundant vegetation, the absence of records in forests is very frequent, especially between the ground and the base of the tree crown [41]. The occlusion of the points that record the tree trunk in the process of recording naridal photogrammetry means that the SfM model, using a UAV, cannot reconstruct its position. Therefore, in the open dolmen records in the field, occlusions may appear, as shown in Figure 13. The advantage of using a mobile phone LiDAR light detection and ranging sensor is the possibility of getting closer to the details of the morphology of the environment of the archaeological site, which, in the case of the SfM using a UAV, if the flight plan is not processed correctly, can affect the final results. The traditional way of characterising an archaeological complex through TLS involves certain difficulties [42], especially in horizontal planes of the ground, such as eliminating nondetailed parts, inaccessibility and bias in the collection of overhead data.

Figure 13. Overlap of the two sets of SfM points using UAV and iPhone 13 Pro LiDAR Light Detection and Ranging Sensor from Dolmen no. 3.

Once the differential between both point clouds has been processed, it is possible to extract information from matching points, where the CloudCompare software [39] determines the C2C deviation on a colour scale that represents the deviation of the compared model.

The distance between the point sets was calculated through a comparison between SfM_{UAV} and HMLS, Figure 14 analysis, and Figure 15 histogram of the previous figure. The parameters that yield the results of the comparison process they expressed in Table 2: are the root mean square error (RMS), the standard deviation, the minimum distance, the maximum distance, the mean distance and the maximum estimation error. The deviation between sets of similar points has two characteristics: (i) the high presence of points in values close to zero in relation to other values that are especially far from the value 0.04 m and the rest of the intervals; and (ii) the high standard deviation according to the formulation of the points along those intervals (Equation (4)).

$$\sigma = \sqrt{\frac{1}{n-1}\sum_{i=1}^{n}(x_i - \overline{x})^2} \quad (4)$$

where n is the sample size, x_i are the points in the intervals and $\overline{x}$ is the average sample value.

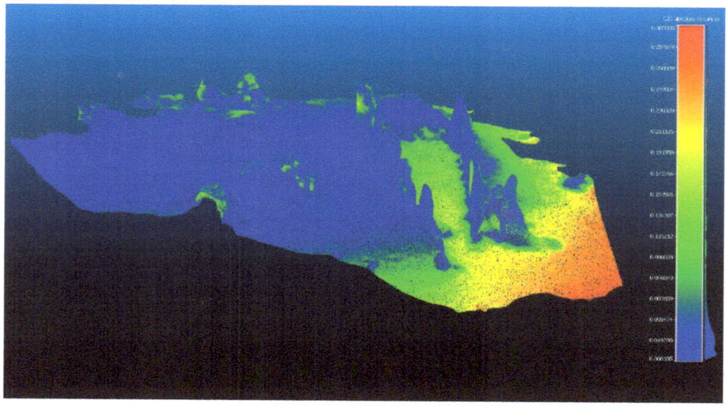

Figure 14. Analysis of the difference between SfM_{UAV} and HMLS. Result of applying the C2C algorithm. Colour scale determines the distance range. In the colour scale, red represents the maximum distance, expressed in metres.

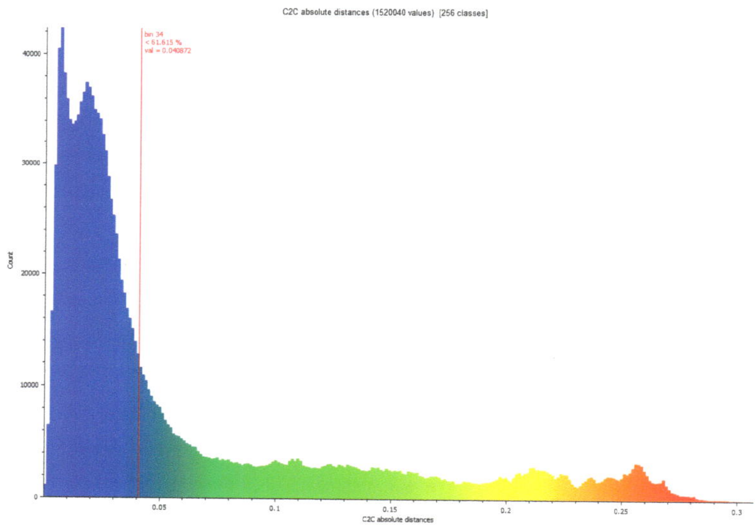

Figure 15. Histogram of Figure 14. Histogram units: metres (*X* axis) and number of points (*Y* axis).

Table 2. Table of deviation between the SfM_{UAV} and HMLS sets of points from the survey in Figure 14.

Comparison between SfMUAV and HMLS	Standard Deviation (σ) (m)	Min. Distance (m)	Max. Distance (m)	Average Distance (m)	Estimated Standard Error (m)
STEP 1	0.0889	0	0.386	0.0403	0.09651

5.3. Analysis of the Deviation between the Cloud of Points Obtained from SfM_{UAV} Using GCP by HMLS and TLS

It is evident that the direct processing that is carried out can bring knowledge of the results closer, taking into account that the main objective of the research work is to know the limitations of the experimental methods in relation to the precision of the use of the iPhone 13 Pro as a measurement element to establish control points on the ground with the help of photogrammetry using unmanned aerial vehicles (UAV) in archaeological sites.

For this case, a point cloud was processed using UAV image capture with ground control points, in which the "Scala Bars" parameters were entered with the dimensions obtained by the LiDAR light detection and ranging sensor of the mobile phone. For both analyses of the sets of points, a partial part of the area surrounding the archaeological site was selected, segmenting both previously aligned sets.

According to studies carried out in other works, remote sensing-based image capture (SfM) using remote sensing is reaching great popularity in the characterisation of landscapes with centimetre-level precision. For example, in landslide studies [43] (where it was shown that the use of multi-rotor UAVs improves the point cloud by 16% compared to the use of fixed-wing UAVs) or in the heritage preservation field [44] (in a study on a Monastery, the quality parameters of the point cloud were evaluated using what the author calls the Remotely Piloted Aircraft System (RPAS) and the TLS). However, in the field of archaeology, there are studies [45] that only approach this investigation using SfM without UAV. Hence, it can be deduced from the previous approaches that the deviation analysis between SfM_{UAV} by means of GCP by HMLS and TLS is innovative.

Figure 16 shows the deviation between aerial photogrammetry using intelligent flight at 24 m and TLS developed on four BLK360 stands using LiDAR technology. The results show a standard deviation of 0.066 m and a maximum distance of 0.83 m.

Figure 16. Analysis of the difference between SfM_{UAV} and GCP by HMLS and TLS. Result of applying the C2C algorithm. Distance distribution colour map. In the colour scale, red represents the maximum distance, expressed in metres, and blue represents the minimum distance.

Certainly, from the point of view of geometric understanding, the highest values are found in the registration points of the treetops, but these data are not representative. The mostly green area that corresponds to values of 0.31 m refers to occlusion points on the floor of the Dolmen recorded by the TLS.

Figure 17 shows the histogram of the study, where 90% of all points have a coincidence of less than 0.032 m. The coincidence between Figures 16 and 17 is consistent since most of the points are in the consolidated values in the blue colour range. The quantitative data provided by the analysis software is represented in Table 3.

Analysing the morphological structure of archaeological sites and the geomorphology of the environment requires accurate topographic information, but when deposits are inaccessible, operators must resort to technologies that revolutionise the current digitalisation market. The presence of trees and shrubs should not be considered an obstacle to characterising these structures. In the processing and analysis stages, it was observed that the SfM_{UAV} point cloud at 12 m behaved differently from the SfM_{UAV} point cloud at 24 m. For this reason, it was decided to open a new analysis section.

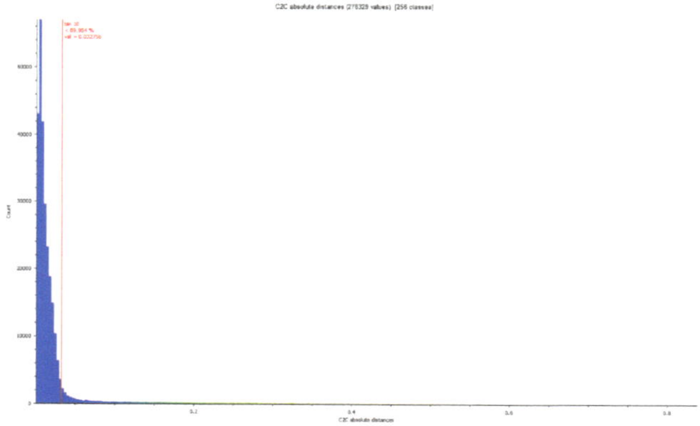

Figure 17. Histogram of Figure 16. Histogram units: metres (*X* axis) and number of points (*Y* axis).

Table 3. Table of deviation between the SfM_{UAV} and TLS sets of points from the survey in Figure 16.

Comparison between SfMUAV and TLS	Standard Deviation (σ) (m)	Min. Distance (m)	Max. Distance (m)	Average Distance (m)	Estimated Standard Error (m)
STEP 1	0.06658	0	0.8194	0.01769	0.0359

5.4. Analysis of the Deviation between Photogrammetry Point Clouds Using a UAV in Intelligent Flight at a Height of 12 m and 24 m

Very recent studies emphasise the study of processing through image capture records using UAVs at different distances on the *Z* axis. These studies have focused on forestry and agriculture to analyse the morphology of trees, essentially volume analysis. In other words, most studies [46] fall within the framework of forestry analysis and geology studies [25], in which different flight scenarios are exposed with different angles of the UAV camera in complex landscapes with high tree density and structured layered morphology, while other works fall within the framework of knowledge about the accuracy and reconstruction of extensive topographical environments [47]. Thus, it seems interesting that, in the context of archaeological sites, the difficulties that we can find in processing field work in different intelligent flight plans are exposed. For this, a study including the processing within the context of the dolmen was carried out, taking the set of SfM_{UAV} points at 24 and 12 m, as can be determined in Figure 18.

Figure 18. Analysis of the difference between SfM_{UAV} at 12 and 24 m. Result of applying the C2C algorithm. Distance distribution colour map. In the colour scale, red represents the maximum distance, expressed in metres, and blue represents the minimum distance.

The data revealed that 71% of the points are coincident at values lower than 0.122 m. This is an important deviation, considering that many points are in the green zone with deviations of 40 cm. This analysis can be seen in the histogram of Figure 19.

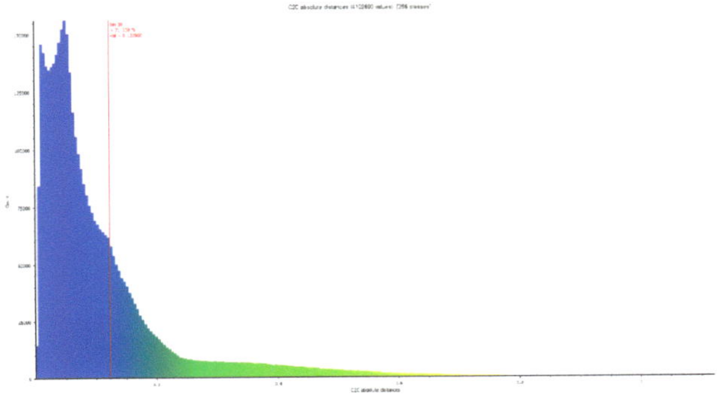

Figure 19. Histogram of Figure 18. Histogram units: metres (X axis) and number of points (Y axis).

The presence of deviations around forty centimetres that represent the majority of the treetops, shrubs and, especially, the upper part of the orthostats, that is, the stone blocks that make up the course of the wall of the dolmen, can be observed. From this line, it can be inferred that there is a presence of systematic vertical deviations. Such errors, which in typical UAV usage data are expressed as a vertical "dome" of the surface, are the result of a combination of nearly parallel imaging directions and an imprecise correction for the distortion of the radial lens [48].

The error in the systematic vertical deviations can be corrected with the additional capture of oblique images. By separating the context from the site and focusing on the dolmen itself, the evidence for deviations is more significant, and a greater appreciation of the results can be obtained. In this second data output, a segmentation of a rectangular prism is carried out on Dolmen no. 3, in which the set of points of the 24 m SfM_{UAV} is taken as a reference.

In Figure 20, it can be seen that the systematic vertical deviations referred to in the studies by James and Robson [48] appear in red.

Figure 20. Analysis of the difference between SfM_{UAV} at 12 and 24 m. Result of applying the C2C algorithm. Distance distribution colour map. In the colour scale, red represents the maximum distance, expressed in metres.

The resulting histogram (Figure 21) shows the number of points per distance. This is counted in absolute values and in metric units. The distribution parameters corresponding to the maximum, minimum, standard deviation and mean values are not really significant here, but the most relevant piece of data is that the area of points in red in the range of colours has a range between 0.15 and 0.25 m.

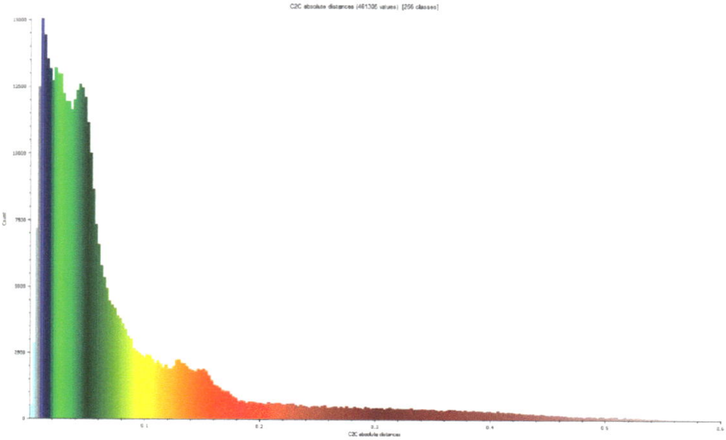

Figure 21. Histogram of Figure 20. Histogram units: metres (*X* axis) and number of points (*Y* axis).

6. Scanning for BIM in an Archaeological Context

The use of HBIM for archaeology arises in analogy and continuity with research and applications on BIM for historical buildings, a subject on which the scientific community has been debating for many years, with different regulatory repercussions from country to country. Nevertheless, there are several specificities in the use of BIM models for archaeology that call for a separate discussion.

As it is well known, BIM is, according to ISO 19650, a series of standards whose content is intended to support agents and operators that participate in any phase of the building cycle or infrastructure pathways through BIM methodology [49], which includes a part of process management. In the case of archaeology, this tool could help not only in the management of all the activities related to preservation, conservation and restoration but also in the "interpretation, presentation, access and public use of the material remains of the past" through virtual archaeology. In the international document "Principles of Seville" [50], aimed at providing practical applications of the principles of the "London Charter" [51], virtual archaeology is defined as "the scientific discipline that seeks to research and develop ways of using computer-based visualisations for the comprehensive management of archaeological heritage" [52], including "virtual restoration, anastylosys, reconstruction, and recreation" [53,54]. To this aim, "inventories, surveys, excavation work, documentation, and research" become fundamental knowledge moments to create two- and three-dimensional representations of the real objects, guaranteeing their "authenticity", in the meaning expressed in the Sevilla Charter, to be obtained through a rigorous and faithful process of representation of their real consistency, also using parametric software, which is fundamental to being true to the geometric and morphological complexity that characterises the stratified ancient heterogeneous structures [55]. All this makes the construction of an HBIM particularly complex [56]. However, research and experimentation in the field of three-dimensional and parametric modelling of these kinds of objects, especially through Scan-to-BIM and photogrammetric processes, have led to the identification of modelling workflows shared by the scientific community [57].

From this point of view, the use of HBIM models represents a valid aid in terms of integrated management, in a single three-dimensional work environment, of different types of data: geometric, stratigraphic, historical, material, environmental, geological, etc. The HBIM model could be considered a 3D database that can be queried in different ways [58] during the design, construction and valorisation phases. The possibility of linking parametric objects, more or less simplified according to different levels of geometric and informative detail, depending on the purpose, with relational databases to be implemented and updated continuously, guarantees that we have a tool for various users to exchange and share information, from archaeologists to conservation offices to scholars. Operators involved in a restoration project can consult the model to access related documents, such as degradation mapping, to use them directly on the archaeological site and plan the interventions to be carried out [59]. In the same way, the model can be consulted by experts interested in accessing iconographic material or archival sources inserted within the HBIM, allowing for new spatial correlations to be established and not only documentary ones. At the same time, there is the possibility of the models being consulted by tourists or simply curious people who want to live immersive experiences through new modes of digital narration, using, for example, augmented and virtual reality [60]. Finally, an HBIM could represent a valid support for public administrations, taking advantage of a valid and reliable verification and control system that, in the long term, provides a mapping of works on the territory and their state of conservation [61], thanks to the possibility of BIM/GIS integration.

It is for this reason that, in recent years, many researchers, at an international level, have dealt with different topics related to BIM for archaeology. The most frequent ones tackle the issues of parametric three-dimensional modelling [62,63], starting from complex data surveying [64], ontologies and sharable vocabularies [65], interoperability between different software [66] and stratigraphic analysis [67].

Currently, there is an evaluation method that combines the creation of a geometric model from a cloud of points with the set of points obtained from area photogrammetry. This method, known as C2M and implemented through CloudCompare, has been used by several researchers [42,68,69] in the context of Cultural Heritage. The direct geometric model is the result of the process of creating static 3D models with simple or complex surfaces [70]. In the field of archaeology, most of the time complex or very complex surfaces are involved, which means that field tests for their evaluation require more complete processes.

7. Discussion

New data capture systems for three-dimensional representation are changing the way geomatics operators and topography experts work. Newer and easier-to-use tools appear, but they can call into question their precision compared to other, more experimental equipment.

The incorporation of LiDAR techniques, such as sensors integrated into mobile phones, is revolutionising the world of 3D scanning. One of the most positive aspects in terms of LiDAR smartphones is their working conditions, among which one can find the possibility of scanning different surfaces from a medium-range distance, including materials in the same scenario, and the processing of data in the "app" of the system lasting a few minutes.

But these record systems are not exempt from serious problems. For example, Costantino et al. [14], for the evaluation of an iPhone 12 Pro, showed that small-scale variations in objects of up to 2 cm of difference in different materials and the architectural space used with somewhat larger dimensions, such as the transept of a church, were detected with precisions of the order of 1 to 3 cm of variation. Some inertial navigation system drift and loss of planarity issues were also mentioned.

Luetzenburg et al. [12], who made measurements of a 130 m by 15 m wide cliff, compared the SfM-MVS with the iPhone 12 Pro point cloud, and the results obtained showed that for 80% of all recorded points, the maximum distance was 15 cm. Through the analysis of the graph of the percentage of precision with respect to the three coordinate axes, it should be emphasised in this work that Z (height) yields quite imprecise values.

The comparison between the integrated sensor of an iPad Pro and TLS was measured by Spreafico et al. [3], who determined that the precision was 2 cm and the accuracy was 4 cm.

One of the most interesting studies developed in the field of interior spaces is the work of Chase et al. [15], where the results in workspaces with a length between 1 and 13 m determined an average error of 3 cm in coordinate (X) and 7 mm in coordinate (Z) in fifteen different shots.

The main difference in the results between Luetzenburg et al. and Chase et al. [15] is the variability of the precision in coordinate (Z), and we believe that Chase's work occurs in conditions where the surface is slightly flat, so the iPhone sensor behaves extremely precisely in these measurements. On the other hand, when the morphology is abrupt and highly variable, precision changes are important.

One of the study samples that made him reevaluate the behaviour of the mobile phone sensor was the result that he showed in the work of Chase et al. [15]. Thus, we selected the results from Table 3 of his work and proceeded to graph them, observing the existence of a linear correlation between the equipment error and the distance. This process was carried out by validating the total station equipment as a reference on the iPhone. This graph is shown in Figure 22, in which it can be seen that there is no correlation between distance and relative precision, so it is determined that the differences may be related to other factors.

We can assume that the relationship between the distance between target points and their precision can be maintained in increasing order and thus predict the behaviour of the deviation between the values of length and precision of the LiDAR system from the linear regression line.

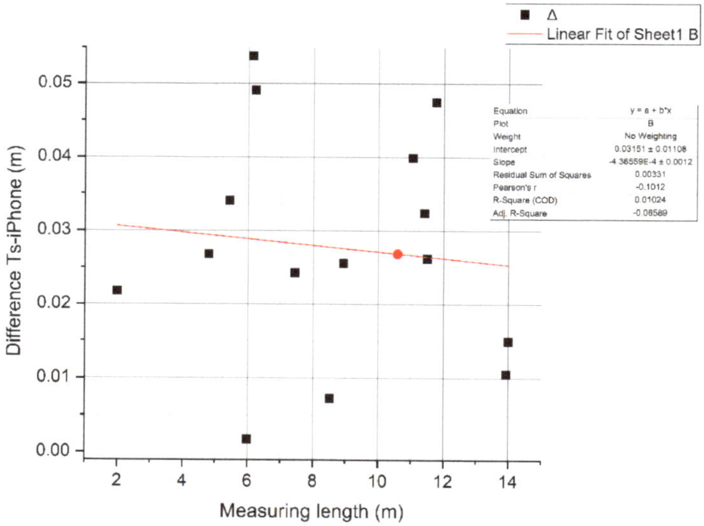

Figure 22. Linear correlation diagram of the deviation in graphing of the results from Table 3 of the work of Chase et al. [15].

On the basis of these studies, we carried out a section on the uncertainty associated with the measurement method, which has been called "residual uncertainty", by studying the density of points on the object to be measured. Interestingly, the method that registered the highest density of points was TLS, followed by the LiDAR light detection and ranging sensor of the mobile phone. Regarding the dispersion of the points in the measurements, the iPhone behaved quite stably; that is, in short measurements, the set of points is very distributive, reaching uncertainty values of a range of 0.005 m, that is, values of 5 millimetres.

Next, we generated an experimental methodology consisting of four sections. The first one had to do with the errors of the LiDAR light detection and ranging sensor of the mobile phone compared to the data obtained by the total station (TS) using the LEICA Flexline TS02 equipment. On the basis of this analysis, we proceeded to determine the existence of a linear correlation between the distance of the GCPs and the deviation that occurs with respect to the measurements with the TS. The statistical results of the comparison revealed that there is a slight relationship between the parameters analysed (R2) in Figures 7 and 8, with very low values of 0.04 and 0.09 m, respectively. The graph that presents some correlation is Figure 11, with values of 0.24.

One must think that the differences between the graphs of Chase et al. [15] and those elaborated in this investigation are justified in that the author takes distances as absolute values. From the subsequent analysis (Figure 12), there are three cases that comprise a standard deviation between 0.021 and 0.069 m. In the case of the station's field work study, the deviation value was quite satisfactory with only a 21 mm difference, but the rest of the studies presented highly significant variations.

Within the development of the field work, the case studies of the School and the Station were studied on slightly flat surfaces, which was not the case in the Dolmen case study, where the Z axis was more variable. But it is not ruled out that the analysis surface in a variable geomorphological environment increases the error value of the LiDAR light detection and ranging sensor of the mobile phone.

In the next study, an analysis of the difference between photogrammetry using UAV and HMLS was carried out. The results showed a standard deviation of 0.0889 m. However, this deviation was estimated to be due to non-alignment values since most of the scattered points were at the end of the CGPs, as shown in Figure 14. A significant value was the high presence of points in values close to 0 in relation to other values that were especially far from the value of 0.04 m and the rest of the intervals; that is to say, most of the points presented a differential of 4 centimetres. This would comprise an average interval of the results obtained in the previous investigation [15].

Regarding the analysis of the difference between SfM_{UAV} and GCP by HMLS and TLS, there was a good degree of agreement. In other words, after meeting the requirements of the objectives set in this part of the study, the results were positive. In this case, the standard deviation parameter was lower than in the previous one, with a result of 0.0666 m. A significant value was the high presence of points at values close to 0, in which 90% of all points had a coincidence of less than 0.032 m. These values were lower than in previous studies. Therefore, when we process photogrammetry using a UAV at a height of 24 m with intelligent flight and nadiral and oblique images and use the iPhone 13 Pro to measure ground control points, we obtain an error against the TLS using the BLK360 32 mm. Thus, these are some results that can be assumed for Scan-to-BIM environments.

Also, the use of aerial photogrammetry at different levels and ways of proceeding is essential to knowing how an optimal 3D model can be obtained. Thus, an analysis of the difference between SfM_{UAV} at 12 and 24 m with the intelligent flights provided by the drone used was determined with the next field work test. In this work, it was possible to determine the presence of systematic vertical deviations, in addition to the fact that these errors, in the typical data of the use of the UAV, are expressed as a vertical "dome" of the surface in the absence of oblique images. In this document, it was possible to quantify the deviation value of the orthostats, establishing a range between 0.15 and 0.25 m.

To provide comprehensive knowledge in the context of Scan-to-BIM applied to archaeological sites, all results have been graphed in Figure 23, where the representation in the range of millimetres of the slight accuracy (LOA) of the independent standard deviation is combined for each method used to acquire the rank of these measurements in USIBD [71]. The green points are sets of points randomly established to represent the record of a surface from a point cloud, where the actual surface is established as an imaginary line close to a value of zero. The second overlaid graph represents the mean deviation of four researchers and the results of the four experiments conducted in this work in a range expressed in metres. The equivalence is evident between the coordinates of the ordinate axes in red and black, although both are expressed in different units. It should be mentioned that the difference between the last survey and the rest of the tests is that each and every one of the tests was conducted with the LiDAR light detection and ranging sensor equipment of the mobile phone except for Survey_4, which determines the deviation between photogrammetry areas with heights of 12 and 24 m.

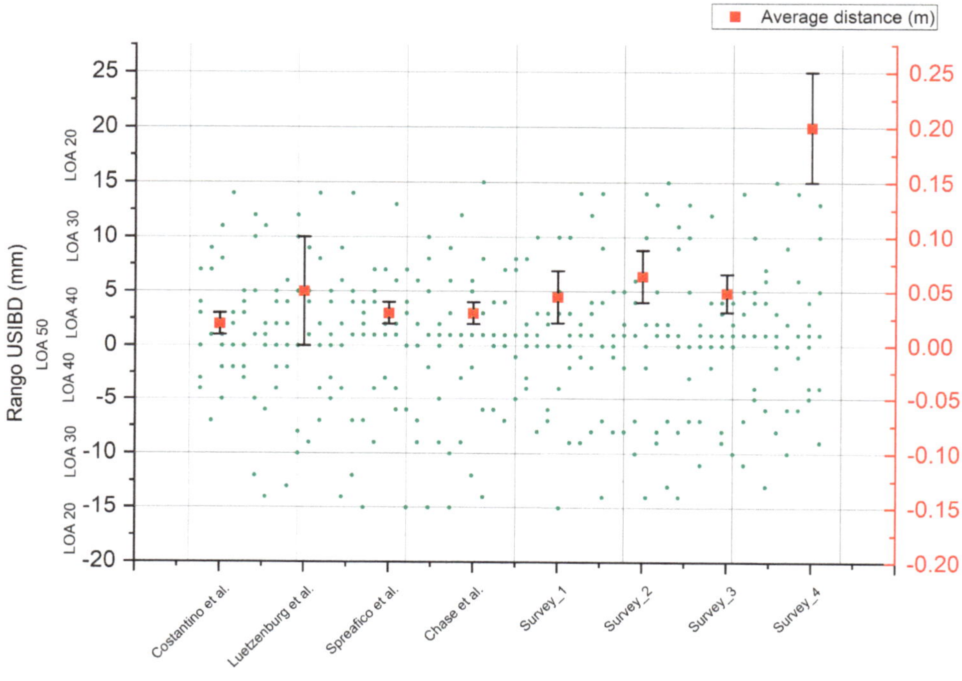

Figure 23. Average values with their errors in a graph with LOA range in the USIBD guide comparing the authors: Costantino et al. [14]; Luetzenburg et al. [12]; Spreafico et al. [3] and Chase et al. [15] with surveys carried out.

8. Conclusions

Disruptive techniques of massive data acquisition are presented as a new opportunity for work that completes the records of 3D digitisation and the implementation of Scan-to-BIM nowadays. Each method requires different data processing and is generally associated with specific software that processes the data until the point cloud is ready to take it to BIM. With this study, we have had the possibility of knowing the deviations shown by the use of the LiDAR light detection and ranging sensor on the iPhone 13 Pro.

Thus, in the data processing phase of this work, four comparison studies were carried out, along with a section in which an approximation to the Scan-to-BIM of archaeological sites was investigated. Through the study of aerial photogrammetry data obtained through two intelligent flight plans, the deviations generated by the measurements of a mobile phone could be compared, and its applicability to capture data from GCPs (ground control points) that serve to improve photogrammetry was assessed.

It must be taken into account that the analysis of the morphological structure of archaeological sites and their geomorphology is carried out in a complex context, such as on open sites where abrupt vegetation and groves abound. Most of these deposits are difficult to access, so having a drone and a mobile phone with these characteristics in one's backpack means that any expert can carry out 3D survey work with certain guarantees of precision.

With the set of data obtained, data on the precision of this equipment can be inferred from these investigations, although the door is left open for other investigations that relate to the capture of records made with a LiDAR light detection and ranging sensor of a mobile phone, correlating the results with other parameters to clarify questions about how to reach precisions close to an LOA of 50 (range values between 0 and 1 millimetre).

It must be recognised that applying BIM to heritage is a form of digital registration that is open to different administrations and that, sometimes, the registration can be optimal in its performance. The path that leads us towards digitisation and the construction information model in historical buildings or archaeological sites consists of innumerable stages, one of the first being the generation of geospatial data. Knowing the requirements and specifications of the levels of precision allows professionals in the Architecture, Engineering, Construction and Ownership (AECO) industry to clearly articulate the objectives of the works. Therefore, precisely understanding the scope of the impact of component errors improves the results of the applicability of BIM.

In the future, through experimental work related to point cloud mapping in archaeological pieces and altarpieces, it will be possible to study how certain deformations produced by the LiDAR light detection and ranging sensor of the mobile phone influence linear structures, quantifying the deformation according to image capture.

Finally, it is also considered necessary to indicate that one of the most important objectives of this work has been fulfilled since a new scenario of possibilities for the applicability of new techniques in the management of the acquisition of massive records to produce increasingly accurate 3D models has opened up.

Author Contributions: J.M.: Conceptualization, Data curation, Formal analysis, Funding acquisition, Investigation, Methodology, Software, Supervision, Visualization, Writing—original draft, Writing—review and editing; J.E.N.-J.: Data curation, Software, Supervision, Validation; M.F.-A.: Visualization, Writing—original draft, Writing; D.O.: Data curation, Resources, Visualization, Writing—original draft, Writing, Project administration; R.E.-P.: Data curation, Software, Resources. All authors have read and agreed to the published version of the manuscript.

Funding: This research was funded by Universidad de Sevilla through VII Plan Propio de Investigación y Transferencia (VIIPPIT).

Data Availability Statement: Not applicable.

Conflicts of Interest: The authors declare no conflict of interest.

References

1. Daniotti, B.; Pavan, A.; Lupica Spagnolo, S.; Caffi, V.; Pasini, D.; Mirarchi, C. Collaborative Working in a BIM Environment (BIM Platform). In *BIM-Based Collaborative Building Process Management*; Springer Tracts in Civil Engineering; Springer: Cham, Swizerland, 2020; pp. 71–102. [CrossRef]
2. Moyano Campos, J.J. *Desde la Forma al Modelo Digital: Teoría de la Morfogénesis en el Ejercicio de la Forma Natural*; de Sevilla, U., Ed.; Universidad de Sevilla: Sevilla, Spain, 2022; ISBN 9788447223800.
3. Spreafico, A.; Chiabrando, F.; Teppati Losè, L.; Giulio Tonolo, F.; Spreafico, A.; Chiabrando, F.; Teppati Losè, L.; Giulio Tonolo, F. The Ipad Pro Built-In LIDAR Sensor: 3d Rapid Mapping Tests and Quality Assessment. *Int. Arch. Photogramm. Remote Sens. Spat. Inf. Sci.* **2021**, *43*, 63–69. [CrossRef]
4. Borrero, M.; Stroth, L.R. A Proposal for the Standardized Reporting of Error and Paradata Regarding Structure from Motion (SfM) 3D Models Used in Recording and Consolidating Archaeological Architecture. *Adv. Archaeol. Pract.* **2020**, *8*, 376–388. [CrossRef]
5. Jones, C.A.; Church, E. Photogrammetry is for everyone: Structure-from-motion software user experiences in archaeology. *J. Archaeol. Sci. Rep.* **2020**, *30*, 102261. [CrossRef]
6. Colosi, F.; Malinverni, E.S.; Leon Trujillo, F.J.; Pierdicca, R.; Orazi, R.; Di Stefano, F. Exploiting HBIM for Historical Mud Architecture: The Huaca Arco Iris in Chan Chan (Peru). *Heritage* **2022**, *5*, 2062–2082. [CrossRef]
7. Duró, G.; Crosato, A.; Kleinhans, M.G.; Uijttewaal, W.S.J. Bank erosion processes measured with UAV-SfM along complex banklines of a straight mid-sized river reach. *Earth Surf. Dyn.* **2018**, *6*, 933–953. [CrossRef]

8. Turner, D.; Lucieer, A.; de Jong, S.M. Time series analysis of landslide dynamics using an Unmanned Aerial Vehicle (UAV). *Remote Sens.* **2015**, *7*, 1736–1757. [CrossRef]
9. de Melo, R.R.S.; Costa, D.B.; Álvares, J.S.; Irizarry, J. Applicability of unmanned aerial system (UAS) for safety inspection on construction sites. *Saf. Sci.* **2017**, *98*, 174–185. [CrossRef]
10. Elkhrachy, I. Accuracy Assessment of Low-Cost Unmanned Aerial Vehicle (UAV) Photogrammetry. *Alex. Eng. J.* **2021**, *60*, 5579–5590. [CrossRef]
11. Moyano, J.; Justo-Estebaranz, Á.; Nieto-Julián, J.E.; Barrera, A.O.; Fernández-Alconchel, M. Evaluation of records using terrestrial laser scanner in architectural heritage for information modeling in HBIM construction: The case study of the La Anunciación church (Seville). *J. Build. Eng.* **2022**, *62*, 105190. [CrossRef]
12. Luetzenburg, G.; Kroon, A.; Bjørk, A.A. Evaluation of the Apple iPhone 12 Pro LiDAR for an Application in Geosciences. *Sci. Rep.* **2021**, *11*, 22221. [CrossRef]
13. Łabędź, P.; Skabek, K.; Ozimek, P.; Rola, D.; Ozimek, A.; Ostrowska, K. Accuracy Verification of Surface Models of Architectural Objects from the iPad LiDAR in the Context of Photogrammetry Methods. *Sensors* **2022**, *22*, 8504. [CrossRef]
14. Costantino, D.; Vozza, G.; Pepe, M.; Alfio, V.S. Smartphone LiDAR Technologies for Surveying and Reality Modelling in Urban Scenarios: Evaluation Methods, Performance and Challenges. *Appl. Syst. Innov.* **2022**, *5*, 63. [CrossRef]
15. Chase, P.P.C.; Clarke, K.H.; Hawkes, A.J.; Jabari, S.; Jakus, J.S. Apple IPhone 13 Pro LiDAR Accuracy Assessment for Engineering Applications. *Transform. Constr. Real. Capture Technol.* **2022**, 1–10. [CrossRef]
16. Kršák, B.; Blišťan, P.; Pauliková, A.; Puškárová, P.; Kovanič, L.; Palková, J.; Zelizňaková, V. Use of low-cost UAV photogrammetry to analyze the accuracy of a digital elevation model in a case study. *Meas. J. Int. Meas. Confed.* **2016**, *91*, 276–287. [CrossRef]
17. Lazarich, M.; Briceño, E.; Ramos, A.; Carreras, A.; Fernández, J.V.; Jenkins, V.; Feliu, M.J.; Versaci, M.; Torres, F.; Richarte, M.J.; et al. Análisis arquitectónico y territorial de los conjuntos megalíticos de Los Gabrieles (Valverde del Camino) y El Gallego-Hornueco (Berrocal-Madroño): El megalitismo en el Andévalo Oriental (Huelva). In *IV Encuentro de Arqueología del Suroeste Peninsular [Recurso Electrónico]*; Universidad de Huelva: Huelva, Spain, 2010; pp. 209–248, ISBN 978-84-92679-59-1.
18. Qguis Project Descarga QGIS. Available online: https://qgis.org/es/site/forusers/download.html (accessed on 29 May 2023).
19. Tian, J.; Dai, T.; Li, H.; Liao, C.; Teng, W.; Hu, Q.; Ma, W.; Xu, Y. A Novel Tree Height Extraction Approach for Individual Trees by Combining TLS and UAV Image-Based Point Cloud Integration. *Forests* **2019**, *10*, 537. [CrossRef]
20. Agisoft PhotoScan Software. Agisoft Metashape. Available online: https://www.agisoft.com/ (accessed on 30 January 2020).
21. Romero, N.L.; Gimenez Chornet, V.V.G.C.; Cobos, J.S.; Selles Carot, A.A.S.C.; Centellas, F.C.; Mendez, M.C. Recovery of descriptive information in images from digital libraries by means of EXIF metadata. *Libr. Hi Tech* **2008**, *26*, 302–315. [CrossRef]
22. Westoby, M.J.; Brasington, J.; Glasser, N.F.; Hambrey, M.J.; Reynolds, J.M. 'Structure-from-Motion' photogrammetry: A low-cost, effective tool for geoscience applications. *Geomorphology* **2012**, *179*, 300–314. [CrossRef]
23. Park, H.; Lee, D. Comparison between point cloud and mesh models using images from an unmanned aerial vehicle. *Measurement* **2019**, *138*, 461–466. [CrossRef]
24. dos Santos, D.R.; Dal Poz, A.P.; Khoshelham, K. Indirect Georeferencing of Terrestrial Laser Scanning Data using Control Lines. *Photogramm. Rec.* **2013**, *28*, 276–292. [CrossRef]
25. Sanz-Ablanedo, E.; Chandler, J.H.; Rodríguez-Pérez, J.R.; Ordóñez, C. Accuracy of Unmanned Aerial Vehicle (UAV) and SfM Photogrammetry Survey as a Function of the Number and Location of Ground Control Points Used. *Remote Sens.* **2018**, *10*, 1606. [CrossRef]
26. Geosystems Leica Geosystems. Leica FlexLine TS02/TS06/TS09 User Manual. 2008. Available online: https://leica-geosystems.com/ (accessed on 16 March 2020).
27. Moyano, J.; Nieto-Julián, J.E.; Antón, D.; Cabrera, E.; Bienvenido-Huertas, D.; Sánchez, N. Suitability Study of Structure-from-Motion for the Digitisation of Architectural (Heritage) Spaces to Apply Divergent Photograph Collection. *Symmetry* **2020**, *12*, 1981. [CrossRef]
28. Moyano, J.; Fernández-Alconchel, M.; Nieto-Julián, J.E.; Carretero-Ayuso, M.J. Methodologies to Determine Geometrical Similarity Patterns as Experimental Models for Shapes in Architectural Heritage. *Symmetry* **2022**, *14*, 1893. [CrossRef]
29. Monsalve, A.; Yager, E.M.; Tonina, D. Evaluating Apple iPhone LiDAR measurements of topography and roughness elements in coarse bedded streams. *J. Ecohydraulics* **2023**, 1–11. [CrossRef]
30. Carter, N.; Hashemian, A.; Rose, N.A.; Neale, W.T.C. Evaluation of the Accuracy of Image Based Scanning as a Basis for Photogrammetric Reconstruction of Physical Evidence. *SAE Tech. Pap.* **2016**. [CrossRef]
31. James, M.R.; Chandler, J.H.; Eltner, A.; Fraser, C.; Miller, P.E.; Mills, J.P.; Noble, T.; Robson, S.; Lane, S.N. Guidelines on the use of structure-from-motion photogrammetry in geomorphic research. *Earth Surf. Process. Landf.* **2019**, *44*, 2081–2084. [CrossRef]
32. Girardeau-Montaut, D. Cloud-to-Mesh Distance. Available online: http://www.cloudcompare.org/doc/wiki/index.php?title=Cloud-to-Mesh_Distance (accessed on 25 February 2023).
33. Razali, M.I.; Idris, A.N.; Razali, M.H.; Syafuan, W.M. Quality Assessment of 3D Point Clouds on the Different Surface Materials Generated from iPhone LiDAR Sensor. *Int. J. Geoinform.* **2022**, *18*, 51–59. [CrossRef]
34. Bernardini, F.; Mittleman, J.; Rushmeier, H.; Silva, C.; Taubin, G. The ball-pivoting algorithm for surface reconstruction. *IEEE Trans. Vis. Comput. Graph.* **1999**, *5*, 349–359. [CrossRef]
35. Kazhdan, M.; Bolitho, M.; Hoppe, H. Poisson Surface Reconstruction. In Proceedings of the Eurographics Symposium on Geometry Processing 2006, Sardinia, Italy, 26–28 June 2006.

36. Mukhlisin, M.; Astuti, H.W.; Kusumawardani, R.; Wardihani, E.D.; Supriyo, B. Rapid and low cost ground displacement mapping using UAV photogrammetry. *Phys. Chem. Earth* **2023**, *130*, 103367. [CrossRef]
37. Gindraux, S.; Boesch, R.; Farinotti, D. Accuracy assessment of digital surface models from Unmanned Aerial Vehicles' imagery on glaciers. *Remote Sens.* **2017**, *9*, 186. [CrossRef]
38. Harwin, S.; Lucieer, A. Assessing the accuracy of georeferenced point clouds produced via multi-view stereopsis from Unmanned Aerial Vehicle (UAV) imagery. *Remote Sens.* **2012**, *4*, 1573–1599. [CrossRef]
39. Girardeau-Montaut, D. CloudCompare Point Cloud Processing Workshop. Available online: https://docplayer.net/184042061-Cloudcompare-point-cloud-processing-workshop.html (accessed on 28 January 2021).
40. Moyano, J.; Cabrera-Revuelta, E.; Nieto-Julián, J.E.; Fernández-Alconchel, M.; Fernández-Valderrama, P. Evaluation of Geometric Data Registration of Small Objects from Non-Invasive Techniques: Applicability to the HBIM Field. *Sensors* **2023**, *23*, 1730. [CrossRef] [PubMed]
41. Webster, C.; Westoby, M.; Rutter, N.; Jonas, T. Three-dimensional thermal characterization of forest canopies using UAV photogrammetry. *Remote Sens. Environ.* **2018**, *209*, 835–847. [CrossRef]
42. Moyano, J.; Odriozola, C.P.; Nieto-Julián, J.E.; Vargas, J.M.; Barrera, J.A.; León, J. Bringing BIM to archaeological heritage: Interdisciplinary method/strategy and accuracy applied to a megalithic monument of the Copper Age. *J. Cult. Herit.* **2020**, *45*, 303–314. [CrossRef]
43. Ruggles, S.; Clark, J.; Franke, K.W.; Wolfe, D.; Reimschiissel, B.; Martin, R.A.; Okeson, T.J.; Hedengren, J.D. Comparison of sfm computer vision point clouds of a landslide derived from multiple small uav platforms and sensors to a tls-based model. *J. Unmanned Veh. Syst.* **2016**, *4*, 246–265. [CrossRef]
44. Deliry, S.I.; Ugur, A. Accuracy evaluation of UAS photogrammetry and structure from motion in 3D modeling and volumetric calculations. *J. Appl. Remote Sens.* **2023**, *17*, 024515. [CrossRef]
45. Marín-Buzón, C.; Pérez-Romero, A.M.; León-Bonillo, M.J.; Martínez-álvarez, R.; Mejías-García, J.C.; Manzano-Agugliaro, F. Photogrammetry (SfM) vs. Terrestrial Laser Scanning (TLS) for Archaeological Excavations: Mosaic of Cantillana (Spain) as a Case Study. *Appl. Sci.* **2021**, *11*, 11994. [CrossRef]
46. Mlambo, R.; Woodhouse, I.H.; Gerard, F.; Anderson, K. Structure from motion (SfM) photogrammetry with drone data: A low cost method for monitoring greenhouse gas emissions from forests in developing countries. *Forests* **2017**, *8*, 68. [CrossRef]
47. Del Savio, A.A.; Luna Torres, A.; Chicchón Apaza, M.A.; Vergara Olivera, M.A.; Llimpe Rojas, S.R.; Urday Ibarra, G.T.; Reyes Ñique, J.L.; Macedo Arevalo, R.I. Integrating a LiDAR Sensor in a UAV Platform to Obtain a Georeferenced Point Cloud. *Appl. Sci.* **2022**, *12*, 12838. [CrossRef]
48. James, M.R.; Robson, S. Mitigating systematic error in topographic models derived from UAV and ground-based image networks. *Earth Surf. Process. Landf.* **2014**, *39*, 1413–1420. [CrossRef]
49. Almeida, R.; Chaves, L.; Silva, M.; Carvalho, M.; Caldas, L. Integration between BIM and EPDs: Evaluation of the main difficulties and proposal of a framework based on ISO 19650:2018. *J. Build. Eng.* **2023**, *68*, 106091. [CrossRef]
50. ICOMOS Principles of Seville. International Priciples of Virtual Archeology. Available online: http://www.sevilleprinciples.com/ (accessed on 15 February 2020).
51. Londoncharter, L. London Charter for the Computer-Based Visualisation of Cultural Heritage. Available online: https://www.london-charter.org/downloads.html (accessed on 30 May 2023).
52. Valente, R.; Brumana, R.; Oreni, D.; Banfi, F.; Barazzetti, L.; Previtali, M.; Valente, R.; Brumana, R.; Oreni, D.; Banfi, F.; et al. Object-Oriented Approach for 3d Archaeological Documentation. *Int. Arch. Photogramm. Remote Sens. Spat. Inf. Sci.* **2017**, *42*, 707–712. [CrossRef]
53. Fiorillo, F.; Rossi, C. 3D Parametric Modelling based on Point Cloud for the Interpretation of the Archaeological Remains. *Disegnare Con* **2021**, *14*, 1–11. [CrossRef]
54. Demetrescu, E. Archaeological stratigraphy as a formal language for virtual reconstruction. Theory and practice. *J. Archaeol. Sci.* **2015**, *57*, 42–55. [CrossRef]
55. Banfi, F.; Brumana, R.; Landi, A.G.; Previtali, M.; Roncoroni, F.; Stanga, C. Building archaeology informative modelling turned into 3D volume stratigraphy and extended reality time-lapse communication. *Virtual Archaeol. Rev.* **2022**, *13*, 1–21. [CrossRef]
56. Barazzetti, L. Parametric as-built model generation of complex shapes from point clouds. *Adv. Eng. Inform.* **2016**, *30*, 298–311. [CrossRef]
57. Banfi, F. HBIM, 3D drawing and virtual reality for archaeological sites and ancient ruins. *Virtual Archaeol. Rev.* **2020**, *11*, 16–33. [CrossRef]
58. Diara, F.; Rinaudo, F. Building Archaeology Documentation and Analysis Through Open Source Hbim Solutions via Nurbs Modelling. *Int. Arch. Photogramm. Remote Sens. Spat. Inf. Sci.* **2020**, *43*, 1381–1388. [CrossRef]
59. Trizio, I.; Savini, F.; Giannangeli, A.; Boccabella, R.; Petrucci, G.; Trizio, I.; Savini, F.; Giannangeli, A.; Boccabella, R.; Petrucci, G. The Archaeological Analysis of Masonry for the Restoration Project in Hbim. *Int. Arch. Photogramm. Remote Sens. Spat. Inf. Sci.* **2019**, *42*, 715–722. [CrossRef]
60. Trizio, I.; Savini, F. Archaeology of buildings and HBIM methodology: Integrated tools for documentation and knowledge management of architectural heritage. In Proceedings of the IMEKO International Conference on Metrology for Archaeology and Cultural Heritage, TC4 MetroArchaeo, Bergamo, Italy, 22–24 October 2020; pp. 84–89.

61. Chiabrando, F.; Lo Turco, M.; Rinaudo, F.; Chiabrando, F.; Lo Turco, M.; Rinaudo, F. Modeling the Decay in AN Hbim Starting from 3d Point Clouds. A Followed Approach for Cultural Heritage Knowledge. *Int. Arch. Photogramm. Remote Sens. Spat. Inf. Sci.* **2017**, *42*, 605–612. [CrossRef]
62. Dore, C.; Murphy, M.; McCarthy, S.; Brechin, F.; Casidy, C.; Dirix, E. Structural simulations and conservation analysis-historic building information model (HBIM). *Int. Arch. Photogramm. Remote Sens. Spat. Inf. Sci.* **2015**, *40*, 351–357. [CrossRef]
63. Brumana, R.; Banfi, F.; Cantini, L.; Previtali, M.; Della Torre, S. Hbim level of detail-geometry-Accuracy and survey analysis for architectural preservation. *ISPRS Ann. Photogramm. Remote Sens. Spat. Inf. Sci.* **2019**, *42*, 293–299. [CrossRef]
64. Adami, A. (Ed.) Geomatica e HBIM per i beni culturali. In *Geomatica e HBIM per i Beni Culturali*; Franco Angeli: Milan, Italy, 2021; pp. 1–233.
65. Acierno, M.; Cursi, S.; Simeone, D.; Fiorani, D. Architectural heritage knowledge modelling: An ontology-based framework for conservation process. *J. Cult. Herit.* **2017**, *24*, 124–133. [CrossRef]
66. Daniotti, B.; Gianinetto, M.; Della Torre, S. (Eds.) *Digital Transformation of the Design, Construction and Management Processes of the Built Environment*; Research for Development; Springer Nature: Cham, Swizerland, 2020; p. 400, ISBN 978-3-030-33569-4.
67. Brusaporci, S.; Trizio, I.; Ruggeri, G.; Maiezza, P.; Tata, A.; Giannangeli, A. AHBIM per l'analisi stratigrafica dell'architettura storica. *Restauro Archeol.* **2018**, *26*, 112–131. [CrossRef]
68. Fryskowska, A.; Stachelek, J. A no-reference method of geometric content quality analysis of 3D models generated from laser scanning point clouds for hBIM. *J. Cult. Herit.* **2018**, *34*, 95–108. [CrossRef]
69. Bonduel, M.; Bassier, M.; Vergauwen, M.; Pauwels, P.; Klein, R. Scan-to-bim output validation: Towards a standardized geometric quality assessment of building information models based on point clouds. *Int. Arch. Photogramm. Remote Sens. Spat. Inf. Sci.* **2017**, *42*, 45–52. [CrossRef]
70. Barazzetti, L.; Banfi, F.; Brumana, R.; Previtali, M. Creation of Parametric BIM Objects from Point Clouds Using Nurbs. *Photogramm. Rec.* **2015**, *30*, 339–362. [CrossRef]
71. U.S. Institute of BUILDING Documentation USIBD Level of Accuracy (LOA) Specification Guide Document C120 TM [Guide] Version 2.0-2016 Guide for USIBD Document C220 TM: Level of Accuracy (LOA) Specification for Building Documentation. Available online: https://cdn.ymaws.com/www.nysapls.org/resource/resmgr/2019_conference/handouts/hale-g_bim_loa_guide_c120_v2.pdf (accessed on 25 February 2020).

drones

MDPI

Article

Integrated Surveying, from Laser Scanning to UAV Systems, for Detailed Documentation of Architectural and Archeological Heritage

Daniele Calisi *, Stefano Botta and Alessandro Cannata

Department of Architecture, Università degli Studi Roma Tre, 00153 Rome, Italy; stefano.botta@uniroma3.it (S.B.); ale.cannata1@stud.uniroma3.it (A.C.)
* Correspondence: daniele.calisi@uniroma3.it

Abstract: Nowadays, the study and digitization of historical, architectural, and archaeological heritage are extremely important, covering the creation of digital twins—virtual replicas of real spaces and environments. Such reconstructions can be achieved using technologies with passive or active light sensors: laser scanners as light emitters, or photogrammetry through the creation of photographic images. As for the latter case, a distinction must be made between terrestrial and aerial shots, increasingly facilitated by the spread of UAV systems. Point clouds are aligned using georeferenced points measured with a total station. To create a faithful virtual model of the subjects, dense point clouds from a laser scanner are used to generate meshes, which are textured in high resolution from aerial and terrestrial photographs. All techniques can be integrated with each other, as demonstrated through the experiences of two case studies, each serving different purposes. The first is a detailed survey conducted for CAD representation of certain areas of Rocca Farnese in Capodimonte. The second is an instrumental survey for the creation of a realistic digital twin, aimed at providing an immersive VR experience of the archaeological area of Santa Croce in Gerusalemme in Rome.

Keywords: survey; photogrammetry; UAV; laser scanner; virtual reality; digital twin; restoration; archeology; architecture

Citation: Calisi, D.; Botta, S.; Cannata, A. Integrated Surveying, from Laser Scanning to UAV Systems, for Detailed Documentation of Architectural and Archeological Heritage. *Drones* **2023**, *7*, 568. https://doi.org/10.3390/drones7090568

Academic Editors: Daniela Oreni, Fabrizio Banfi and Davide Mezzino

Received: 10 August 2023
Revised: 30 August 2023
Accepted: 1 September 2023
Published: 4 September 2023
Corrected: 24 February 2024

1. Introduction

The development of survey techniques has ancient origins, evolving in response to the needs of its time. It began with the earliest land surveying practices to define agricultural areas, progressing through tools such as pretorian tables, which were used from the late 16th century for surveys even on a large scale by targeting lines through the instrument. This evolution continues to the present day, encompassing the latest digitalization of archaeological and architectural artifacts using both terrestrial and aerial instruments. This progression has occurred in constant dialogue with developments in other scientific fields, such as optics and geometry, resulting in continuous refinements and innovations. Notable studies include the development of stereoscopy from the 17th century and the construction of lasers in the 1950s [1].

To date, numerous methods and tools for digital surveying have been developed, each with varying costs and characteristics. Their particularity, compared with traditional techniques, lies in their ability to produce 3D models of captured subjects as outputs, characterized by a high fidelity to the original. Many advanced surveying tools fall into the category of contactless devices with the surveyed subject (Figure 1), a feature which makes them suitable for the preservation needs of areas such as archaeological areas. In particular, photogrammetry belongs to systems which acquire and process data from the passive reception of light radiation, in contrast to laser scanning devices like LiDAR, which actively emit radiation and then analyze the reflected echo data for surveying purposes [2].

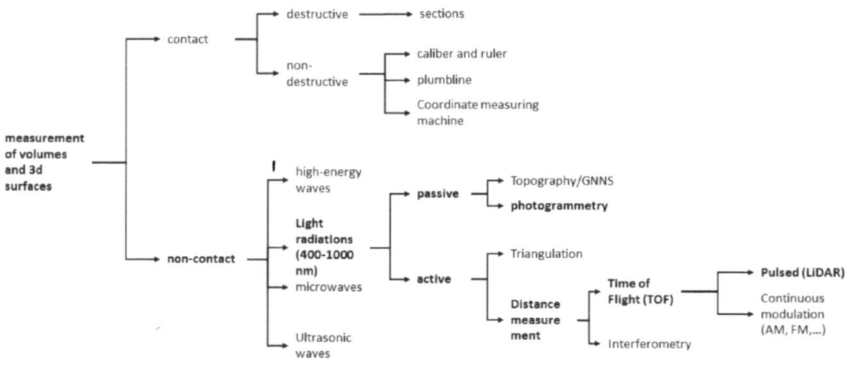

Figure 1. Taxonomy of measuring instruments in traditional and technological surveys with contact and non-contact techniques (image by the authors).

The ability to directly digitize architectural and artistic artifacts makes it possible to work with multi-scalar information, ranging from the smallest details to urban complexity. This has become an essential tool for archiving, preserving, and studying cultural heritage, as it facilitates the creation of databases of digital replicas capable of enhancing and assisting their analysis, dissemination, and management [3].

Among these techniques, photogrammetry refers to the principles of stereoscopy and the perceptual reconstruction of three-dimensional structures based on motion, supporting spatial detection and reconstruction. The photogrammetric surveying technique makes it possible to determine the shapes, sizes, and positions of objects in space through analysis, currently carried out by specific software, of at least two photographs of the subjects taken from two distinct points (known as a stereo pair). This process allows for the recognition of corresponding points, from which the scene can be reconstructed from different angles. The movement of the camera in relation to the object is crucial for deducing its three-dimensionality, which is why the technique is defined as structure from motion (SfM). For this reason, photogrammetric captures must occur in sequences of shots with uniform characteristics, depicting the object with a certain overlap percentage (at least 50%, though it might be variable) to allow for the recognition of homologous points within the set of images [4,5].

The idea of using photography as a measuring tool began to take shape around 1850. It started with Arago, the Director of the Paris Observatory, who requested specific resources from the French Chamber of Deputies to develop photographic techniques for producing increasingly accurate topographic maps [6,7]. In 1849, Colonel Aime Laussedat demonstrated the usefulness of photography for creating topographic maps. The first shots for this purpose were taken from the tops of mountains or the roofs of particularly tall buildings. Later, equipment was installed on early hot air balloons (Figure 2). The technique of obtaining measurements from a photograph was initially called "metrophotography" and was later referred to as photogrammetry, a term coined by the German engineer Meydenbaur in 1870 [8]. Having already applied intersection calculations to perspective views using the "camera lucida" (an optical device used for accurate drawing of viewed subjects), Lussedat later applied it to photographs of urban and mountain landscapes. His demonstrations in Paris and Buc, near Versailles, in 1861, convinced the French army to establish a specialized brigade for the development of photogrammetry [9].

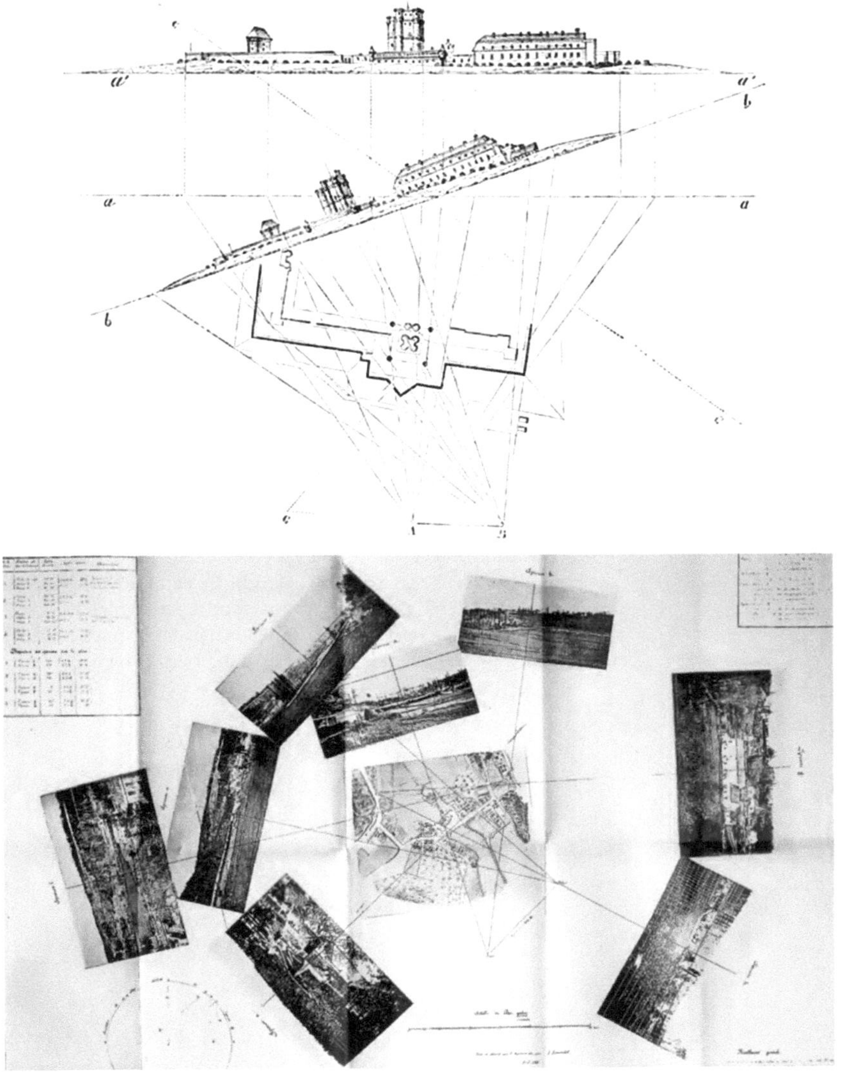

Figure 2. From the top: Laussedat, reconstruction of the facades of the Vincennes Fort in Paris carried out in 1850; Laussedat, photographic survey of Buc near Versailles carried out in May 1861. The eight constituent photographs of the survey, placed near the recording stations, with the developing map in the center [8].

Exploited by numerous scientific disciplines, photogrammetric techniques can be divided into two main categories: aerial and terrestrial photogrammetry. The distinction lies in the position of the camera and the distance of the sensor from the surveyed object, although the quality of graphical results obtained from the two photogrammetric modes can sometimes be similar.

Terrestrial photogrammetry employs primarily static devices through which sequential shots are taken using different capturing techniques, mostly keeping a close distance from the subject. There are a range of devices with different uses and levels of precision. This includes metric cameras, characterized by specialized lenses to control distortions and internal orientation systems, as well as amateur cameras, which can still achieve good resolutions.

Aerial photogrammetry, on the other hand, exploits devices such as drones to capture areas hardly accessible from the ground, and also optimizes results for subjects typically reconstructed by terrestrial survey. The advantages of unmanned aerial vehicles (UAVs) also lie in reducing acquisition time and costs, minimizing the number of shots required, and lowering risks for operators, who can control the flight remotely from safe areas, even in situations of inaccessibility or other hazards [10]. UAVs can be equipped with various types of sensors, making them versatile surveying tools across multiple sectors. The propulsion systems of the flying device are equally crucial in executing an aerial photogrammetric campaign. Fixed-wing drones, for instance, are more suitable for large open areas, requiring more takeoff and landing space. They cannot move at low speeds or get too close to the subject being surveyed. Helicopters and multirotor drones are commonly employed in this field, as they are easy to maneuver in all directions and can stably hover in place. There are also less common hybrid systems which allow for conversion between vertical and horizontal propulsion, making them suitable for various situations [11].

It is possible to determine the achievable level of detail with a device in advance, based on its technical specifications: the pixel size of the sensor and the focal length used for capturing make it possible to calculate the minimum visible size of an object in the image, depending on the distance from the object. This is known as ground sample distance (GSD) [5]. Using a simple proportion (Figure 3), it is easy to determine the maximum acceptable distance between the device and the surveyed scene to achieve a specific representation scale. In these cases, photographic datasets are composed by planning the flights over the analyzed areas, which is extensively useful for surveying large surfaces, where it is important to plan the percentage of overlap between photographs based on the altitude and the spacing between parallel flight paths executed by the drone.

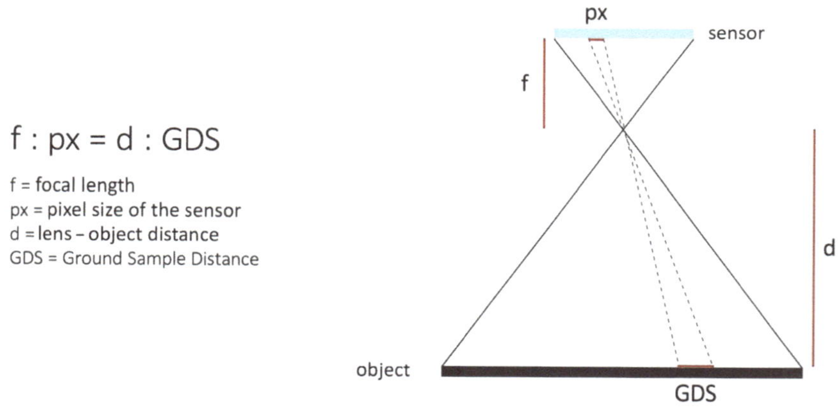

Figure 3. Definition of GSD (ground sample distance) in relation to the characteristics of the camera and its distance from the detected object. Based on the GSD it is possible to understand up to which scale of representation it is possible to represent the architectural object so that the perspective error of the restitution is acceptable (image by the authors).

Both techniques employ control points, which are positioned on the ground. In the case of aerial photogrammetry, these points are spatially identified through geographic coordinates and are usually defined using GPS or GNSS equipment, which provide centimeter-level precision in spatial coordinate determination. In terrestrial photogrammetry, these points are placed directly on the object being surveyed to facilitate measurements between different targets. These control points are also crucial as references in the subsequent alignment processes among different types of surveys, allowing the algorithms of the photomodeling software to recognize and align datasets with each other. This way, it is possible to combine aerial and terrestrial photogrammetric techniques, achieving not only complete reconstructions of artifacts but also maintaining different scales; this is particularly useful for structuring detailed models of extensive areas [12].

Digital acquisition can lead to various two-dimensional and three-dimensional graphical elaborations (georeferenced or not), including point clouds, meshes, and orthophotos [13]. For instance, digital terrain models (DTMs) and digital surface models (DSMs) are products of this technology: the former focuses on the land surface with information related to the topography of the ground plane, excluding all objects in the area; the latter represents the surface as captured, including vegetation, urban structures, and all anthropic interventions [14].

The advantages of both aerial and terrestrial photogrammetry lie in the speed and efficiency of data acquisition and processing. Typically, information supporting territorial and urban planning is updated every 10–15 years. Using photogrammetry, cartographic databases for local-scale projects can be updated in a shorter time frame [15].

New integrated survey technologies have allowed the analysis and study of archaeological, historical, and architectural heritage much more accurately. Public administrations and institutions have gradually begun a process of virtual digitization of assets under their responsibility. They understand the importance of having access to easily consultable digital twins for the management, visualization, cataloging, and programmatic planning of interventions for the conservation of the heritage itself. The case studies are numerous, and, at an international level, each case obviously has its specificity and the approach to each site is always different based on the geo-morphological characteristics of the artefact to be surveyed. Furthermore, it is also necessary to diversify the case studies based on the ultimate purpose of the survey itself: from the mere creation of a digital copy, to the mapping of materials and degradations for the purpose of conservative restoration, to visualization in a virtual environment through XR, up to immersivity in the site and in its context [16–20].

The presented research proposes an analysis of integrated digital survey processes and methodologies involving both aerial and terrestrial photogrammetric techniques in the detailed reconstruction of architectural and archaeological artifacts, in conjunction with other surveying tools like laser scanners. The aim is to obtain faithful replicas of reality, which can serve as the basis for further bidimensional and three-dimensional studies of the surveyed subjects. The purpose is not only to evaluate geometric and material qualities, which are essential in fields like restoration, but also to address data management issues arising from different types of instrumental surveys. Despite their different goals, the applied methodology is retraced through the two presented case studies. This encompasses the stages of survey and data capture, as well as the development and alignment of point clouds, leading to the production of the final models. The first case study focuses on Rocca Farnese in Capodimonte (VT, Italy), a 13th-century fortress on Lake Bolsena. The integrated survey aims to provide 2D–3D documentation for the study and restoration of the artifact. The second case study concerns the archaeological area of Santa Croce in Gerusalemme in Rome (RM, Italy). Aside from its analytical purposes, the project aims to develop an interactive and immersive virtual reality experience.

2. Materials and Methods

The first case study is the result of a collaboration with architect Anelinda Di Muzio on behalf of the Soprintendenza Archeologia Belle Arti e Paesaggio per la Provincia di Viterbo e per l'Etruria Meridionale. The subject is Rocca Farnese in Capodimonte, located in the province of Viterbo, Italy (Figure 4). Situated in the center of the Tuscia, in an area which preserves the memories of the Etruscans and the Renaissance residences of Caprarola, Vasanello, Vignanello, Bomarzo, Villa Lante, and Capodimonte, Rocca Farnese plays a significant role as a testament to the region's rich cultural heritage. Originally constructed as a fortress in the 13th century by the ancient nobility of Bisenzio, it was later transformed into a palace in the 16th century, for which the project was assigned to Antonio da Sangallo il Giovane, who emphasized its quadrangular plan. The building is situated on top of the historic center, on a volcanic cliff overlooking Lake Bolsena. In the past, the Rocca was the residence of the noble Farnese family and hosted some of the most illustrious popes in history. I also became a representative residency of Duchy of Castro, until its decline in 1649. Nowadays, Rocca Farnese is a private property of multiple owners but it's still partially visitable.

Figure 4. Aerial view of Rocca Farnese in Capodimonte on the Bolsena Lake. The small village is dominated by the Rocca which greatly dominates the entire town. The defensive walls containing the palace gardens are also significantly important (image by the authors).

The restoration of the Rocca Farnese in Capodimonte and all related activities, including the survey of the monument, are funded by the European Union, PNRR-CAPUT MUNDI, NEXT GENERATION EU IN ROME, Mission 1 Component C3, Ministry of Tourism subject holder of the funding, Soprintendenza Archeologia Belle Arti e Paesaggio per la Provincia di Viterbo e per l'Etruria Meridionale subject implementing the funding.

The survey of the Rocca was carried out using two different methodologies: one conducted using a laser scanner, the other involving aerial photogrammetry with various drones, different in sizes and characteristics.

During the initial phase of the campaign, which was crucial for the execution of the integrated survey, various types of targets were positioned. These targets served as reference points for the topographic survey, aerial photogrammetry with UAVs, and the laser survey. These targets played an essential role in ensuring the accuracy and consistency of geospatial data collected during the entire surveying process. The activities related to the topographic and drone surveys required distinct targets, considering the different techniques used and the specific goals of each operation. For the topographic survey, small metal plates were strategically placed in the surrounding environment. These plates were attached to structural elements, such as walls, pillars, and poles. This allowed for the establishment of a reference coordinate system and ensured a precise georeferencing of collected images and data. For the drone survey, a total of 124 targets were used, known as photogrammetric control points. To maximize operational efficiency, predefined targets were directly exported from Agisoft Metashape Professional 1.7.2 and then physically placed in the site. This choice was made to facilitate recognition by the same software, used later in the process to generate the point cloud. These targets consisted of reference images with specific geometric patterns. The shape and distribution of these patterns allowed the photogrammetric software to easily recognize them in the photos acquired by the drone. Accuracy in positioning of the reference points significantly influenced the overall precision of the survey and the subsequent cartographic processing. Furthermore, it was crucial to create detailed documentation regarding the position of each target, including coordinates and information about the surrounding environment. This facilitated the post-processing phase of the data. The topographic survey was then conducted to georeference characteristic points and determine their spatial coordinates (X, Y, Z) within a national reference system by a total station connected to a GNSS. Subsequently, this information was applied to the aerial photogrammetric survey to identify and associate common points between the two datasets.

During this phase, a survey campaign was also conducted using the Z+F IMAGER 5010 laser scanner (Figure 5) for the spaces in front of the main facade, the bridge and the area beneath it, the courtyard, and all interior rooms on the ground floor, including access points to adjacent levels.

Figure 5. Photo taken during the survey campaign, showing the laser scanner employed and one of the targets located on the wall next to the entrance arch. Different quality levels were used during the scanning process, mainly based on the distance between the artifact and the device. In particular, external scans were set to a higher-resolution quality to obtain greater accuracy of the data (image by the authors).

The laser scanner survey project was previously planned on a historical floor plan provided by the client, aiming at minimizing blind areas and adjusting the number of stations to ensure visibility between different positions for an appropriate alignment of point clouds, with minimal and acceptable margins of error. A total of 103 scans were required, mostly at medium resolution due to the relatively small size of the interior spaces (Figure 6). The presence of furniture and a significant number of windows resulted in numerous blind spots and noise caused by laser refraction and reflection on surfaces such as glass and mirrors. However, this did not affect processing, as the resulting point cloud was only meant for 2D redesign of the interiors (minimizing blind areas due to occlusions) and for understanding wall thicknesses; there was no need to fully recreate the furnished interior in the final output.

Photography played a crucial role in the architectural survey conducted with the assistance of drones. Exploiting modern technologies and the flight capabilities of these devices, it was possible to capture unique and distant angles, hardly accessible otherwise. The drones used during the survey activities included the DJI Mavic 3 Cine with an L2D-20c camera, the DJI Inspire 2 with an HG310 camera, and the DJI Mini 3 Pro with an FC3582 camera. The use of these tools allowed for the integration of information acquired during terrestrial survey campaigns, capturing the complexity of this particularly wide architecture, with resolutions of up to 5280 × 3956 pixels. Simultaneously, a photographic campaign was conducted for the interior rooms of the ground floor, the courtyard, the outdoor spaces above and beneath the bridge, and in front of the main facade. The dataset was acquired with a Canon EOS 6D camera with a Canon EF 24–105 mm f/4 L IS USM lens (mostly set at a focal length of 24 mm, with variable zoom for detailed shots of decorations). A total of 1270 RAW photos were taken and cataloged according to 27 areas to be easily checked during the CAD redrawing phase. This resulted in a wide archive of references, including an additional 160 photos taken by smartphone (Huawei 30 Pro, Leica Quad Camera, with a Super Spectrum CMOS main sensor and stabilized 27 mm f/1.6 OIS lens), as well as around 20 videos in MP4 and AVI by drone.

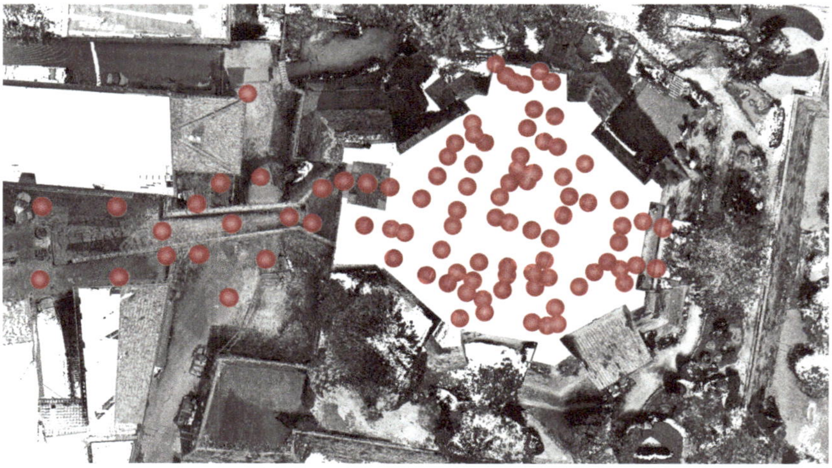

Figure 6. In red, the positions of the laser scans according to the subsequently processed point cloud in Autodesk ReCap. The laser scanner survey included the ground floor, the main facade on the north, the access bridge, and the surrounding context (image by the authors).

The availability of a wide set of photographs allowed for a comprehensive view and a deep control of Rocca Farnese, including details which were difficult to reach. To ensure optimal results in photogrammetry, a meticulous calibration of the cameras was necessary, considering lighting conditions and reflections in the surrounding environment (Figure 7). Additionally, ground-based reference points and targets were used for image georeferencing.

The entire drone survey operation involved a total of 8567 photographs, of which only 3602 were later included on Agisoft Metashape Professional 1.7.2. Among these 3602 images, 1371 shots were taken by the L2D-20c camera, characterized by a focal length of 12.29 mm, a resolution of 5280 × 3956 pixels, and a pixel size of 3.36 × 3.36 μm. An additional 97 photographs were acquired using the HG310 camera, featuring a focal length of 3.61 mm, a resolution of 4000 × 2250 pixels, and a pixel size of 1.7 × 1.7 μm. Lastly, 2134 images were captured by the FC3582 camera, with a focal length of 6.72 mm, a resolution of 4032 × 3024 pixels, and a pixel size of 2.4 × 2.4 μm. During the photograph alignment phase, some images were not recognized by the software, resulting in a final count of 3204 successfully aligned cameras.

To begin with the alignment process, the initial step involved the manual identification and recognition of targets from the topographic survey in all the drone photographs. By assigning spatial coordinates (X, Y, Z) to each of these targets, a rigid grid of fixed topographic points was effectively established, which was later considered by the software during the construction of the model. Subsequently, certain technical parameters were set, including the accuracy level being set to "High" to ensure the best result available in aligning photographs, fully exploiting the capabilities of the software. The accuracy parameter is directly proportional to processing time, but inversely proportional to the desired margin of error during this phase. Processing time is influenced by the PC specifications: in this case, a laptop with a 64-bit Windows operating system, equipped with 127.77 GB RAM, a 13th Gen Intel(R) Core(TM) i9-13900K CPU, and an NVIDIA GeForce RTX 4090 GPU.

Camera Calibration

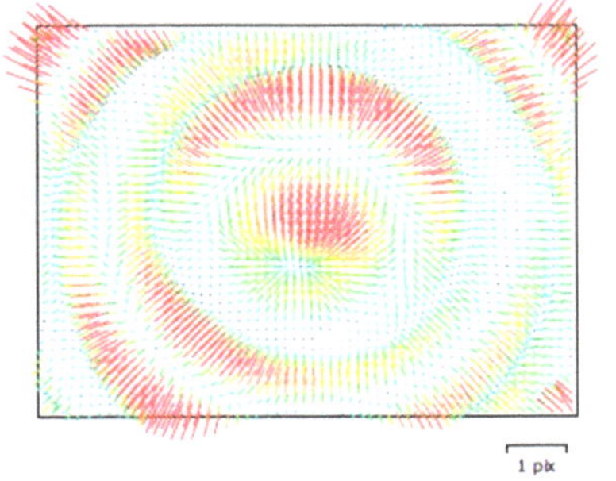

L2D-20c (12.29mm)
1371 images

Type	Resolution	Focal Length	Pixel Size
Frame	**5280 x 3956**	**12.29 mm**	**3.36 x 3.36 µm**

	Value	Error	F	Cx	Cy	B1	B2	K1	K2	K3	K4	P1	P2
F	**3728.31**	0.011	1.00	0.03	-0.20	-0.24	-0.00	-0.48	0.46	-0.42	0.38	0.02	-0.22
Cx	**-6.868**	0.022		1.00	0.01	-0.04	0.04	0.00	-0.00	-0.00	0.00	0.95	0.01
Cy	**20.86**	0.02			1.00	-0.02	-0.04	-0.01	0.00	0.00	-0.00	0.02	0.93
B1	**0.417082**	0.0065				1.00	-0.01	-0.00	-0.00	0.00	-0.00	-0.03	0.09
B2	**0.312897**	0.0062					1.00	-0.00	0.00	-0.00	0.00	-0.05	-0.01
K1	**0.0366065**	2.9e-05						1.00	-0.97	0.93	-0.88	0.00	-0.01
K2	**-0.187408**	0.00016							1.00	-0.99	0.96	-0.00	-0.00
K3	**0.405141**	0.00036								1.00	-0.99	0.00	0.00
K4	**-0.264526**	0.00027									1.00	-0.00	-0.00
P1	**-0.000817715**	1.7e-06										1.00	0.02
P2	**0.000725016**	1.5e-06											1.00

Figure 7. Calibration of the drone camera. In the table below, the calibration coefficients and correlation matrix (image by the authors).

Regarding the initial phase of the alignment process, called matching, it took 16 min and 47 s, with a memory usage of 3.42 GB. The second phase of the process, alignment, required a longer time, 1 h and 22 min, with a memory usage of 4.41 GB. The outcome of the alignment phase led the software to recognize the photographs in space and align them to create an initial sparse point cloud, consisting of 1,722,938 points (Figure 8).

The subsequent phase focused on generating the dense point cloud (Figure 9), with careful attention to setting specific parameters, such as configuring the quality to "High" and setting the filtering mode to "Aggressive." The process took a total of 6 h and 58 min, with a memory usage of 71.95 GB, and resulted in a file size of 7.64 GB. Initially, the point cloud consisted of 537,192,463 points, but it was later subjected to a cleaning and optimizing process to remove points which were irrelevant to the context of the surveyed area, as well as dense vegetation and shrubs which covered significant portions of the Rocca. The cleaning operation was carried out on each laser scan, in some cases also using the cloud clipping tool to focus on superfluous or incorrect parts (due to reflections or refractions) to avoid deleting useful parts of the survey. This operation aimed at avoiding issues during two-dimensional reconstruction of the building's facades. As a result, the model was reduced to 181,914,544 points, significantly lowering the size of the project file and making it more manageable.

Figure 8. Sparse point cloud built in Agisoft Metashape. The creation of the sparse point cloud is the first step of the reconstruction, resulting from the alignment of the photographs in space. The alignment process was implemented by recognizing the targets whose real coordinates were previously taken by the total station connected to a GNSS (image by the authors).

Figure 9. Dense point cloud built in Agisoft Metashape. The rigid grid defined by the 124 targets (with real coordinates taken from the total station connected to a GNSS) allowed for the alignment of all the frames of the virtual reconstruction with a minimal margin of error (image by the authors).

A 3D mesh reconstruction was then generated by starting the "Build Mesh" process. This operation led to the creation of a model consisting of 39,526,220 faces and 19,806,251 vertices, ready for the texturing process. Specific parameters were set during this phase, configuring the mapping mode to "Generic" and setting the blending mode to "Mosaic". Defining a texture size of 8192 × 8192 pixels was decisive to achieve an optimal level of detail in the mapping. The options for "Enable hole filling" and "Enable ghosting filter" were both enabled to ensure high-quality results in the process. The initial phase of the texturing process, known as UV mapping, took a total of 24 min and 41 s, with a memory usage of 5.07 GB. The subsequent blending phase took 42 min and 37 s, utilizing 11.13 GB of memory. At the end of the texturing process, the file size increased by 1.90 GB in addition to the previous memory used in the earlier stages. During these processes, a second subgroup of points, called a "chunk", was created in parallel with the one used for processing, in order to preserve an intact copy of the dense point cloud, without any point removal. This operation proved useful during testing in the meshing (or texturing) process, as it made it possible to identify the areas of file optimization based on the final weight of the product undergoing texturing. During a later phase of the process, this mesh model was used, together with the complete point cloud (made by joining UAV and laser scanner models), to elaborate and export the orthophotos needed for the 2D reconstruction. In fact, simultaneously with the processing of the drone-based aerial photogrammetry, the scans obtained by laser scanning were imported and aligned using the Autodesk ReCap 2023 v.22.0 algorithm, obtaining the point cloud of the interior areas of the ground floor and the exterior areas close to the main fronts, above and below the access bridge. This point cloud consisted of a total of around 638,600,000 points (Figure 10).

Figure 10. Point cloud processed in Autodesk ReCap from the laser scanner survey. The alignment of the scans starts from a master station which controls the entire project, to which the subsequent scans begin their alignment. The process is a progression: the second aligns with the first, the third with the second, and so on. Sometimes the source and destination scans need to be set manually, especially if the survey does not follow a chronological order, as often happens. The overlap is done automatically, or by indicating three pairs of points on two different scans (image by the authors).

The dense point cloud generated by the drone was simultaneously exported in E57 format to proceed with the alignment to the point cloud generated by the laser scanner. This model was then imported into Autodesk ReCap 2023 software within the project, where the laser scanner point cloud was already processed. Both point clouds were in real scale (1:1), as the drone-generated point cloud had already integrated the data obtained from the topographic survey (associated to the national reference system), while the laser scanner point cloud was already in real scale through its own reference system. However, the point clouds were in different positions of the virtual environment. Therefore, it was necessary to move and align the dense point cloud generated by the drone to the point cloud from the laser scanner; this was made by selecting the same set of points in both models, choosing three targets positioned within the courtyard. This procedure resulted in the loss of topographic data and the national reference system. However, such data were not essential for the purposes of the research, which included the two-dimensional representations of plans, elevations, and sections. After this operation, a complete point cloud was obtained, in which the information acquired from the laser scans was integrated with the information from the drone survey. Before proceeding further, a verification was carried out to ensure that the divergence between the two point clouds was only 2 mm. Once the accuracy of the result was confirmed, the new complete point cloud was saved in RCP format.

To proceed with the two-dimensional reconstruction phase, the complete point cloud was imported into AutoCAD 2022 as an external reference and centered into the standard origin of the workspace. The next step was to define a section plane for the ground floor and for each elevation to be surveyed and redrawn. As for the ground floor, the reference z = 0.000 elevation of the workspace was taken to create a "slice" type section plane (to control not only the position of the cut but also its thickness, managing both sections and projections). This plane cut through the point cloud at a height of +1.25 m relative to the chosen building reference elevation (0.00), which was a point on the floor of the internal courtyard portico (nearly flat); this data was crucial for dimensioning the drawings, both plans and elevations. Starting from the section obtained from this plane, all vertical section planes ("slice" type) were positioned and sized (Figure 11), with attention to ensure their parallelism with their respective facades. This resulted in a floor plan and 16 elevations: three main external elevations corresponding to the entrance facades, four internal elevations corresponding to the courtyard, and two elevations of the access bridge, for which a 2D redrawing at a 1:50 scale was planned along with material representation by overlaying the orthophotos from the mesh model; seven elevations of the remaining external fronts, for which only the material representation through orthophotos at 1:50 was planned; and a ground floor plan at a 1:100 scale (including the entire garden and access bridge) and a 1:50 scale (focused on the Rocca), for which only the two-dimensional redrawing was planned. Since the work was done in 2D on a sectioned point cloud within a 3D environment, a specific user coordinate system (UCS) was defined for each section plane, placing the XY plane on it and the pointing Z in relation to the plane's reading direction. This approach not only allowed the elevation to be displayed correctly on the orthogonal projection coinciding with the section plane but also allowed projection on the XY plane of any drawn element (with Z elevation = 0 related to the chosen UCS) by activating the "replace Z value with current elevation" setting, thus facilitating and optimizing the reconstruction work. This CAD project was then used as a template for each drawing.

Figure 11. The point cloud produced by laser scanning was subsequently aligned with the point cloud deriving from photo-modeling in Metashape and exported in E57. The complete point cloud, here imported into AutoCAD 2022, is cut by the section plane created for the floor plan. Here, the section plane of one of the elevations is underlined in blue (image by the authors).

As previously mentioned, for 2D redrawing of the facades, an overlay with high-resolution orthophotos was planned to accurately represent material variations on the surveyed surfaces. For this reason, the meshes obtained earlier were not further cleaned or optimized, but directly exploited to create the images using the "Build Orthomosaic" tool in Agisoft Metashape Professional 1.7.2. These orthophotos were developed to be aligned in reference to the section planes previously defined in the CAD environment on the point cloud, so they could later be placed in coincidence with them. This made it possible to work simultaneously on the sectioned point cloud and its corresponding raster image obtained from the mesh, providing a more detailed view of facade material qualities during the 2D redrawing phase; it also enhanced understanding of the structure and improved control over the three-dimensional aspects of the drawing. To obtain each facade, three markers were placed on the mesh to be recognized and used to set the horizontal and vertical axes of the projection plane. Using the mesh as the reference surface and enabling automatic hole filling, all orthophotos were generated (Figure 12), including the facades that did not require redrawing, with a pixel size of approximately 0.027 px/m (considering 300 DPI resolution images to be scaled at 1:50). These were exported in PNG format to maintain background transparency, with dimensions ranging from 3288 × 4096 px to 17391 × 10239 px.

Figure 12. Elaborations showing the two sides of the bridge at the entrance. Here, the orthophotos generated were also integrated with parts from the point cloud where the mesh version of the model was missing (the church in the background of the image on the left, for instance). Then, the bridge was re-drawn in AutoCAD, considering the orthophotos, to convey a full and detailed representation of the manufact and its material characteristics (images by the authors).

Some of the orthophotos processed in Agisoft Metashape had issues related to the quality of the mesh; in particular, the edges of the facades facing the courtyard of the Rocca were unclear and difficult to read due to an inability to get the drone close to the internal corners to avoid the risk of collision. Therefore, it was decided to attempt a reconstruction of the interior orthophotos using another SfM program, specifically, Reality Capture Beta 1.0, to evaluate other results using a different algorithm. Following the same workflow, textured mesh models of the courtyard facades were processed and orthophotos were generated.

Each orthophoto was then imported as a raster reference into the respective file, placed on the correct section plane, carefully scaled, and aligned to the orthogonal view from the point cloud. For each redrawn facade, two versions of the same orthophoto were elaborated: the first at 100% opacity, to be overlaid by CAD drawing; the second with a white background under the redrawn areas, leaving the remaining context at 60% opacity to highlight the two-dimensional representation (Figure 13).

The final phase of the process was the 2D reconstruction of the surveyed floor plans and elevations (Figures 14 and 15), working in detail by integrating information derived from orthophotos, point clouds, and high-resolution photographs obtained from aerial and terrestrial campaigns. These were essential to understand details such as decorations, which were redrawn geometrically where possible, or in their current state due to the widespread degradation of certain architectural elements.

Figure 13. The two versions of the facade: on the left, the 2D representation overlaying the orthophoto at 100% opacity; on the right, the redrawn facade with a white background and the orthophoto at 60% opacity that completes the rest of the facade. The double representation was necessary to appreciate the quality of the details of the 2D drawings, which were the basis for an initial mapping of materials for the subsequent restoration project requested by the client, which analyzes degradations and interventions. At the same time, the full-opacity representation also makes it possible to appreciate the detail of the orthophoto in its material details (images by the authors).

Figure 14. The 1:100 scale plan of the ground floor of Rocca Farnese, elaborated in AutoCAD 2022. The restitution process is complex, and it is also the result of a cross-check between the point cloud and the detailed photographic documentation, which allows for a better understanding of the various moldings and their generating geometries (images by the authors).

Figure 15. Parts of the elevation on the side of the main facade, redrawn from the integrated digital survey, overlaid on the orthophoto. On the right, a detailed focus of the same elaboration (images by the authors).

The second case study involves an integrated digital survey with different purposes, carried out at the archaeological site of Santa Croce in Gerusalemme in Rome (Figure 16). Located in the Rione Esquilino, the studied area is enclosed to the north by the Aurelian Walls and the Claudian Aqueduct. It includes a complex system of archaeological structures spanning various periods and typologies: the homonymous early Christian basilica, built between the 1st and 2nd centuries AD but completed only after the Baroque restorations in the mid-18th century; the Circo Variano, also from the 1st century AD, discovered and identified only through excavations in the 20th century; the Amphitheatrum Castrense, probably built by Elagabalus in the 2nd–3rd centuries AD on the highest point of the area; the Domus Costantiniane, with imperial-era masonry structures excavated during mid-20th century works; and the Civil Basilica of the Sessorian Palace, also known as the Temple of Venus and Cupid (circa 4th century AD) [21].

The project was conducted in collaboration with and with the permission of the Ministry of Culture (MiC) and the *Soprintendenza Speciale di Roma Archeologia Belle Arti Paesaggio*, aiming not only at an in-depth analysis of the archaeological area of Santa Croce but primarily at the development of an interactive and immersive experience in virtual reality. This is an experience to allow future users to explore the evolution of these places in an alternative way, exploiting the active and emotional nature of a multisensory simulation. An integrated survey was therefore necessary to create a digital twin of the site, an accurate replica of reality used as a realistic digital environment for a freely explorable virtual museum. This exhibition would be characterized by forms and behaviors intentionally discordant with the actual location, blending high-tech and metaphysical elements.

The laser scanning survey was planned according to a series of different criteria. The scanning was conducted in two phases during September 2022 (Figure 17), using the morning hours to achieve uniform lighting. In the first phase, special attention was given to scanning the Claudian Aqueduct, the Aurelian Walls, and the archaeological remains of the Circo Variano, as these elements were essential to the project; therefore, a high level of detail was necessary for their scanning. In the second phase, the scanning initially focused on the Domus Costantiniane, including a section of the adjacent aqueduct. Subsequently, the scanning campaign proceeded to the area beyond the tree-lined path, intended to serve as a background for the project. However, a lower level of detail was chosen to avoid excessive data overload The laser scanner used for the survey was the Z+F IMAGER 5010C.

Figure 16. Aerial image of the Santa Croce in the Gerusalemme site, captured during the survey campaign. The area is particularly complex due to the historical–archaeological stratifications that distinguish the Horti Variani area: the Claudian Aqueduct, the Aurelian Walls, the archaeological remains of the Circo Variano, the various Domus Costantiniane, the church and the annexed Amphitheatrum Castrense, the cupid temple, and more (image by the authors).

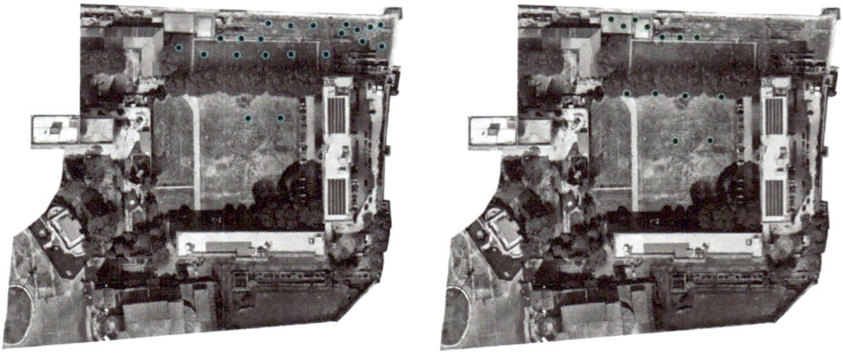

Figure 17. Plan of the first and second phases of the laser scanner survey campaign. The survey project was set up to avoid possible blind areas due to the arches of the aqueduct and the numerous wall ruins which could occlude the laser scanner. Therefore, stations were programmed on paper, calibrated on site where necessary, and all performed at a high resolution, to create a dense point cloud capable of expressing the material details of the survey (image by the authors).

Initial operations included tool calibration, setting the resolution, scanning speed, and accuracy. Survey points were then taken along the defined internal and external paths of the designed campaign. Special attention was given to collimation points between interconnected spaces that were not completely visible, ensuring control over possible blind areas of the point cloud. After completing all the scans for both phases, the two point clouds were generated and merged using Autodesk Recap 2022. The merging was achieved by recognizing common points between the models within the project, resulting in a unified dense point cloud. The result included a total of 30 scans from various positions within the Santa Croce in the Gerusalemme archaeological area. Due to the high level of detail used during the first phase, the resulting point cloud was extremely heavy, causing file management issues. Therefore, it was decided to optimize the number of points to reduce its size. However, a different approach was taken for the area of the domus and the aqueduct, whose number of points remained high to ensure greater accuracy. Through this process, defective areas were identified in the coloring of the point cloud, caused by the interference of sunlight during data acquisition. However, this issue was anticipated from the beginning, and a second survey with a drone was planned to integrate panoramic data acquired by the laser scanner.

Once the generation of the dense point cloud obtained from the laser scanner was completed, the model was exported in E57 format and subsequently imported into the structure-from-motion software Agisoft Metashape Professional 1.7.2. This process was necessary to divide the dense point cloud into four distinct chunks using selection and point removal tools. This division was carried out to optimize operations and set each chunk specifically based on the role the scanned subject would have within the project, considering the final goal. The four chunks corresponded to different areas: the Domus Costantiniane, the Claudian Aqueduct, the Aurelian Walls and Circo Variano, and the urban context.

The survey by UAV, planned in collaboration with an external certified technician, was also divided into two phases (Figure 18). During the first phase, the focus was on the Claudian Aqueduct and the Aurelian Walls, fundamental for the project and therefore captured with a high level of detail. The second phase was dedicated to a survey of the domus, protected by a shelter about two meters tall, which created some flying difficulties due to the limited space; the operation was nonetheless successfully completed. Subsequently, the surrounding urban context of the archaeological area of Santa Croce in Gerusalemme was scanned to be used as a background for the virtual experience. To initiate the survey process on the Claudian Aqueduct and the Aurelian Walls, targets were positioned and later used as reference points for the composition and processing of the point cloud using Agisoft Metashape. Different types of drones were used as surveying tools, including the DJI Mavic 2 Pro (Claudian Aqueduct), Mavic 3 Cinema (Claudian Aqueduct, Aurelian Walls e Circo Variano, Background), and DJI Mini 2 (Domus, Claudian Aqueduct, Background). For example, the DJI Mini 2, smaller and lighter than the others but equipped with a good camera, was used for the photogrammetric survey of the Domus Costantiniane, covered by a shelter, which required special attention in piloting to avoid damaging the artifacts.

Once the survey phase was completed (Table 1), the process of importing and aligning the images acquired through UAVs was carried out using the SfM software Agisoft Metashape Professional 1.7.2. A sparse point cloud was generated first, followed by the creation of a highly dense point cloud. Afterwards, the same process of division of the point cloud into four chunks was performed, similar to what was done for the one derived from the laser scanner. This operation was carried out with special attention to ensure that the chunks were delimited the same way as those previously obtained.

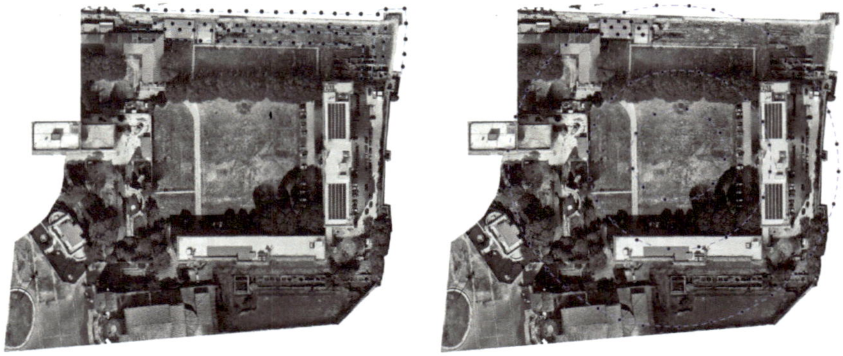

Figure 18. Plan of the first and second phase of the aerial photogrammetric survey campaign by drone. Acquisition of the photos was accomplished following the two main shooting techniques for reality-based models: with parallel axes and with converging axes. For some areas, such as for the Domus Costantiniane, hidden by an opaque roof, a precision flight was performed manually to avoid obstacles like supporting pillars (image by the authors).

Table 1. Summary of survey parameters and output data, divided according to the chunks.

	Domus. Point Cloud Laser Scanner/Drone + Texture Drone	Claudian Aqueduct. Point Cloud Laser Scanner/Drone + Texture Drone			Aurelian Walls + Circo Variano. Point Cloud Laser Scanner/Drone + Texture Drone	Background. Point Cloud Drone + Texture Drone	
Drone	DJI Mini 2	DJI Mavic 3 Cine	DJI Mavic 2 Pro	DJI Mini 2	DJI Mavic 3 Cine	DJI Mavic 3 Cine	DJI Mini 2
Images	21	556	4	21	190	643	5
Camera	FC7303	L2D-20c	L1D-20c	FC7303	L2D-20c	L2D-20c	FC7303
Resolution	4000 × 2250 pixel	5280 × 3956 pixel	5472 × 3648 pixel	4000 × 2250 pixel	5280 × 3956 pixel	5280 × 3956 pixel	4000 × 2250 pixel
Focal length	4.49 mm	12.29 mm	10.26 mm	4.49 mm	12.29 mm	12.29 mm	4.49 mm
Pixel size	1.76 × 1.76 µm	3.36 × 3.36 µm	2.41 × 2.41 µm	1.76 × 1.76 µm	3.36 × 3.36 µm	3.36 × 3.36 µm	1.76 × 1.76 µm
Aligned cameras	578	552			190	648	
Coordinate system	WGS 84 (EPSG::4326).	WGS 84 (EPSG::4326).			WGS 84 (EPSG::4326).	WGS 84 (EPSG::4326).	
Point cloud	445,755 points	180,164 points			107,946 points	435,376 points	
Accuracy	Highest	High			Highest	Highest	
Matching time	10 min 28 s	7 min 12 s			1 min 42 s	5 min 57 s	
Matching memory usage	670.27 MB	2.55 GB			957.68 MB	2.94 GB	
Alignment time	12 min 46 s	11 min 30 s			3 min 27 s	20 min 18 s	
Alignment memory usage	271.74 MB	283.58 MB			93.32 MB	520.86 MB	
File size	53.60 MB	41.35 MB			18.85 MB	63.85 MB	
Dense point cloud	19,534,249 points	142,211,824 points			78,761,296 points	198,642,533 points	
Processing time	20 min 32 s	53 min 25 s			5 h 32 min	2 h 25 min	
File size	300.82 MB	2.09 GB			3.80 GB	2.66 GB	
Faces	43,323,442	10,535,455			24,999,999	4,075,419	
Vertices	21,716,129	5,281,503			17,030,491	4,370,785	
Texture size	12.288 × 12.88 × 4	4.096 × 4.096 × 12			4.096 × 4.096 × 4	4.096 × 4.096 × 4	
Reconstruction time	1 h 29 min	4 min 8 s			4 h 2 min	1 h 38 min	
Uv mapping time	28 min 51 s	7 min 57 s			2 h 6 min	26 min 42 s	
Uv mapping memory usage	9.81 GB	3.82 GB			3.91 GB	8.75 GB	
Blending time	11 min 18 s	17 min 17 s			2 min 22 s	9 min 26 s	
Blending memory usage	23.99 GB	14.97 GB			6.82 GB	21.41 GB	
File size	2.65 GB	681.17 MB			1.23 GB	128.88 MB	

This provides an overview of the settings used for individual monuments, as well as the quality of the outputs developed, such as point clouds and meshes (image by the authors).

A careful evaluation of the data obtained from both campaigns indicated some key considerations for outlining an effective methodology for the following model development and texturing. The remarkable accuracy of the data acquired during the laser scanner survey allowed the generation of a dense point cloud that was very true to reality; however, the use of only panoramic photos proved to be less effective for a high-quality reconstruction of the textures of the mesh model. On the other hand, the use of aerial photogrammetry by UAVs made it possible to obtain more detailed and precise textural information, despite a sparser point cloud compared with the laser scanner. The reconstruction algorithm in the software, working with many shots taken from various positions, selected portions of the photos that best approximated a direction perpendicular to the polygon normal. These portions were combined and reprojected onto the modeled object as an image mosaic, generating the final texture. Therefore, the decision was to use the point cloud from the laser scanner to generate the mesh and then apply the texture developed through the drone survey.

The two dense point clouds from the distinct surveys were processed separately using Agisoft Metashape and later exported in E57 format. A unique reference coordinate system (RDN2008/Italy zone (N-E)—EPSG: 6875) was chosen to align them using the CloudCompare software, given the absence of georeferencing in the laser scanner data compared with the data acquired from the drone (Figure 19).

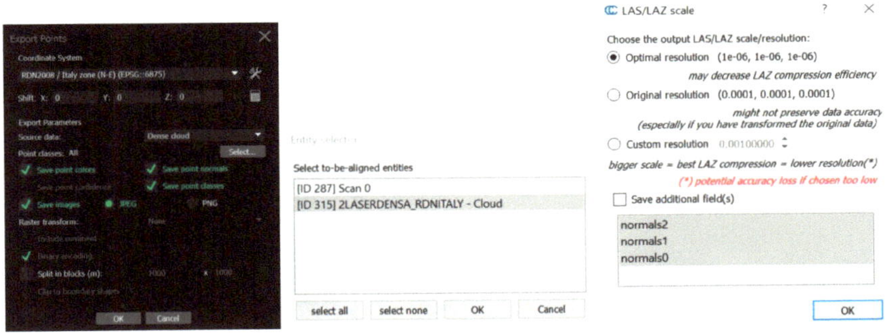

Figure 19. From left: export settings screen for the dense point clouds from laser scanners and drones in Agisoft Metashape; import screens in CloudCompare for the previously exported files (image by the authors).

Once the data was imported into the software, a set of matching pairs of points was identified between the two point clouds, which was used for alignment. The model obtained from the drone was used as the reference base for aligning the one from the laser scanner. The selection of common points between the clouds was performed with great care, as the precision in defining them directly affected the coherence of the overlap between the dense point clouds. The goal was to keep the displacement error between the two models constantly below 0.01 m to obtain an accurate alignment, finally achieving a value of 0.001 m (Figure 20).

The software provided the option to adjust the scale between the point clouds during the alignment process (Figure 21), transforming the second point cloud based on the primary one to achieve a perfect fit. In this operation, the larger the displacement error, the more significant the scale adjustment. Once aligned, the two point clouds were kept in two manageable chunks. Subsequently, the aligned point clouds were exported again in LAS format, using an optimal resolution from CloudCompare and preserving the new coordinates.

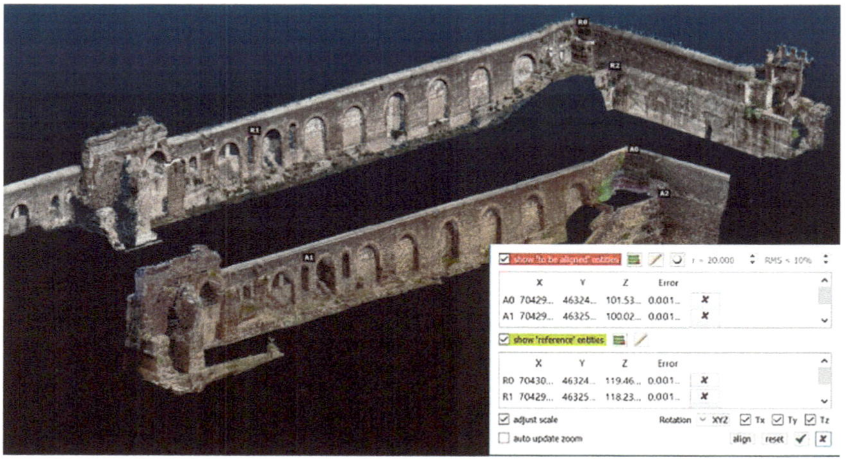

Figure 20. Selection of common points between the two imported point clouds in CloudCompare. The choice of targets for the alignment of the two point clouds must be such as to guarantee some characteristics: the points must be rather distant, to avoid a specific scaling and rototranslation in favor of a general one (in this case, for example, the length of the aqueduct); the points must belong to at least two different and/or orthogonal planes, preferably on the three Cartesian planes (image by the authors).

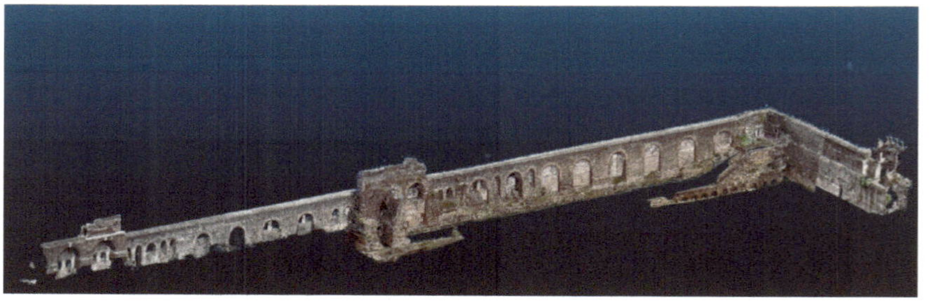

Figure 21. Aligned point clouds from drone survey and laser scanning in CloudCompare. The overlap error between the two clouds is 0.001 m, optimal for proceeding to the next stage by returning to the Metashape software (image by the authors).

The next step was to reintroduce the two aligned point clouds into Agisoft Metashape, divided into two separate chunks, to proceed with the construction of the mesh model. During the meshing phase, the laser scanner point cloud was processed with a high level of detail, disabling the "face count" option to avoid potential reductions in the number of surface polygons (Figure 22).

The obtained mesh was then subjected to the texturing phase using photos captured by the drone, applying the "Mosaic" blending mode to obtain the final high-resolution JPEG texture (12,288 × 12,228 pixels). This process, from the alignment phase in CloudCompare to the texturing phase in Agisoft Metashape, was replicated for each of the four areas that constituted the entire project. This resulted in the creation of four distinct files, each configured specifically based on the expected final output (Figure 23).

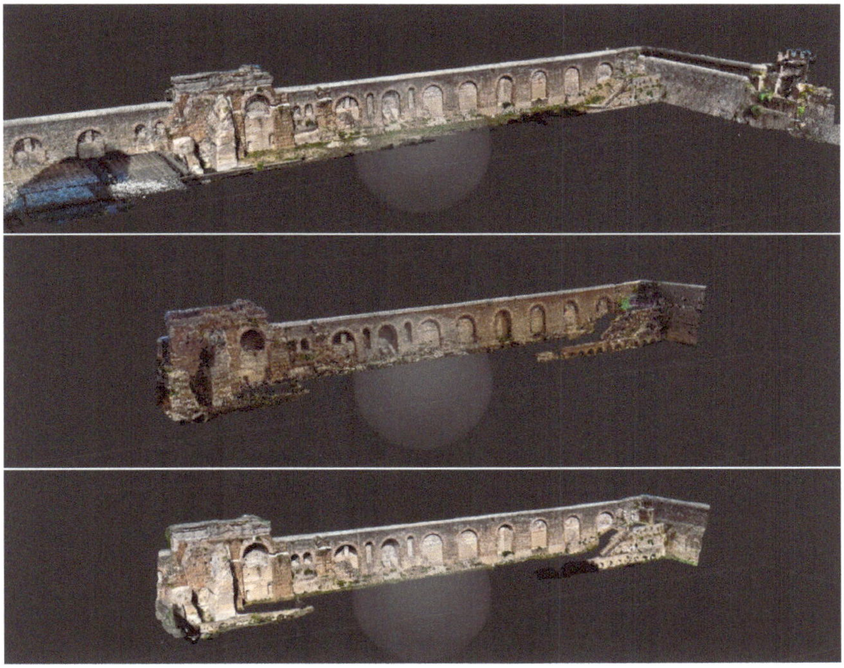

Figure 22. The dense laser scanner point cloud was exported from CloudCompare with the new Cartesian position coincident with the lower-quality drone point cloud. This way, it was possible to import the laser cloud into the Metashape source file, in the same position, and proceed to the creation of a mesh model more detailed than one from SfM. The 3D model finally underwent the texturing process, exploiting the photos taken by drone. From the top: dense point cloud from drone survey; dense point cloud from laser scanning; laser-scanner-derived mesh, textured by the drone survey (images by the authors).

Figure 23. From left to right: focus on the dense point cloud from drone survey; focus on the dense point cloud from laser scanning; focus on the mesh from laser scanning textured by the drone survey. The number of points in the laser scanner cloud made it possible to mesh with a high resolution. The perfect coincidence of the two models, laser and drone, has allowed the creation of a high-resolution texture from the set of photographs derived from the flights. This allowed for the creation of a coherent digital twin close to reality, on which the immersive VR experience could be based (images by the authors).

Once the processes of meshing and texturing for the entire archaeological area included in the project were completed, it was necessary to further divide each file into four chunks to achieve a proper export in OBJ format to be imported into Unreal Engine 5.0.3, the software chosen to develop the virtual experience. This procedure was carried out with the aim of optimizing the graphics performance of the project in virtual reality. Considering that the maximum size limit for importing into UE5 is around 1.5 GB per OBJ file, numerous exports were made from Agisoft Metashape, keeping each group of meshes below this threshold Furthermore, each exported mesh group was divided to have a small overlapping area with adjacent ones, to ensure a correct overlap when placed into UE5 environment, avoiding gaps in the final model. During the export process, the same local reference system was consistently used, devoid of georeferencing, as the final project output did not require such information. Given that the overall project included four distinct files, four different local reference systems were employed. Once imported into Unreal Engine, this resulted in a different arrangement of the archaeological artifacts compared with their real-world localization. To facilitate the correct placement of the parts, a complete version of the archaeological area was also included as a reference, to be then removed at the end of the process. This complete version was obtained by meshing the drone-generated point cloud at a lower quality to keep its weight below 1.5 GB. At this point, the original positions of individual archaeological artifacts in high quality were recomposed by adjusting the position and rotation coordinates of the parts.

The development of the VR experience on UE5 was conducted on a workstation equipped with a Windows 10 Pro 64-bit OS, 48 GB of RAM, an Intel(R) Core(TM) i7-8700 CPU, and a NVIDIA GeForce RTX 3090 Ti GPU. In order to handle the highly detailed meshes produced earlier, which had a high poly count, these were imported while activating the dedicated Nanite geometry visualization system (Figure 24). Recently developed and implemented in Unreal Engine 5, Nanite allows the real-time rendering of elements with a very high object count based on the observer's distance, allowing continuous internal optimization without necessarily simplifying assets in advance through other programs. Nanite is crucial for using digital twins with high levels of detail (LOD), especially when combined in a complex scene like Santa Croce in Gerusalemme [22,23]. Otherwise, hardware resources would struggle to manage the enormous amount of data. The Nanite mesh essentially remains a triangle mesh, which is automatically analyzed and segmented into clusters of triangles. These clusters dynamically adjust their LOD, increasing or decreasing it based on the camera's position, allocating more memory to visible details [24]. Nanite does not support objects without opaque and single-sided materials yet, but this is not an issue for the imported digital twin, as opaque and structurally simple materials were applied to the surfaces.

The VR Museum experience of the archaeological site was developed using a standard VR template provided by the game engine. This template was specifically designed to simplify the initial scene setup and already included useful scene elements, coding components, interactions, and animations. The entire experience was composed of three interconnected scenarios, created by blending scanned artifacts, hand-modeled elements in Rhinoceros 7, and high-resolution objects and materials sourced directly from the Megascans library by Quixel, accessible within the UE5 editor. The library facilitated the employment of digital vegetation to enhance the environments, chosen according to the species existing at the actual sites. The main scene, representing the archaeological area, was enriched with abstract designed artifacts to guide the user's experience. Additionally, a museum/control room was created within the scene, providing access to other environments (Figure 25).

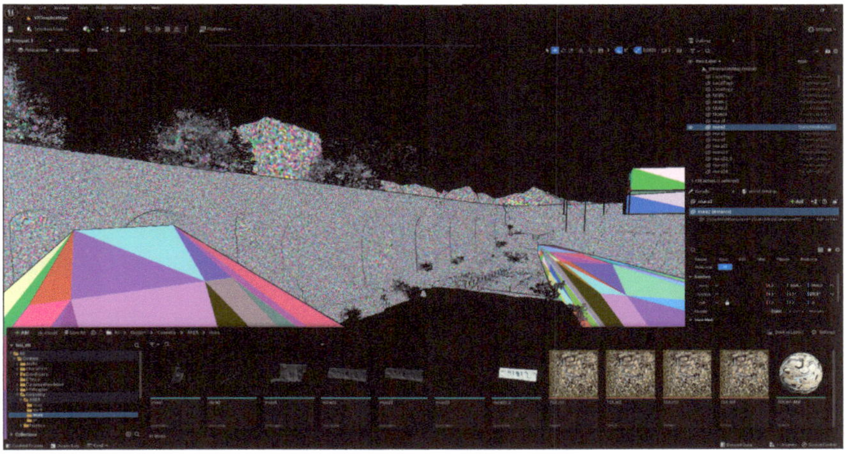

Figure 24. Nanite geometry visualization in Unreal Engine 5.0.3: the polygon count of the Aqueduct is visibly higher than for the structure in the background, while the foreground walkways appear much simpler as they were manually modeled and were already less complex (image by the authors).

Figure 25. A screenshot captured from within the virtual experience of the main scenario. In the background, the digital twin of the Claudian Aqueduct can be seen, while, in the foreground, hand-modeled elements have been placed to create evocative scenery, along with access to the virtual museum (image by the authors).

The other two scenes, the Domus Costantiniane and the Circo Variano, represent two focuses that reuse the meshes produced from the integrated survey to create isolated exhibition rooms. For performance reasons and to create the sensation of a surreal and abstract environment, the transition between scenarios (called maps in UE5) is achieved through a specially coded teleportation system. Other interactive elements, such as popup info points, particle effects, dynamic lights, and music, have been carefully placed in the area and programmed using the internal blueprint coding system to enrich the scenes and make them more engaging for the users. After wearing the headset and exploring a site which blends reality with metaphysics, they can discover and learn about the history of the place (Figures 26 and 27).

Figure 26. Screen from within the virtual experience, showing the digital hands the user exploits to interact with the domus (**on the left**) and the teleportation system (**on the right**) that allows movement within the environment (images by the authors).

Figure 27. Selections of some planimetric elaborations created for the digital exhibition within the virtual museum. On the left, the orthophoto of the domus floors; on the right, a focus on the redrawing of the remaining mosaics (images by the authors).

3. Results

Integrated digital surveying through UAV aerial photogrammetry and laser scanning allowed for the creation of highly accurate models of the two research sites, utilizing inherently similar methodologies but adapted for very different purposes. On the one hand, the 2D restitution of Rocca Farnese was carried out for subsequent analysis and restoration interventions, requiring careful consideration of dimensions and materials to accurately represent the current state of the site. On the other hand, the 3D reconstruction of the archeological site of Santa Croce in Gerusalemme aimed at creating an engaging environment in virtual reality, in which mesh surface optimization and integrity were combined with a need for detail and realism to construct an immersive experience of the locations. This reconstruction was also presented at the *Biennale dello Spazio Pubblico 2023*, held by the Department of Architecture, University of Roma Tre, where visitors could immerse themselves in this and other interactive virtual environments using an Oculus Rift S head-mounted display, freely navigating the museum with appropriate controllers and experiencing the archaeological site in an alternative way.

In both case studies, the methodology followed led to the elaboration of point clouds from different surveys, which were successfully combined with a divergence of only 2 mm between the two models for Rocca Farnese (Table 2a) and only 1 mm for those of Santa Croce in Gerusalemme (Table 2b). Once cleaned and refined, the final point clouds allowed the generation of high-density meshes, which were used without the need for further simplification or retopology due to the ultimate goals of the research and the exploiting of systems such as Nanite.

Table 2. Final summary of the results obtained from the models elaborated in the two case studies.

	(a)	
	Rocca Farnese	
UAV point cloud	n. points	181,914,544
Laser scanner point cloud	n. points	638,600,000
Final point cloud model	n. points	820,500,000
	final divergence between joined point clouds (m)	0.002
Mesh model	faces	39,526,220
	vertices	19,806,251
	(b)	
	Santa Croce in Gerusalemme	
UAV point cloud	n. points	251,961,595
Laser scanner point cloud	n. points	187,188,307
Final point cloud model	n. points	439,149,902
	final divergence between joined point clouds (m)	0.001
Mesh model	faces	82,934,315
	vertices	48,398,908

4. Discussion

Other than allowing the survey of spaces inaccessible by drone (such as the interiors of the Rocca), the implementation of laser scanning provides greater geometric detail of the areas captured through aerial photogrammetry. This implementation allowed full completion of the architectural volumes of the sites. The use of UAVs also facilitated the reconstruction processes, especially in the case of Rocca Farnese, where dense vegetation and numerous garden levels would have posed further challenges and required more time if attempted only through terrestrial photogrammetry and laser scanning. However, the vegetation, particularly the tree canopies near the artifacts, did affect the acquisition of certain fronts, creating gaps and necessitating the integration of terrestrial data. In the case of Santa Croce in Gerusalemme, tall vegetation did not present the same issue as it was generally situated at a distance from the archaeological site. Low vegetation and shrubs close to the ground connections of the artifacts, however, required meticulous work during mesh cleaning to preserve the captured data. During the VR environment development, efforts were made to fill and conceal any gaps with digital plant assets.

The adoption of an integrated survey methodology, exploiting various indirect techniques, significantly optimized this 2D–3D reconstruction work in architecture and archaeology. It allowed the combination of different levels of detail and addressed obstacles (such as volumetric articulations and vegetation) that are challenging to overcome through direct measurement. The use of orthophotos as a base for reconstructions increased the level of detail in redrawing and streamlined the process compared with traditional redrawing.

The two case studies described here are just examples of possible applications of this methodology. The approach needs to be adapted for cases with different goals, surveyed subjects, and contexts. Although the instruments may be similar and compatible, a survey campaign must always be designed according to the result to be obtained, avoiding waste of energy, resources, and costs. For Rocca Farnese, the client required a two-dimensional survey that would highlight the dimensional and material aspects as the basis for a subsequent map of materials and degradations to plan conservative restoration interventions. In this case, the effort was made to obtain a sufficiently detailed point cloud for CAD redrawing and the creation of strongly detailed orthophotos to ensure clear reading of surfaces. In the case of Santa Croce in Gerusalemme, the purpose of the survey was the creation of an immersive virtual museum, as established with the *Soprintendenza Speciale di Roma*. This required exploitation of the archaeological remains in the area to create a digital twin with optimal quality for an experience of a virtual space true to reality. Therefore the density of the laser scanner point cloud (whose textures were low in quality) was exploited to generate a mesh to be integrated with high-resolution textures obtained by UAV photographs.

Nonetheless, integrated surveys may have very different purposes. Many institutions are now digitizing architectural and landscape heritage, creating archives with many goals: from the simple consultation of the asset in a virtual environment, to the analysis or management of the digital copy, to the systematic programming of interventions, up to the analysis of the temporal decay of an asset subjected to many surveys through time. It is possible to structure a survey in HBIM to plan conservation or evaluate possible risks (for example, for buildings damaged by destructive earthquakes). Thanks to the speed of data acquisition, integrated surveys can be exploited to document the progressive stratifications during archaeological excavations, to allow a better understanding of the various covered levels in a virtual environment. In some cases, virtual models deduced from the surveys become the basis for experiences in XR. Especially after the pandemic crisis, museums and institutes are gradually starting to provide virtual tours of portions of their galleries, with various degrees of immersion. [25]. These alternatives have improved digital tourism, especially in a period in which circulation has been prohibited, reaching new frontiers in the creation of interactive digital twins. This process can be followed for those archaeological sites which, for various reasons, are not accessible to the public.

Ultimately, the integrated survey is just a new tool for replicating in a digital environment the same information that was typically represented by analogue survey: architectural or material plans, elevations, and sections in orthogonal projection, which are useful for representations at canonical scales, measurement, and analysis of architecture. These elaborations are still among the most requested by institutions and administrations.

The methodology adopted proved effective in diverse contexts with numerous purposes. While campaigns required various types of integrations during the process, this highlights the importance of an approach that integrates multiple tools and techniques to adapt to the specific needs of each case. The need to balance performance and quality, especially in projects requiring optimization, like those in VR, is an interesting research area that may need further critical exploration, both of the experiences described, and in further research in agreement with the *Soprintendenza Speciale di Roma* or with the *Parco dell'Appia Antica*. Surely, one of the research areas on which the research group is focused is the use of XR to enhance the understanding of places, whether they are real or just designed and virtually reconstructed [26].

The research carried out and the experiments on open conventions allow for the improvement of a process which still has issues, especially related to the significant sizes of files and their management. Therefore, alternative forms of experimentation could deepen the investigation into integrated systems using data, including working on software with innovative algorithms (such as Nanite for Unreal Engine).

From a social and perceptual point of view, VR (or XR in general) is a subject for wider interdisciplinary research, which attempts to study the different effects of experiencing these virtual environments on the personalities and development of new generations. Many sociologists have long criticized the educational power of some role-playing video games, but new studies are taking place on so-called serious games, immersive virtual experiences aimed at educating and teaching. The digital twins created and described in this article are part of these virtual reconstructions, which become navigable environments for the "role player" and which require a realistic and plausible level of detail.

These sectors are now at the center of global media attention, and it is the mission of universities to develop these issues and provide research ideas, investigate solutions, and create interdisciplinary relationships with the aim of improving the reconstruction, management, and perception of digital twins.

Author Contributions: Conceptualization, D.C., S.B. and A.C.; methodology, D.C., S.B. and A.C.; software, D.C., S.B. and A.C.; validation, D.C., S.B. and A.C.; formal analysis, D.C., S.B. and A.C.; investigation, D.C., S.B. and A.C.; resources, D.C., S.B. and A.C.; data curation, D.C., S.B. and A.C.; writing—original draft preparation, D.C., S.B. and A.C.; writing—review and editing, D.C., S.B. and A.C.; visualization, D.C., S.B. and A.C.; supervision, D.C.; project administration, D.C.; funding acquisition, D.C. All authors have read and agreed to the published version of the manuscript.

Funding: The research concerning the survey of Rocca Farnese was funded by Arch. Anelinda Di Muzio, Ph.D, specialist in monument restoration, on behalf of the *Soprintendenza Archeologia Belle Arti e Paesaggio per la Provincia di Viterbo e per l'Etruria Meridionale*, by scientific collaboration agreement *Analisi, dimensionamento e restituzione materica delle superfici della Rocca Farnese a Capodimonte (VT) attraverso tecnologie avanzate di misurazione e rappresentazione*—Dipartimento di Architettura Università Roma Tre.

Data Availability Statement: All data included in this paper is strictly connected to the research, is not yet published, and is the result of personal studies and analysis carried out by the research group.

Acknowledgments: This research was made possible by collaboration with Arch. Anelinda Di Muzio, *Soprintendenza Archeologia Belle Arti e Paesaggio per la Provincia di Viterbo e per l'Etruria Meridionale*, and Alessandro D'Accolti for Rocca Farnese, and with *Ministero della Cultura* (MiC), Arch. Alessandra Centroni, and Dott.ssa Simona Morretta on behalf of the *Soprintendenza Speciale di Roma Archeologia Belle Arti Paesaggio* for the archeological site of Santa Croce in Gerusalemme.

Conflicts of Interest: The authors declare no conflict of interest.

References

1. Docci, M.; Maestri, D. *Manuale di Rilevamento Architettonico e Urbano*; GLF Editori Laterza: Bari, Italy, 2020; ISBN 8842090689.
2. Bianchini, C.; Ippolito, A.; Bartolomei, C. The surveying and representation process applied to architecture: Non-contact methods for the documentation of cultural heritage. In *Handbook of Research on Emerging Digital Tools for Architectural Surveying, Modeling, and Representation*; IGI Global: Hershey, PA, USA, 2015; pp. 44–93. [CrossRef]
3. Pepe, M.; Alfio, V.S.; Costantino, D. UAV Platforms and the SfM-MVS Approach in the 3D Surveys and Modelling: A Review in the Cultural Heritage Field. *Appl. Sci.* **2022**, *12*, 12886. [CrossRef]
4. De Luca, L. *La fotomodellazione Architettonica*; Flaccovio Editore: Palermo, Italy, 2011; ISBN 9788857900704.
5. Cannarozzo, R.; Cucchiarini, L.; Meschieri, W. *Misure, Rilievo, Progetto per Costruzioni, Ambiente e Territorio*, 5th ed.; Zanichelli: Bologna, Italy, 2010; Volume III, ISBN 9788808123817.
6. Brunet, F. Inventing the literary prehistory of photography: From François Arago to Helmut Gernsheim. *Hist. Photogr.* **2010**, *34*, 368–372. [CrossRef]
7. Struffolino, D. *La Città Misurata. Tecniche e Strumenti di Rilevamento nei Trattati a Stampa del Cinquecento*, 1st ed.; Salerno Editrice: Roma, Italy, 1999; ISBN 8884022886.
8. Granshaw, S.I. Laussedat bicentenary: Origins of photogrammetry. *Photogramm. Rec.* **2019**, *34*, 128–147. [CrossRef]
9. Polidori, L. On Laussedat's contribution to the emergence of photogrammetry. *ISPRS* **2020**, *43*, 893–899. [CrossRef]
10. Ulvi, A. Using UAV Photogrammetric Technique for Monitoring, Change Detection, and Analysis of Archeological Exc vation Sites. *J. Comput. Cult. Herit.* **2022**, *15*, 1–19. [CrossRef]
11. Grassi, S.; Santarsiero, D.; Sirigu, M. *Guida Rapida alla Fotogrammetria 3d Con i Droni*, 1st ed.; myGeo Edizioni: Empoli, Italy, 2019; pp. 14–16, ISBN 9781701313026.
12. Barba, S.; Limongello, M.; Parrinello, S.; Dell'Amico, A.; di Filippo, A. *D-SITE. Drones. Systems of Information on Cultural Heritage for a Spatial and Social Investigation*; Pavia University Press: Pavia, Italy, 2020; Volume 1.
13. Vacca, G.; Dessi, A.; Sacco, A. Utilizzo delle immagini UAV a supporto della pianificazione territoriale. In Proceedings of the 21st National Conference ASITA, Salerno, Italy, 21–23 November 2017.
14. Liu, R.; Zhu, Y.; Luo, Y.; Liu, X. Construction urban infrastructure based on core techniques of digital photogrammetry and remote sensing. In Proceedings of the 2009 International Forum on Information Technology and Applications, Chengdu, China, 15–17 May 2009. [CrossRef]
15. Caroti, G.; Piemonte, A.; Nespoli, R. UAV-Borne photogrammetry: A low-cost 3D surveying methodology for cartographic update. In Proceedings of the MATEC Web of Conferences, Sharjah, United Arab Emirates, 18–20 April 2017. [CrossRef]
16. Herrero-Tejedor, T.R.; Maté-González, M.Á.; Pérez-Martín, E.; López-Cuervo, S.; López de Herrera, J.; Sánchez-Aparicio, L.J.; Villanueva Llauradó, P. Documentation and Virtualisation of Vernacular Cultural Heritage: The Case of UndergroundWine Cellars in Atauta (Soria). *Heritage* **2023**, *6*, 5130–5150. [CrossRef]
17. Elbshbeshi, A.; Gomaa, A.; Mohamed, A.; Othman, A.; Ghazala, H. Applying geomatics techniques for documenting heritage buildings in Aswan region, Egypt: A Case study of the Temple of Abu Simbel. *Heritage* **2023**, *6*, 742–761. [CrossRef]
18. Calisi, D.; Botta, S. Virtual Reality and Captured Reality for Cultural Landscape Communication. In Proceedings of the 9th International Workshop 3D-ARCH "3D Virtual Reconstruction and Visualization of Complex Architectures", Mantua, Italy, 2–4 March 2022; pp. 113–120. [CrossRef]

19. Zhang, M.; Yaoxin, Z.; Xue, F.; Xiaoqi, W. A 3D Spatial Diagnostic Framework of Sustainable Historic and Cultural District Preservation: A Case Study in Henan, China. *Building* **2023**, *13*, 1344. [CrossRef]
20. Wang, J.; Yi, T.; Liang, X.; Ueda, T. Application of 3D Laser Scanning Technology Using Laser Radar System to Error Analysis in the Curtain Wall Construction. *Remote Sens.* **2023**, *15*, 64. [CrossRef]
21. Colli, D. Il palazzo Sessoriano nell'area archeologica di S. Croce in Gerusalemme: Ultima sede imperiale a Roma? *Mélanges De L'école Française De Rome. Antiq.* **1996**, *108*, 771–815. [CrossRef]
22. Bevilacqua, M.G.; Russo, M.; Giordano, A.; Spallone, R. 3D Reconstruction, Digital Twinning, and Virtual Reality: Architectural Heritage Applications. In Proceedings of the 2022 IEEE Conference on Virtual Reality and 3D User Interfaces Abstracts and Workshops (VRW), Christchurch, New Zealand, 12–16 March 2022; pp. 92–96. [CrossRef]
23. Giordano, A.; Russo, M.; Spallone, R. Representation Challenges: Searching for New Frontiers of AR and AI Research. In *Representation Challenges—New Frontiers of AR and AI Research for Cultural Heritage and Innovative Design*; Giordano, A., Russo, M., Spallone, R., Eds.; Franco Angeli: Milano, Italy, 2022; pp. 9–18. [CrossRef]
24. Karis, B. Nanite—A deep dive. In Proceedings of the Real-Time Rendering in Games course at SIGGRAPH2021, Virtual, 9–13 August 2021.
25. Botta, S.; Calisi, D. Using Virtual Reality as a Tool to Research, Analyze and Learn: The Competition for Palazzo del Littorio in Rome. In Proceedings of the 3rd International and Interdisciplinary Conference on Image and Imagination, Milan, Italy, 25–26 November 2021; pp. 552–561. [CrossRef]
26. Mostra del Dipartimento di Architettura. Available online: https://architettura.uniroma3.it/mostra-architettura (accessed on 22 August 2023).

drones

Article

Enhancing Building Archaeology: Drawing, UAV Photogrammetry and Scan-to-BIM-to-VR Process of Ancient Roman Ruins

Chiara Stanga [1,*], Fabrizio Banfi [1] and Stefano Roascio [2]

1 ABClab GIcarus, Department of Architecture, Built Environment and Construction Engineering, Politecnico di Milano, Via Ponzio 31, 20133 Milan, Italy; fabrizio.banfi@polimi.it
2 Ministero della Cultura Parco Archeologico dell'Appia Antica, Piazza delle Finanze 1, 00185 Roma, Italy; stefano.roascio@cultura.gov.it
* Correspondence: chiara.stanga@polimi.it

Abstract: This research investigates the utilisation of the scan-to-HBIM-to-XR process and unmanned aerial vehicle (UAV) photogrammetry to improve the depiction of archaeological ruins, specifically focusing on the Claudius Anio Novus aqueduct in Tor Fiscale Park, Rome. UAV photogrammetry is vital in capturing detailed aerial imagery of the aqueduct and its surroundings. Drones with high-resolution cameras acquire precise and accurate data from multiple perspectives. Subsequently, the acquired data are processed to generate orthophotos, drawings and historic building information modelling (HBIM) of the aqueduct, contributing to the future development of a digital twin. Virtual and augmented reality (VR-AR) technology is then employed to create an immersive experience for users. By leveraging XR, individuals can virtually explore and interact with the aqueduct, providing realistic and captivating visualisation of the archaeological site. The successful application of the scan-to-HBIM-to-XR process and UAV photogrammetry demonstrates their potential to enhance the representation of building archaeology. This approach contributes to the conservation of cultural heritage, enables educational and tourism opportunities and fosters novel research avenues for the comprehension and experience of ancient structures.

Keywords: Roman aqueduct; historic building information modelling (HBIM); UAV photogrammetry; representation; building archaeology; extended reality (XR)

Citation: Stanga, C.; Banfi, F.; Roascio, S. Enhancing Building Archaeology: Drawing, UAV Photogrammetry and Scan-to-BIM-to-VR Process of Ancient Roman Ruins. *Drones* **2023**, *7*, 521. https://doi.org/10.3390/drones7080521

Academic Editor: Geert Verhoeven

Received: 22 June 2023
Revised: 3 August 2023
Accepted: 7 August 2023
Published: 9 August 2023

1. Introduction

In addition to its international recognition for its significant role in documenting and preserving cultural heritage, building archaeology has emerged as a discipline extensively utilised by conservation architects and archaeologists in the study and practical implementation of building conservation history. Furthermore, current research in this field is focused on advancing the integration of recording methods and techniques, addressing interpretative issues such as the application of stratigraphic analysis to standing buildings, exploring the impact of theoretical paradigm shifts on the discipline and harnessing the potential of new virtual and visualisation technologies to revolutionise both methodological and interpretative approaches in the future.

In this context, integrating advanced survey techniques, including laser scanners and photogrammetry, enables the acquisition of increasingly precise point clouds and models as a foundation for building archaeology studies. Particularly for large-scale archaeological sites or structures situated at elevated locations, the use of UAV photogrammetry allows for the documentation and identification of sites that are otherwise difficult to access. Our case study is the Claudius Anio Novus aqueduct. The aqueduct, especially the portion that runs through the Tor Fiscale Park, is a discontinuous structure stratified by successive transformations and is challenging to interpret. For this reason, building archaeology has been employed to guide analyses and elaborations.

The paper intends to offer a contribution to the ongoing research on the application of UAV photogrammetry in archaeological sites and monuments, highlighting the potential benefits of integrating drones into a comprehensive survey strategy that integrates topographic networking, laser scanning, terrestrial photogrammetry with HBIM and extended reality (XR). In recent years XR has gained renewed attention and interest in the fields of architecture and archaeology. The benefits of XR applied to built heritage are manifold in different areas, such as the documentation of historical sites and the possibility to experience immersive tours, together with the virtual reconstruction and visualisation of lost heritage and data sharing thanks to platforms that facilitate collaboration among experts involved in site analysis.

Furthermore, the 3D model has been used to realise the first approach to a digital twin (DT) of the aqueduct, which is meant to capture, store and monitor diverse information about the object of interest within a spatial and semantic database.

By employing UAV photogrammetry and a scan-to-HBIM-to-XR process, the paper shows the potential of these technologies in capturing detailed data, enabling precise documentation and supporting heritage interpretation and informed decision-making processes in archaeological and complex contexts.

2. Building Archaeology to Study Historical Buildings

2.1. Short Note about Building Archaeology in Italy

In Italy, the foundations of building archaeology, in terms of both practical and theoretical aspects, were established around 1987. During the 1990s, the discipline shifted towards various building archaeology and restoration topics, such as materials, construction techniques, building types, art history, historical interpretation and urban planning [1]. Initially, the term "building archaeology" aimed to encompass diverse experiences from multiple research centres and universities across Italy (Siena, Genova, Venezia, Brescia, Roma, Milano). However, there is still a lack of shared tools and practices in the field. Nevertheless, despite differing perspectives, building archaeology is widely recognised as a valuable tool for knowledge and an essential guide in responsible preservation projects.

The tools of building archaeology include building stratigraphy, chrono-typology, archaeometry, dendrochronology, the analysis of materials, surface finishes, construction techniques and non-destructive diagnostic investigations [2]. Building stratigraphy, borrowed from archaeology, aids in understanding the vertical sequence and interpreting the construction phases. It relies on the concept of stratigraphic units (SUs), which represent portions of a building that were likely constructed together due to a single constructive action. Therefore, knowledge of the historical construction techniques and processes specific to the building's cultural and geographical context is necessary [3,4]. However, a building's lifespan is also marked by further processes of addition (positive SUs) or removal (negative SUs) of parts to repair damages or fulfil new needs. SUs can also be categorised as surfacing units (wall masonry) or covering units (finishing). Studying unit borders and the "contact points" between different SUs becomes crucial in understanding their relationships and reconstructing the chronological sequence. Examining mortar joints is a fundamental step in the case of exposed masonry facades.

Building archaeology is a path at the boundary between two disciplines—architecture and archaeology—that complement and enrich each other. It is a multidisciplinary approach in which the archaeologist expands their observation to artefacts different from those they typically approach. At the same time, the architect borrows investigative tools from archaeology to analyse the structure they intend to restore or trace its historical evolution. Within this logic, the process of understanding a structure is a dual-lane road, where the first lane is occupied by bibliographic sources and the second by reading materials and construction techniques. Historical documentary research provides a starting point, but written documents and drawings are indirect sources, not always easy to interpret. The "material source," on the other hand, speaks for itself directly through its forms, colours and signs of workmanship or deterioration, even if layered. However, there are limits to interpretation, even in this case [5].

For this latter aspect, the survey has served as a starting point, providing knowledge of the aqueduct's geometric, dimensional and constructive characteristics. It serves as the graphic foundation for any hypothesis' verification. The three-dimensional model has also contributed to the understanding, allowing the visualisation of the relationship between the parts of the structure and volumetric conjectures. Indirect and direct sources are complementary: their dialectical relationship enables the understanding of the complexity of the architectural object and the verification of the hypotheses formulated for various analysed aspects.

The aim of building archaeology is to understand historical phases, the succession of specific actions carried out on the structure (including the history of the construction site) and the dating of certain elements or parts of the building, and to determine the construction methods, material treatment and supply of the construction site.

To "predict the past," archaeology utilises an evidential method that, starting from visible phenomena, seeks to trace back to their causes based on deductive and inductive reasoning, from the general to the particular and vice versa. Visible phenomena can be observed in the "signs," specific characteristics of the structure through which architecture communicates its history. They provide clues and evidence for the archaeologist and should be sought in the traces of the following [6]:

- the interfaces of stratigraphic units: the boundary, the meeting point between different units, becomes essential in establishing any existing relationships between two stratigraphic units;
- mortar joints and beds of masonry: this helps to verify the constructional continuity;
- employed materials: their surface treatment, typology and characteristics provide valuable information in understanding the material culture of a specific historical phase;
- construction materials: they allow the measurement of valuable elements—for example, for mensiochronology.

Obviously, what may serve as evidence for one construction may not be applicable to another, and each aspect can be a "sign." What is currently not understood may have significance in the future, perhaps with the discovery of a new investigative system.

Especially in the field of archaeology, the outcome of an investigation is never exhaustive because it depends on many variables, such as the specific preparations of the scholar (perspective, preliminary knowledge) and the development of new methods and analysis techniques that are continually updated over time to understand certain phenomena better. Rather than providing definitive answers to the numerous existing questions, these investigations open up new research scenarios that could enrich our knowledge of such a significant monument in Rome's construction history.

2.2. *Building Archaeology into HBIM in Italy: A Brief State of the Art*

Research centres have developed strategies to incorporate material and decay analyses into informative platforms to align HBIM with knowledge and preservation practices. Initially, geographic information systems (GIS) were employed, providing advantages in quantitative management and data analysis compared to AutoCAD, but still limited to two-dimensional representations [7].

As BIM methods have expanded to encompass built heritage, their potential has been directed towards diverse analyses for maintenance and preservation purposes. In Italy, the Basilica of Collemaggio served as a pioneering project [8] that consolidated decades of experimental practices. Following the devastating earthquake in L'Aquila in April 2009, the basilica underwent a complex preservation endeavour involving local superintendence offices and a design support group comprising research teams from the University of L'Aquila, Sapienza University of Rome and Politecnico di Milano. ENI coordinated the project, providing funding and technical expertise and overseeing the design and construction phases. This preservation initiative stood out by incorporating innovative technologies, such as the HBIM model, to manage various phases of preservation.

Additionally, the HBIM was utilised to generate historically significant architectural elements that were damaged during the earthquake. The model also facilitated construction site management, including formulating hypotheses regarding reconstructing the partially collapsed chevet roof and mapping materials and decay. BIM, functioning as an information system in a three-dimensional graphic space, proves to be a valuable tool for the historical-critical process of understanding built heritage.

Recently, research centres have focused on the transition from the two-dimensional to the three-dimensional visualisation of building archaeology analysis using HBIM. Some studies concentrate on field archaeological work, employing photogrammetric surveys to create 3D models of excavated stratigraphic units [9]. Archaeological stratigraphy analysis is also crucial for virtual reconstruction purposes, as explored by Demetrescu (2015) [10].

The preservation and rehabilitation plan for Castel Masegra, situated atop a rocky hillock overlooking the city of Sondrio, has been a significant undertaking since its acquisition by the Municipality in 2012 [11]. The BIM approach, in this case, aimed to represent the "volume stratigraphy" identified by dividing the model's elements based on different construction phases. Additionally, the BIM model served as a tool for the sharing of the work among the professionals involved in the project and for tourists and virtual visitors. Consequently, the study explored the potential utilisation of BIM for finite element analysis and more interactive sharing methods. Various applications were compared to leverage the model, even remotely.

In 2018, the Department of Civil Engineering at the University of L'Aquila, led by Stefano Brusaporci [12], proposed a BIM-based analysis of building archaeology for the San Vittorino complex near L'Aquila. The study focused on the wall of the old church of San Michele Arcangelo, dating back to the 4th century, to verify construction hypotheses. Two strategies were employed for the external façade: the first involved creating a parametric wall for each stratigraphic unit, comprising external facing, a core and internal facing layers. The second strategy entailed generating parametric walls for each stratigraphic unit, corresponding to the layers constituting the wall. This resulted in a parametric wall for the central core and multiple walls representing the identified stratigraphic units on the external and internal facades. Parameters such as the description, wall typology, construction phase and state of conservation were assigned to each wall. The creation of schedules in Autodesk Revit served to extract information associated with each element in tabular form. The corresponding construction phase was assigned to each stratigraphic unit, allowing for the visualisation of construction units for each historical period. In 2020, the National Research Center [13] conducted a study on the creation of an HBIM for the church of S. Francesco in Rocca Calascio, near L'Aquila. This rural building contains traces of 16th-century frescoes, and modelling of the stratigraphic units proved to be crucial. Two methods were employed: directly modelling each stratigraphic unit as a parametric family in Autodesk Revit, starting from the laser scanner-derived point cloud, and modelling the units within SketchUp and then importing them as a "mass" into Revit to transform them into parametric walls. In both cases, the parametric data for each entity were customised and updated based on the stratigraphic analysis.

Similarly, the approach of modelling directly in Revit after importing the point cloud was utilised in 2019 for the elaboration of Castello di Fossa, an ancient settlement on Monte Circolo [14]. In the same year, Diara and Rinaudo published a study on the realisation of HBIM focused on building archaeology using open-source software such as FreeCAD, which combines CAD and BIM programs. The software was implemented on multiple levels, including libraries, material databases and IFC classification [15].

These various studies represent initial attempts to incorporate building archaeology into BIM. Given the increasing interest in this field [16–18], it is crucial to establish discussion forums and develop shared tools and methods.

3. Case Study: The Claudius Anio Novus Aqueduct in Tor Fiscale Park, Rome

The Aqua Claudia is indicated by Frontinus as the eighth aqueduct of Rome, along with the Appia, the Anio Vetus, the Marcia, the Tepula, the Julia, the Virgo and the Alsietina [19]. The ninth is the Anio Novus. The Claudius Anio Novus aqueduct is therefore composed of the Aqua Claudia, on which the specus (channel) of the Anio Novus was built almost simultaneously.

The waters were collected from the upper Anio Valley, above Subiaco. The route of the two aqueducts overlaps, starting from the limaria pools, which are located at the height of the seventh mile of the Via Latina (now between Via Lucrezia Romana and Via Anagnina). After crossing Via delle Capannelle, the two channels rise first on low arches and then on increasingly higher ones, following the natural terrain and ensuring the necessary slope of the conduits. This is the most well-known stretch of the aqueduct, depicted in numerous paintings portraying the Roman countryside since the 18th century, such as the well-known engraving by Giovan Battista Piranesi. The two aqueducts run northwest towards Roma Vecchia, where the Aqua Marcia also begins its elevated path. From this point onwards, the two aqueducts run parallel to each other. After Villa delle Vignacce, they intersect twice, creating a diamond-shaped space, the so-called Campo Barbarico, near the Tor Fiscale Park, which is the focus of the research. Then, they turn north and, near Porta Furba, they follow the current Via del Mandrione until Casilina Vecchia, heading west. After passing Piazza Lodi, they reach the area of the ancient Sessorian Palace, an imperial residential complex built by Septimius Severus. The section of the Aqua Claudia adjacent to the Palace was incorporated by Aurelian into the construction of the walls, including those sections that crossed Via Labicana and Prenestina and now form Porta Maggiore. The terminal castle of the aqueduct was located near the gate, close to the Temple of Minerva Medicea, but it was demolished during the 19th century [20].

The section of the Claudius Anio Novus aqueduct located in the Tor Fiscale Park required the construction of more visible and continuous reinforcing structures, such as double brick arches (Figure 1). These arches are the only remaining and still visible constructions. The best-preserved part of this section, situated towards the north and closer to the city, consists of two large arches composed of square stone blocks with brickwork reinforcements (section D). In 2008–2009, safety and consolidation works were carried out on this section. Additionally, in 2007–2008, partial remains of the caementa cores of nearby arches (section C) were restored. The longest stretch of the aqueduct includes thirteen spans, which are defined alternately by double-arch load-bearing structures and the remaining caementa core (section G).

The *opus latericium* masonry from the Hadrianic era is well preserved for approximately three double arches. This section shows that peperino and tuff blocks were removed from the pillars and repurposed, as their imprints can still be seen on the *caementa*. Over time, the structures have been significantly altered and weakened due to spoliation, leading to prolonged exposure. Furthermore, from a structural standpoint, only the original consolidation structures of the monument are still functional. The most recent restoration of this section took place in 1964, as a plaque affixed to the structure shows.

In addition to the mentioned thirteen arches, there are other smaller monumental sections, including two peperino arches to the north (section B), which are now privately owned and were not restored in the 2000s. There are also two subsequent double arches (sections A1 and A2), with the latter located within private property. Three isolated *caementa* pillars (sections E, F, H) are separate from the rest of the structure. Only two of these pillars have undergone limited interventions (sections F, H), while section E has never been restored.

Figure 1. Detail of the Tor Fiscale Park plan and the sections of the Claudius Anio Novo aqueduct.

This section of the aqueduct is significant as it highlights the interaction between human construction and the geomorphological characteristics of the environment. These interactions have necessitated rapid structural interventions. The thickening of the piers or brickwork walls around the aqueduct (section D) during the Late Antique period indicates that the early Hadrianic interventions did not resolve the structural issues. On the other hand, the minimal intervention on the long stretch of the Claudius aqueduct in the Aqueduct Park suggests that restoration work was not required due to design deficiencies but rather other location-related problems. Roman architects were typically skilled at identifying the most suitable foundation soil. However, in this case, hydraulic requirements primarily influenced their choices to ensure a constant and controlled water flow.

The *opus quadratum* masonry of the Claudius aqueduct can be observed in section B, where a pillar and part of an arch have been preserved. It is also prominent in section D, particularly on the west side, where the pillars and arches remain (Figure 2).

Figure 2. The *Claudius aqueduct's opus quadratum* masonry identified in section D.

Another section of the original monument is located beneath the Torre Fiscale, which was built using the ancient structure as its base. In the Tor Fiscale area, the Hadrianic walls were constructed alongside each section, and, in most cases, they are the only surviving walls due to the gradual removal of the peperino blocks over the centuries. The practice of spoliation is particularly evident in section G, where the imprints of the removed stone blocks can be seen on the *caementa* walls. The Hadrianic brickwork is mostly preserved, encircling the four sides of the piers, except for the sections necessary to remove the stone blocks. The last construction phases, traditionally attributed to the Honorius period and likely dating from the Late Antique period, utilised masonry that reused Roman bricks with some courses of limestone or marble blocks.

During this phase, particularly in section D, the original pillars were dismantled and reassembled, and the opus quadratum's spoliation is noticeable (Figure 3) [21]. The peperino blocks were repositioned, sometimes reshaped to create smaller elements and placed using stone wedges and ample mortar joints. This indicates that the reshaping of the pillars no longer adhered to the ancient techniques of skilled stonemasons who could produce perfectly uniform ashlars. Specifically, on the east side of section D, the practice of spoliation deviated from the original pillar's morphology as it required strengthening and widening the base. The literature does not give much attention to the Claudius Aqueduct in the Tor Fiscale Park due to the extensive removal of the stone blocks, resulting in the near-total disappearance of the original structure. At first glance, it is challenging to identify the preserved elements of the aqueduct compared to the majestic arches in the Aqueduct Park. However, analysis through building archaeology [22] reveals extensive maintenance activities, indicating the continuous attention of Roman institutions, at least until late antiquity, towards an infrastructure considered vital for the city's functioning and the well-being of its residents.

Figure 3. The *opus quadratum's* spoliation in section D.

4. Materials and Methods

The research was based on a multi-sensor survey campaign, using a total station, laser scanner and terrestrial and aerial photogrammetry to acquire the area's dimensional, geometric and material data, including the park and aqueduct sections.

4.1. Multi-Sensor Survey Campaign

The survey activities, conducted in June and August 2021, focused on the Tor Fiscale Park area and its relation to the sectors of the Claudius aqueduct. The survey objectives included creating a plan of the entire park area with the precise positions of different aqueduct sections in a local and national cartographic reference and conducting a detailed survey of each portion of the aqueduct for the scan-to-BIM process. Overall, the survey operations aimed to accurately map the Tor Fiscale Park area, including the aqueduct sections, paths and park accesses, using geodetic techniques, UAV surveying and image processing.

The survey strategy comprised four main operations (Figure 4).

- Local geodetic network to georeference the survey in a national reference system and provide a stable reference for the combination of different survey campaigns. The network vertices were measured using a Global Navigation Satellite System (GNSS) and topographic techniques. Ground control points were used for photogrammetry and laser scanning.
- Network vertices were measured using a combination of GNSS and topographic techniques. Two different approaches were used to materialise the benchmarks for this study. The first approach, the "Order 0" benchmarks—fundamental vertexes of the network—involved placing topographic benchmarks fixed on the ground. A cast-iron covering protected these benchmarks. The second approach, the "Order I" benchmarks, utilised inox screws fixed within existing structures in the park area, such

as brick and concrete pathways and curbs. Retroreflective targets were positioned corresponding to stable points to facilitate total station repositioning. This allowed for resection strategies or repositioning on a network benchmark, using retroreflective targets to fix the azimuth angle. For the "Order 0" benchmarks (named 1000, 2000, 3000), their positions were determined using the GNSS technique through a static acquisition scheme, with each acquisition session lasting approximately 1.5 h. The measurements from these three stations were adjusted using data collected by the permanent GNSS station of the regional GNNS system located in Rome (ROUN), which served as an approximate baseline of 10 kilometres. Three independent sessions were conducted to determine the baselines: ROUN–1000 and ROUN–2000 (session 1); 1000–2000 (session 2); ROUN–3000 and 2000–3000 (session 3). The geodetic network, consisting of both "Order 0" and "Order I" benchmarks, was measured using a Leica TPS1200 total station. A least-squares adjustment was performed to compensate for discrepancies, resulting in average precision of approximately 1.5 millimetres for the benchmarks.

- A UAV survey was carried out to create a three-dimensional model of the area. The survey was conducted using the DJI Mavic Mini 2 and Mavic Mini 3 pro drones, employing two different flight altitudes. Firstly, a survey was carried out at a height of 25 m, using a double grid schema with 80% longitudinal overlap between images along the strip and 60% transversal overlap between different strips. Secondly, another survey was conducted at a height of 35 m, utilising a single grid schema with 60% longitudinal and transversal overlap. The selection of flight height was made to achieve a final orthophoto with a ground sampling distance (GSD) of 5.0 millimetres. Throughout the survey, approximately 2400 images were captured and processed into a single block. For the bundle adjustment, 10 points were utilised as ground control points (GCP), and 8 points were used as check points (CPs). The root mean square (RMS) error of the GCPs and CPs was approximately 2.0 centimetres. The DJI Mavic Mini 2 drone was used to carry out the initial photographic survey of the area, aiming to generate the overall orthophoto of the park and the three-dimensional models of the aqueduct sections. This first survey was supplemented by a second acquisition, conducted using the DJI Mavic Mini 3 pro drone, which was capable of capturing the more concealed parts of the structures, such as sub-arches, the interior sides of pillars and the water channel, thanks to the camera's ability to be oriented.

The Mini series was introduced in 2019 with the launch of the Mavic Mini, aiming to create a lightweight drone capable of delivering high-quality images and videos. The Mini 2 replaced the original model, which is no longer in production, and brought significant enhancements, especially in video quality and flight safety. The drone features a 12 Mp camera with a 1/2.3″ sensor, and, typical of this type of drone, both the focal length and aperture of the camera are fixed. The camera has a 24 mm equivalent focal length with an f2.8 aperture, and it also comes with fixed focus. Equipped with a 3-axis gimbal, the Mini 2 ensures excellent stabilisation during flights and includes bottom collision detection for added safety. However, it lacks front and rear sensors. The flight autonomy of this model is approximately 30 min under normal conditions (at a flight speed of 4.7 m/s with no wind). The DJI Mavic Mini 3 pro represents a significant advancement over its predecessor, mainly due to the integration of a larger sensor (1/1.3″) with a higher resolution (48 Mp) in the camera. While maintaining the same focal length (24 mm), the camera now boasts a brighter aperture (f1.7) and autofocus capabilities. The drone has improved obstacle detection through front and rear sensors, enhancing flight safety. Additionally, the battery life has been slightly extended to 34 min. An optional controller with its screen can be purchased, eliminating the need for a smartphone during flights. Another notable improvement compared to the previous model is the expanded controllable tilt range of the gimbal, enabling a unique portrait capture mode, which, as mentioned, helped to capture the sub-arches and other parts of the aqueduct. Regarding the camera, DJI has incorporated a Quad Bayer type filter in the sensor. This choice likely serves as a compromise to increase the resolution while minimising noise, especially in video recording. The signal to noise ratio (SNR) influences the noise in an image, with higher values resulting in less visible noise. To improve the SNR, larger sensor cells are preferable to capture more photons and reduce signal variations between cells, thereby reducing image noise. In the Mini 3, despite the larger sensor, the resolution has been quadrupled compared to the Mini 2, leading to smaller cell sizes and potentially decreasing the SNR ratio. However, implementing a Quad Bayer filter helps to control noise when working at lower resolutions, such as 4K video and 12 Mp still photos. Pixel binning, a technique that combines information from four sensor cells for each pixel, is applied in such cases, significantly improving the SNR. This pixel binning technique is present in all analysed drones with resolutions exceeding 12 Mp. On the other hand, the maximum resolution capture mode (48 Mp) can be useful in capturing highly detailed bright scenes, but it may suffer from increased noise levels in low light conditions (Table 1).

Figure 4. Laser scanning (**upper left**) and UAV photogrammetry survey operations (**upper right**). The localisation of the benchmarks "Order 0" (1000–2000–3000) and "Order I" (in red) and retroreflective targets (blue) (**bottom**).

The purpose of the photogrammetric survey was twofold: on the one hand, to reconstruct all non-detectable areas using terrestrial laser scans (such as the upper part of the aqueduct and certain areas in the upper sections of the arches), and, on the other hand, to generate high-resolution textured models and orthophotos of the elevations, top views and sub-arches. The photogrammetric acquisitions were designed to obtain orthophotos at a scale of 1:20. Two different cameras (Canon EOS-1D Mark IV and Nikon D700) were used with varying focal lengths depending on the requirements (Table 2). Generally, a 20 mm lens was used; a 35 mm lens was used for the upper parts, while a 90 mm lens was used for the sub-arches of the highest arches of the aqueduct (section G). Each block was acquired and processed separately. Depending on the size of the blocks, the number of processed photos ranged from 100 to 4000 images. The root mean square (RMS) of the ground control points (GCP) and check points (CP) was approximately 1.0–2.0 cm.

Table 1. Comparison of DJI Mavic Mini 2 and DJI Mavic Mini 3 pro.

	DJI Mini 2	**DJI Mini 3 Pro**
Focal	24 mm	24 mm
Sensor size	1/2.3″	1/1.3″
Image resolution	12 Mp	48 Mp
Aperture	f2.8	f1.7
Raw	yes	yes
Video	4 K—30 fps	4 K—60 fps
Gimbal tilt	−90° 20°	−90° 60°
Collision sensor	Bottom	Bottom Front Rear
Autonomy	32 min	34 min
Range	10 Km	7–12 Km

Table 2. Terrestrial photogrammetric survey: acquisition parameters.

	Terrestrial Camera 1	**Terrestrial Camera 2**
Camera	Canon EOS-1D Mark IV	Nikon D700
Focal	20 mm	20–35–90 mm
Sensor size	27.9×18.6 mm^2	35.8×23.8 mm^2
Image resolution	4896×3264 pix	4256×2832 pix

- A laser scanner survey of the different sections of the aqueduct was conducted using the FARO Focus 3D X 130 HDR (Table 3). The scanning acquisition was designed to ensure the maximum coverage of the aqueduct surfaces, reducing self-occlusions and ensuring an adequate point cloud density (Figure 5). Two different survey campaigns were planned, acquiring 213 scans, georeferenced with average reference precision of approximately 4 mm on the ground control points.

The two-dimensional drawings (Figure 6) were obtained using the point cloud data in Autodesk AutoCAD and McNeel Rhinoceros. For each block, three main horizontal sections were defined at an approximate elevation of 1.20 m from the ground level. These sections corresponded to the elevation of the sub-arches of the first- and second-level reinforcement structures, which varied depending on the different blocks. In some cases, they were also positioned at a higher elevation (section G). Subsequently, a cross-section was created for each block at the key points of the arches. The same section line was used to create the two internal elevations of the pillars. Finally, the top view (obtained through UAV data) and the four main elevations were created. Orthophotos obtained through UAV surveys, including the sub-arches, enriched the various outputs. Simultaneously, three-dimensional models obtained through both ground-based and aerial photogrammetry were processed using the Agisoft Metashape software (version 1.8.0). These models were used to generate the orthophotos. The photogrammetric model, consisting of a textured mesh, was imported into McNeel Rhinoceros to carefully draw certain details, such as the outlines of the peperino stones, which were more challenging to identify in the point cloud. Additionally, other three-dimensional representations were created. A NURBS model in McNeel Rhinoceros was used to study hypotheses and reconstructions of the construction phases of the aqueduct, such as the removed peperino stone blocks of the pillars. An object-based model was created in Autodesk Revit, following the analysis of architectural archaeology. While the former is intended as a visualisation tool within an interchange software that can be used for parametric modelling (Revit) or virtual reality (Twinmotion, Unreal Engine), the latter is aimed at supporting the operational phase of knowledge and the planning of diagnostic surveys and restoration interventions. It assists operators involved in the conservation of the aqueduct.

Table 3. Faro Focus 3D X 130 HDR parameters.

	Laser Scanner
Method	Phase-based method
Error distance	Accuracy up to ±2 mm
Range	From 0.6 m up to 130 m
Noise reduction	50%
HDR	Photo overlay
HD resolution	Up to 165 Mpx colour

Figure 5. Laser scanner point clouds of sections G and F (**top**) and D and E (**bottom**).

Figure 6. The drawings of section D with the dimensions: east and west elevation (**upper**); sections (**bottom left**), and plans at different levels (**bottom right**).

4.2. UAV Photogrammetry of Section D

By delving into the state of the art, various research studies have identified several benefits in digitising built heritage with aerial photogrammetry [23–28]. One of the key advantages is non-invasive data collection. Unlike traditional methods that often require physical access to heritage buildings for surveying and measurements, UAV photogrammetry allows for data collection from a distance. This approach minimises the risks associated with potential damage to delicate or fragile structures. Another benefit is high-resolution documentation [29,30]. Specifically, the drones enabled the documentation of previously inaccessible areas of the Claudius Anio Novus aqueduct, such as the upper parts of the structures and the barrel vaults between each pillar, located approximately 7 m above the ground.

UAV photogrammetry enables the capture of detailed, high-resolution images of heritage buildings from various angles and perspectives. These images serve as valuable records, documenting the current condition of the building, architectural details and historical features [31–33]. UAV photogrammetry also facilitates the comprehensive 3D modelling of heritage buildings [34,35]. Sophisticated approaches demonstrate the possibility of reconstructing buildings in three dimensions by integrating multiple overlapping images captured from different angles.

Additionally, utilising UAV thermal images and small multispectral UAV sensors further enhances the accuracy and richness of the data captured [36–38]. The 3D model enhances the understanding of the building's architecture and spatial layout. Additionally, UAV photogrammetry supports the condition assessment and monitoring of heritage buildings [39,40]. By comparing photogrammetric models created at different time points, changes or deterioration in the building's structure can be identified. These data assist heritage conservation professionals in prioritising restoration efforts and tracking the effectiveness of conservation interventions. The detailed information obtained through UAV photogrammetry is invaluable for preservation and restoration planning. Architects, engineers and conservationists can analyse the 3D models to pinpoint areas needing repair, evaluate structural stability and design appropriate conservation strategies. This information contributes to preserving the historical value of the building while ensuring its long-term sustainability. Lastly, UAV photogrammetry outputs, such as high-resolution images and interactive 3D models, facilitate public engagement and education [41,42]. These resources can be shared with the public through online platforms or exhibitions, providing broader access to heritage buildings. Furthermore, they promote public involvement and education about architectural history, cultural heritage and preservation efforts.

Once the data are acquired, a vital post-processing phase is required, where many photos are processed to obtain 2D and 3D representations, which are essential for subsequent analysis stages [43–47]. During this phase, one of the key elements is the generation of orthomosaics. This involves stitching together multiple high-resolution aerial photographs to create a georeferenced image that provides a detailed and distortion-free overview of the building and its surroundings [48,49]. The orthomosaic serves as a foundation for further analysis and interpretation. Another essential component is the point cloud, which consists of a collection of 3D points accurately representing the building's surface. By extracting and correlating the 3D coordinates of distinctive features captured in multiple images, the point cloud effectively conveys the building's geometry and spatial arrangement. The point cloud data can be processed to create a 3D mesh or polygonal model to enhance the visualisation of the building's architecture. This mesh connects the points, forming triangular or polygonal facets that accurately represent the surfaces and forms of the structure. To achieve a more realistic representation of the building's appearance, the orthomosaic or 3D mesh can be textured using the original aerial imagery. The architectural representation gains visual realism by applying textures and colours to the model's surfaces, facilitating better visualisation and understanding [50–52].

Annotations and measurements also play a crucial role in highlighting specific architectural details or important elements within the representation. Labels, markers, arrows and other annotations can be added to draw attention to key features, while accurate measurements such as distances, areas and volumes can be extracted from the 3D model, supporting analysis and design processes [53–55]. Various software applications are available to visualise and render the architectural representation. These tools allow for the creation of realistic renderings, walkthroughs or virtual tours, providing an immersive experience and a deeper understanding of the building's architectural characteristics.

On the other hand, digital photogrammetry does not provide semantic models, despite the ability to generate a substantial amount of digital outputs such as orthophotos, point clouds and mesh models. Semantic enrichment provides added value and is a prerequisite in historic building information modelling (HBIM) models. As is known, HBIM models, mainly through the scan-to-BIM process, allow the professionals involved to add information. Through bidirectional and parametric logic, each BIM object can have its material type with its physical, mechanical and thermal characteristics, along with other details, such as historical phase dating, construction techniques, degradation pathologies and restoration and repair actions. In this context, understanding the artefact to be digitised and represented in various digital forms becomes the key element for the correct interpretation of each architectural and structural element [56–60]. Drawing and, more specifically, 3D modelling can be directed towards an understanding and digital transposition process in which simple points in space and textured meshes become true digital twins, capable of enhancing the awareness of the historical heritage and the tangible and intangible value of the structure itself [61,62].

All these considerations were observed in the photogrammetric survey of section D of the Claudius Anio Novus aqueduct. The photogrammetric acquisition was performed with the Canon EOS-1D Mark IV and Nikon D700. For the reconstruction of the overall shape of section D, a 20 mm lens was used. The photo-set was then integrated with a 35 mm lens, which was used for the upper parts, and a 90 mm lens was used for the highest arches of the aqueduct. The UAV survey aimed to capture the parts not visible from the ground (top) and the arches (Figure 7). In total, 300 camera and 960 UAV pictures were acquired for a total of 1260. A set of 29 markers was used, measured with the total station. The root mean square (RMS) of the ground control points (GCP) and check points (CP) was approximately 1.0–2.0 cm. The point cloud had 6,801,707 points, while the 3D model had 5,903,747 faces and a texture of 8192 × 4096 × 40.3 bands (file size 545.34 MB).

Figure 7. From point clouds to high-resolution textured models (section D).

5. Results

5.1. Studying Construction Geometries and Techniques

The detailed survey facilitated a better understanding of the construction techniques and phases of the aqueduct, as well as the analysis of material decay, which is crucial in developing a conservation design plan for the monument. Despite the stratification of the different construction phases, the structures and the heterogeneity of the materials, the Claudius Aqueduct is characterised by mostly recognisable construction phases and extensive, distinguishable recent repairs. However, these repairs are difficult to place unless otherwise documented chronologically. Each phase utilises a limited number of materials and construction techniques that are macroscopically homogeneous, usually distinguishable from those employed in preceding or succeeding phases. Specific interventions and localised repairs can also be observed (Figure 8).

The building archaeology analysis of the aqueduct, conducted at a preliminary level, is based on visual observations, bibliographic references and an initial exploration of the archives, particularly regarding 19th-century restorations. The analysis aims to identify the stratigraphic units, which correspond to areas defined by recognisable construction techniques. The analysis aims to reconstruct the history of the restorations and interventions that have affected the aqueduct structure since the early years following its inauguration in 52 AD. Each stratigraphic unit has its own record, summarising the chronological references, initial interpretations, qualitative and visual observations of the materials and possible indications for further analysis. The mapping process was carried out in two dimensions, using drawings created for geometric surveys, and in three dimensions through the development of a dedicated HBIM for the examined sections of the aqueduct. Both masonry stratigraphic units (USM) and surface stratigraphic units (USS) have been identified (Figure 9).

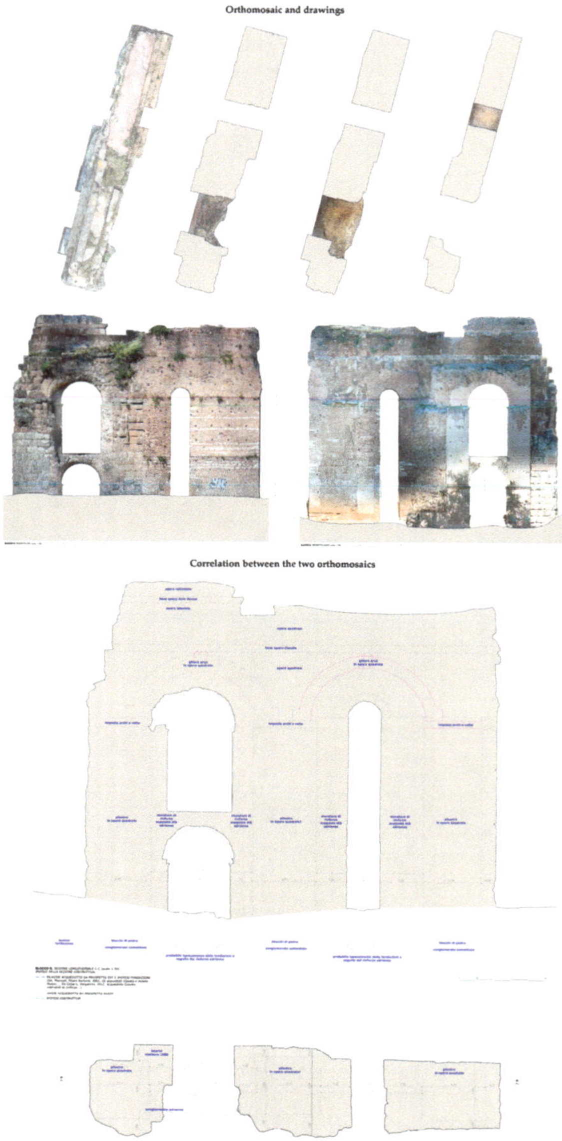

Figure 8. Hypothesis for identification of the presence of pillars and arches in peperino stone in plan and elevation, based on the correlation between the facades (orthomosaics) and the identified stratigraphic units.

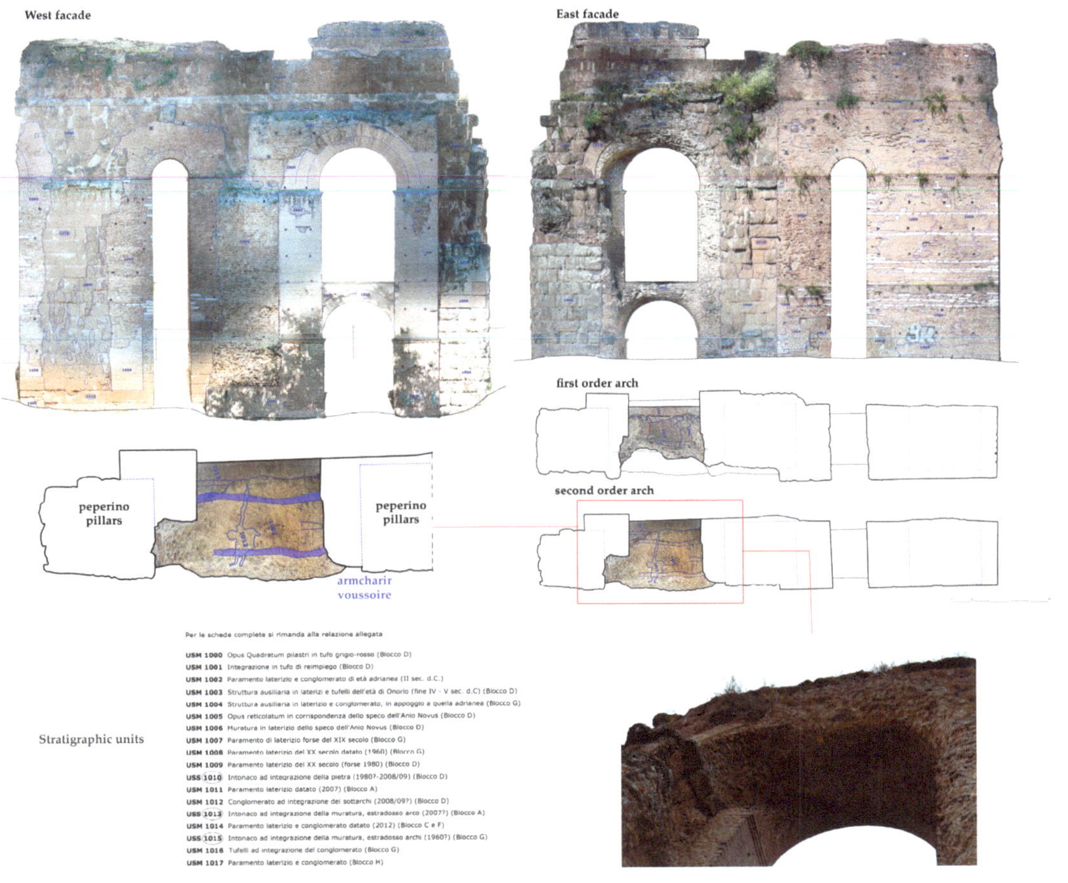

Figure 9. Building archaeology: stratigraphic units of east and west facades and sub-arches.

Section D is the only part of the aqueduct where three pillars and two arches in peperino ashlars are still partially visible. The general length of this portion of the aqueduct is approximately 24 m, with a width of 3.90 m. It consists of a stone-ashlar (peperino) pillar, measuring approximately 3.60 × 3.50 m, located to the south, and another pillar visible on the east elevation, partially incorporated into an *opus vittatum* masonry wall, which probably dates back to Onorio (395–423 d.C.) (Figure 10). A third pillar is partially visible on the west front, incorporated into the *opus vittatum* masonry wall. On both the east and west fronts, the arch in peperino can also be partially recognised. The southern pillar and the one partially incorporated into the late-antiquity wall to the north are reinforced by a double-arch structure from the Hadrianic period.

The distance between the two pillars is approximately 6.16 m. The distance between the second pillar and the third one, located to the north, has been reduced through a masonry intervention that incorporated the last pillar and narrowed the gap between them to approximately 1.80 m by constructing a much narrower arch than the Hadrianic one. The arches rest on imposts of approximately 0.45 m and are composed of a ring of peperino blocks with a trapezoidal keystone, as observed in other sections of the aqueduct, sometimes using tuff blocks. The *specus* of the Claudius aqueduct also rests on a stringcourse. Only one wall remains of the *specus*, on the west side, consisting of four courses of stone. A portion of the *specus* of the Anio Novo aqueduct remains as an *opus reticulatum* masonry wall. The latter was likely covered by a barrel vault, as observed in the Parco degli Acquedotti and the area of the Banca d'Italia.

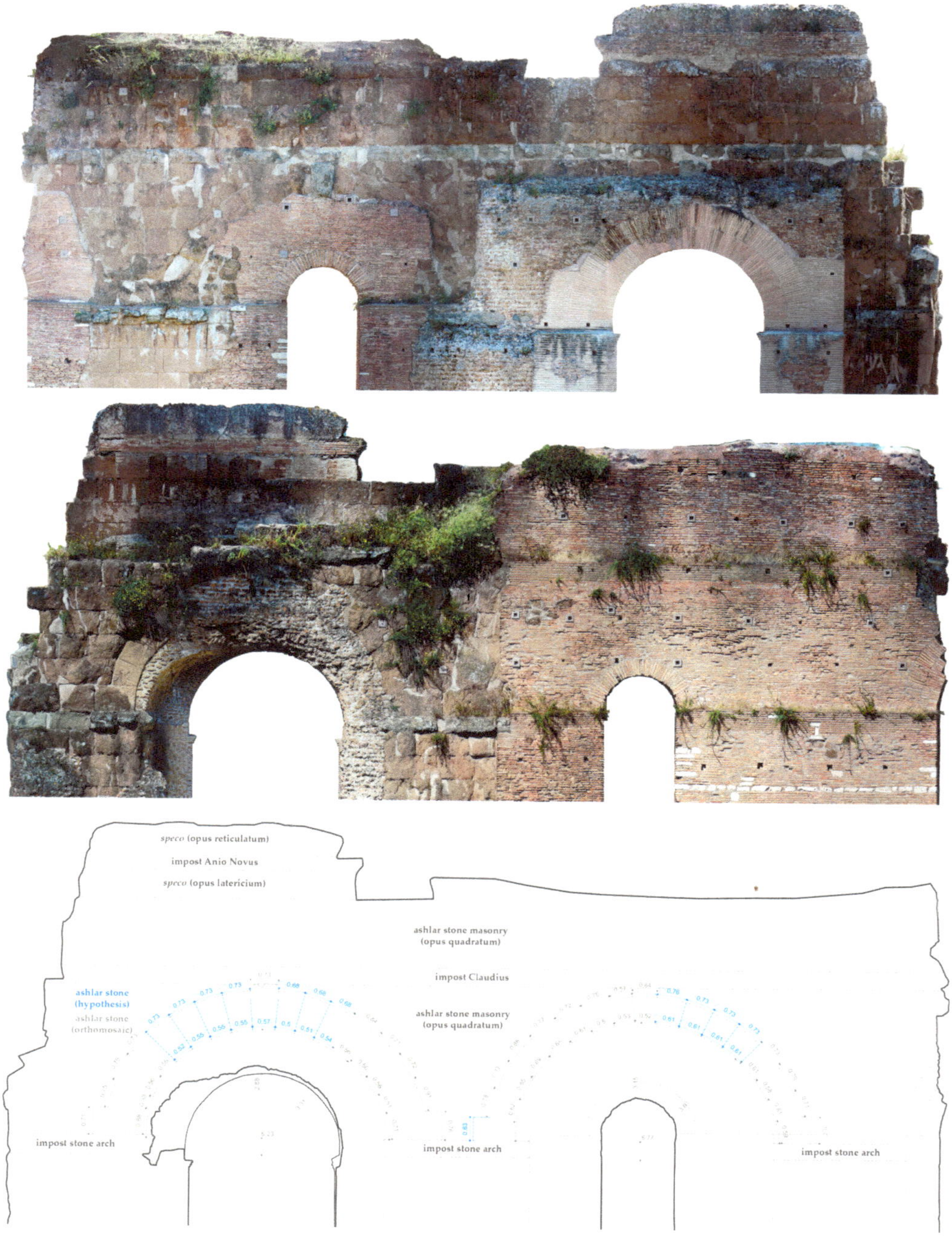

Figure 10. Arch geometries: surbased arch construction and stone dimension.

The current appearance of section D is the result of a restoration conducted in the 1980s. The brickwork was reconstructed, especially on the west side, as well as the *opus caementicium* of the Hadrianic arches, while the tuff stones were consolidated and integrated with mortars to reconstruct the shapes. In 2008, consolidation works were carried out to contain the thrust of the arches and prevent any sliding of wall sections, especially in the area related to the *specus*, and to restore the brick curtains that protected the *opus caementicium*. The additions had to be reversible and compatible with the material of the structure. A longitudinal chain was constructed, embracing the two arches, to which several transverse ties were added at the points of the putlog holes to connect the brickwork wall. It was necessary to compensate for a portion of the southern arch with lithic anchoring devices on the original masonry. Before installing the metallic ties, made of stainless steel, a cannula was inserted to allow the injection of compatible mortars. Inside the tube, the bar is housed, not injected, and therefore easily removable.

Regarding the analysis of the architectural archaeology of the sub-arches during the Hadrianic period, the construction technique of the arches is defined by bipedal brick armchair voussoirs, which also form the arch ring, and a casting of *opus caementicium*. As for the arch of the first order, it consists of two armchair voussoirs located between the two arch rings (total arch thickness: 2.48 m, span: 6.77 m). On the other hand, the arch of the second order consists of five armchair voussoirs, including the arch rings (total arch thickness: 3.70 m, span: 6.25 m). On the arch of the first order, three masonry stratigraphies have been identified: one related to the Hadrianic era (1002, *opus caementicium*) and two related to the restoration works of 1980 (1012, brickwork and *opus caementicium*). The same stratigraphies are found on the arch of the second order.

5.2. HBIM Objects for Building Archaeology Stratigraphic Units

The 3D modelling of the identified stratigraphic units was achieved through the interoperability of various software programs, including Rhinoceros, for pure modelling and parameterisation in the Revit environment [63]. This approach allowed for the accurate and comprehensive representation of the architectural elements, enhancing the site's analysis and interpretation through new information sharing forms, such as virtual and augmented reality (VR-AR).

The steps that led to the creation of the HBIM model involved four crucial moments.

- Data acquisition: laser scanner point cloud data on topographic support and photogrammetric datasets.
- Scan-to-BIM: the creation of two-dimensional drawings from the point cloud (AutoCAD), the enrichment of drawings using orthomosaics of the facades and sub-arches of aqueduct sections and the textured photogrammetric model managed in Rhinoceros. Building archaeology analysis of the elevated structure, including the mapping of stratigraphic units, analysis of alteration and degradation phenomena and creation of the HBIM model based on building archaeology.
- Information mapping: the association of information, documents, texts and images with the parametric objects of the HBIM model and the extraction of a database.
- Sharing of collected data: the transparency of information entered into the model and the possibility of integrating it with virtual or augmented reality programs.

The second point can be better explained with the following observations.

- The plans, elevations and sections of each aqueduct section were created by managing the point cloud within AutoCAD, ensuring proper georeferencing within the coordinate system defined by the topographic network with GNSS support. This ensured interoperability between different software in subsequent work phases.
- The orthomosaics obtained through data processing in Agisoft Metashape were correctly inserted and georeferenced in the CAD environment. Additionally, the CAD drawing of certain details (such as peperino blocks) benefited from the use of pho-

togrammetric models in Rhinoceros, where the three-dimensional representation provided a better understanding and allowed for the better redesign of the elements.

- Photogrammetric models and CAD drawings were georeferenced in Rhinoceros to create a NURBS model. This served two purposes: providing the basis for the parametric model in Revit and moving towards sharing formats accessible to a wider audience, such as augmented reality.
- Brick lining surfaces were parameterised in Revit within Rhinoceros. The cementitious conglomerate required more detailed modelling using the Geomagic software.
- The generated data were imported into Revit to create the parametric model. The individual stratigraphic units were defined based on the analysis of the architectural archaeology, followed by information mapping, including the definition of new parameters.

As mentioned, the modelling of complex morphologies, such as irregularly shaped walls or vaulted structures in historical architecture, including the Claudius Aqueduct conglomerate, within HBIM, using detailed surveys that generate point clouds, faces challenges in expanding the built heritage information system to high levels of detail and information. The main reason is the limited number of BIM modelling tools, which do not allow for the creation complex as-built parametric objects oriented towards built heritage. Therefore, interoperability with different programs becomes necessary to represent and create a semantic model that is as geometrically and dimensionally accurate as possible to reflect the actual state of the surveyed structure.

The challenge posed by the Claudius aqueduct is not only in terms of modelling, recognising "common" elements such as pillars, arches and imposts, which can be difficult and is perhaps less significant for the conglomerate material, but also in terms of conveying information to support the restoration project.

An HBIM supporting the project and restoration site should three-dimensionally reproduce the analyses performed on 2D drawings (or directly create them in 3D) (Figure 11). It should consider the knowledge acquired about the history, construction phases, techniques and materials through observations, diagnostic investigations, assumptions and hypotheses, in case some data have not yet been studied or made available. It should also consider the state of conservation, the mapping of alteration and degradation phenomena and any indications of previous interventions. This integration allows for a synergy between the need for knowledge and the possibilities for simulation.

The final HBIM model thus becomes a true representation not only of the object itself but of the entire construction process of the intervention. It allows for the extraction, management and control of specific aspects of interest for different professionals involved, such as material aspects and degradation. These shared aspects are collected in a single digital database that can be continuously expanded over time and gradually support research and restoration efforts.

Specifically, the implemented parameters for each modelled stratigraphic object, referring to the "Identity Data" section, include the following.

- Stratigraphic unit: This field contains the stratigraphic unit related to the parametric object (e.g., USM 1002).
- Object: Each parametric element is uniquely identified by a code consisting of the first letter corresponding to the aqueduct section (A, B, C, D, E, F, G, H), a second letter indicating the orientation of the stratigraphic unit (N = North, S = South, O = West, E = East), a number representing the stratigraphic unit (from 1000 to 1012), an additional letter indicating the type of stratigraphic unit (M = masonry, S = surface) and a progressive number to differentiate similar elements. Finally, three acronyms, 3D, TX (texture) and BIM, indicate the actual parameterisation and texturing in Revit of the object (e.g., G_O_1002_M_01_3D_TX_BIM).
- Brief description of the stratigraphic unit in Italian and English.
- Construction phase (if known or hypothesised).
- Chronological analysis (for brick linings and peperino blocks if available).

- Reference card: Link to an external repository with stored PDF-format cards for individual stratigraphic units.
- Conservation status: A brief current state description and links to reference cards.

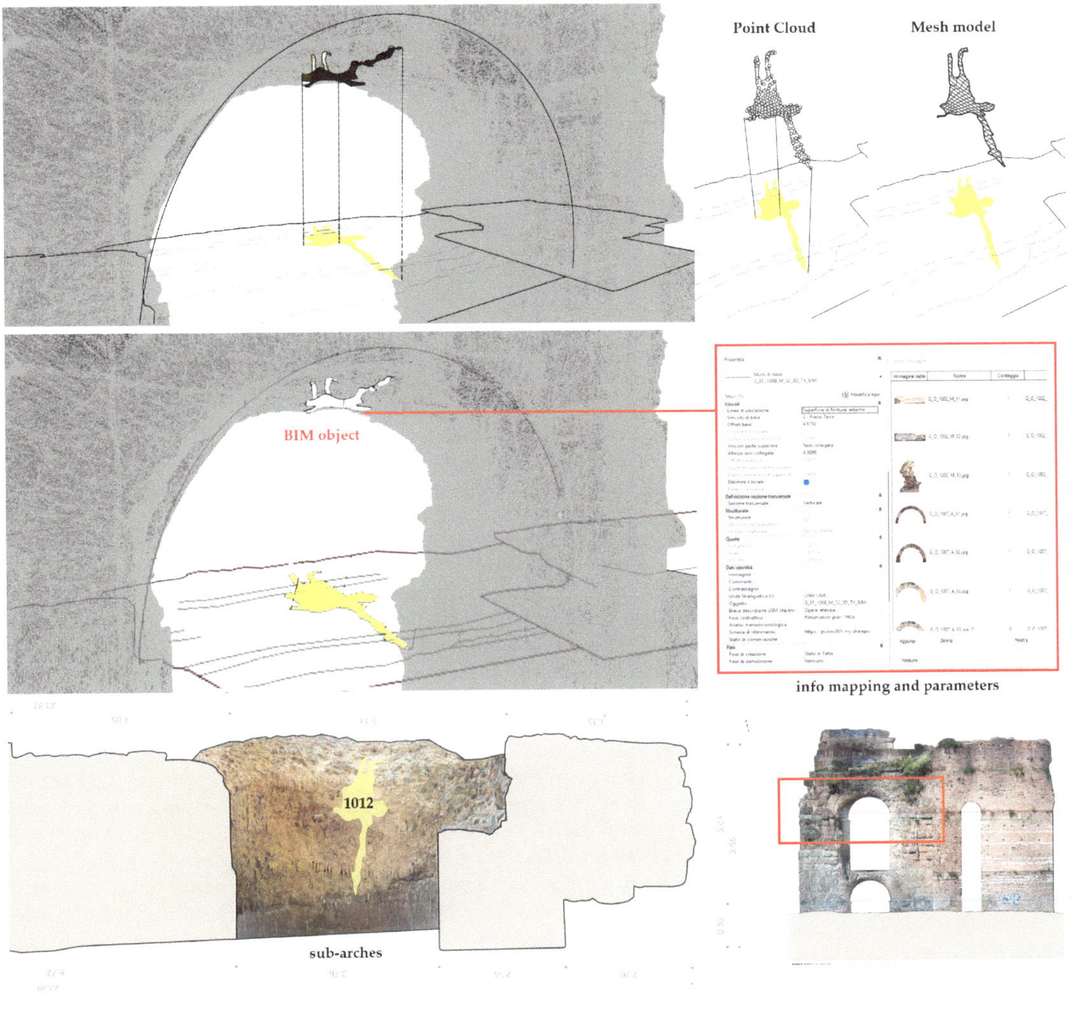

Figure 11. From NURBS to HBIM of complex shapes, such as stratigraphic units and material decay.

Additionally, in the "Image" section, orthomosaics related to each stratigraphic unit have been uploaded. This allows them to be visualised within the BIM environment and accessible as separate consultable documents. The entered data can be exported as an Excel spreadsheet, making them accessible and editable for the involved parties. At the same time, the Excel spreadsheet can serve as a basic grid for the definition of an actual database that can be subsequently linked to the BIM. This database can also consider future diagnostic investigations that need to be carried out.

The different stratigraphic units identified within the sub-arches have been modelled and imported into Revit.

5.3. From Building Archaeology and HBIM to XR

The accessibility of analysis within an HBIM project is frequently confined to a limited group of professionals actively restoring artifacts. The advent of the digital era has showcased the transformative potential of emerging technologies such as extended, virtual and augmented reality, surpassing conventional representation, visualisation and information dissemination methods [64]. In the domain of digital cultural heritage, the advancement of digital twins is still in its nascent stages. Intriguing investigations have demonstrated the feasibility of integrating real-time monitoring data with intricate models, facilitating continuous monitoring and enhancement across various stages of the lifecycle.

However, utilising and interpreting these technologies in the digital milieu necessitates advanced expertise. One must consider the intricate nature of HBIM projects and the intricate interplay between schedules, parametric objects and their associated mapped information. Concurrently, state-of-the-art VR devices necessitate a period of assimilation and adaptation for users to immerse themselves in these digital volumetric reproductions effectively [65]. Notably, the operation of VR headsets mandates specific proficiencies not universally possessed by all users.

Addressing these challenges, this study proposes a user-centric development approach centred on interpreting stratigraphic units, thereby rendering them accessible even to individuals lacking expertise in the domains of HBIM or VR.

The principal objective is to streamline the interpretation of semantic models through a novel form of shared DT facilitated via web-based platforms.

The first step to encompass a web-based platform is the development of an immersive environment wherein end-users can readily delve into the intricacies of the semantic model and its associated analyses by means as simple as a mere click on the desired area of focus. As a result, restoration specialists, archaeologists and even virtual tourists can harness this system to augment their comprehension of the artifact in an intuitive, simplified and interactive manner, with seamless interaction facilitated by touchscreen interfaces. Through targeted development leveraging visual programming languages and implementing a versatile environment accommodating both virtual and augmented reality, the understanding of the conducted analyses has been significantly enhanced on-site and remotely.

The specific development effort entailed the integration of Autodesk Revit into platforms provided by Epic Games (Twinmotion and Unreal Engine). The objective was to offer an interactive representation that was seamlessly integrated with the structural context of the Roman aqueduct while concurrently fostering enriched user interaction through a meticulously reconstructed VR environment. In particular, the VR functionality in Twinmotion supported various VR devices, including virtual reality headsets, allowing users to experience their designs with a high level of realism. By wearing a compatible VR headset, users can interact with their models using intuitive controls and movements, such as walking, teleporting and grabbing objects. This level of interactivity added a new dimension to the representation process, enabling users to test different scenarios and analyses, assess the impacts of decisions and make more informed choices. In addition to the immersive experience, the VR implementation enabled users to capture high-quality VR screenshots and videos, which could be shared with clients, stakeholders and collaborators. These visual assets provided a compelling and realistic representation of the monument, enhancing communication and understanding among all parties involved. These combined solutions facilitated comprehensive analysis directly on the HBIM model. Leveraging the potential of the semantic model, an easily navigable environment was established. Users could interact with the model by clicking on areas of interest, thereby revealing thematic zones wherein pertinent descriptions were associated. In this manner, the accessibility of these analyses surpasses conventional techniques employed for the visualisation of stratigraphic units.

Furthermore, the XR project (Figure 12) illustrates how three-dimensional morphological reproductions of construction techniques facilitate the communication of functional and formal aspects, allowing observers to interpret the detected Sus.

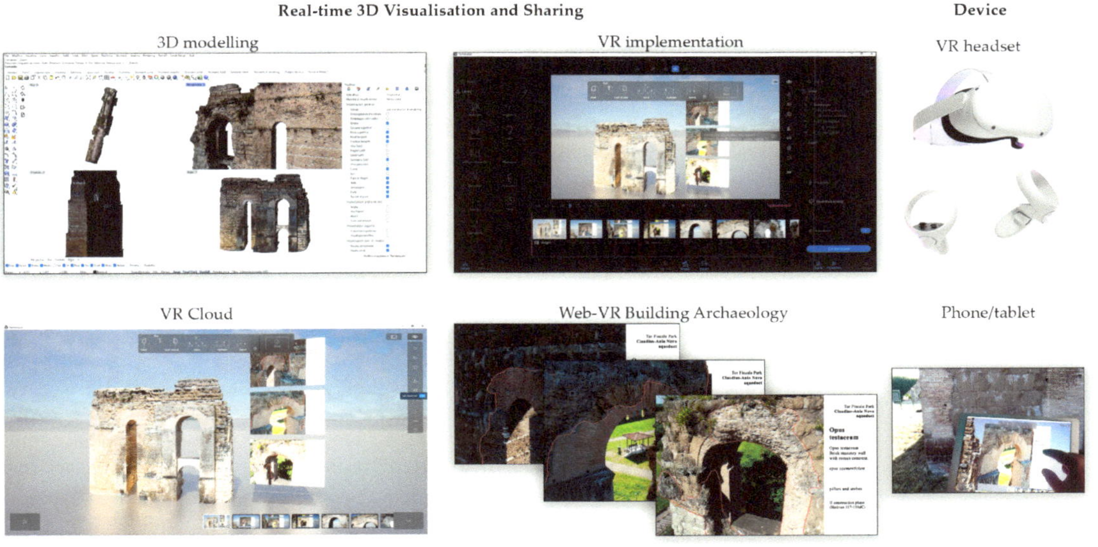

Figure 12. XR implementation for multiple devices and users.

For instance, the process of moving and installing the stone ashlars required technical expertise and construction machinery, as indicated in Plate XXIV of various Vitruvian editions of De Architectura [66].

Throughout history, from Claude Perrault's 1676 edition "Les dix livres d'Architecture de Vitruve corigés et traduits en 1648 for Claude Perrault, 1676" [67] to the 1758 edition "L'Architettura" by Marco Vitruvio Pollione," there have been dedicated chapters on Machines (Book Tenth) and Water (Book Eighth), detailing processing techniques. Table XXIV of Book Tenth illustrates the "L" iron pincer, representing the working technique called the "olive grove," used to lift large stone blocks.

By detecting segments with accuracy of 2–3 mm and representing the monument's complex morphology, the XR platform enables users to examine the three-dimensional finishes of exposed faces. Historical documentation is integrated into the HBIM gradually, including references and links to Vitruvian tables and open-access collection libraries such as Europeana.

6. Discussion and Conclusions

The possibility of integrating UAV photogrammetric surveys for the sub-arches of the Claudius Anio Novus aqueduct has allowed for a comprehensive understanding of the monument, even of the difficult-to-reach upper parts. This is important for two reasons: (i) there was no need to wait for scaffolding to study and analyse the area, and (ii) it was possible to provide preliminary support for the design of the restoration intervention. Certainly, further in-depth investigations will be necessary through observations, close inspections and diagnostic campaigns after the scaffolding is erected. Surveying with a drone has helped to overcome certain problems. The modelling that followed has benefited from the advancements of scan-to-BIM to model complex surfaces, capable of becoming BIM-enabled objects with the addition of parameters, including custom ones. Furthermore, the HBIM model was developed to serve as a foundation for a digital twin.

Building archaeology has benefited from three-dimensional representation to understand the relationships between identified stratigraphic units. Furthermore, the ability to visually represent stratigraphic units in three dimensions will assist specialists involved in restoration in making the data accessible remotely as well. In the future, for example, some of our hypotheses could be tested using instrumental investigation techniques. In fact, ongoing investigations include the analysis of mortar samples for each stratigraphic unit and soil surveys.

Web-VR has several advantages that contributed to its increased accessibility and popularity compared to traditional VR applications. One key advantage was its accessibility, as users could directly access VR experiences through web browsers, without downloading or installing dedicated applications. This eliminated the barriers associated with app installations, making VR content easily accessible to a wider audience. Incorporating textures derived from UAV photogrammetry significantly enhanced the visual fidelity and realism of the virtual environment. By carefully applying textures to the surface of the HBIM model, the colours, patterns and intricate details captured in the original aerial photographs could be faithfully reproduced. This meticulous attention to detail intensified the user's sense of immersion and active participation, creating a captivating and immersive experience.

Another notable benefit of Web-VR was its ease of sharing. Being web-based, users could effortlessly share links or URLs with others, allowing instant access to the VR content. This streamlined sharing process simplified the distribution of VR material and encouraged collaboration and shared engagement among users, facilitating a more interactive and immersive experience.

Web-VR's web-based nature also enabled real-time updates to VR content. Developers could make prompt modifications or publish new content without requiring users to perform separate updates or downloads. This ensured that users always had access to the most up-to-date version of the VR experience, fostering continuous improvement and innovation within the virtual environment.

Furthermore, Web-VR offered compatibility with a wide range of devices, including computers, smartphones, tablets and dedicated VR headsets. This cross-device platform versatility allowed users to access VR experiences through their preferred devices, without the need for specific VR hardware. This flexibility enhanced the accessibility and expanded the potential audience, creating a more inclusive VR ecosystem.

In terms of cost-effectiveness, Web-VR presented notable advantages. By eliminating the need for platform-specific or device-specific application development, the associated costs for development and maintenance were generally reduced. Additionally, users could access VR experiences without the need to invest in dedicated VR hardware, making entry into the VR world more economically viable.

Lastly, the proposed scan-to-BIM-to-XR process adds value to the development of scalable VR experiences that reach diverse audiences while effectively supporting the restoration project. Establishing a common set of tools and facilitating synchronised interventions fostered seamless collaboration among all stakeholders. The web-based infrastructure of the system enabled concurrent access by a multitude of users without any compromise in performance, thereby ensuring exceptional scalability. This aspect proved especially beneficial for immersive shared VR experiences in the research case study, including virtual tours, events and simulations, where multiple participants could interact simultaneously, fostering collaboration and facilitating shared enjoyment.

Author Contributions: Conceptualization, C.S. and F.B.; methodology, C.S. and F.B.; software, C.S. and F.B.; validation, C.S., F.B. and S.R.; formal analysis, C.S., F.B. and S.R.; investigation, C.S., F.B. and S.R.; resources, C.S. and F.B.; data curation, C.S., F.B. and S.R.; writing—original draft preparation, C.S. and F.B.; writing—review and editing, C.S. and F.B.; visualization, C.S. and F.B.; supervision, C.S., F.B. and S.R.; project administration, C.S., F.B. and S.R.; funding acquisition, C.S., F.B. and S.R. All authors have read and agreed to the published version of the manuscript.

Funding: Consultancy program stipulated within the scope of the convention between the Politecnico di Milano and the Appia Antica Archaeological Park for the "Digitization (acquisition, detection, representation, HBIM, XR) aimed at creating an experimental Digital Twin to support knowledge, design, restoration, dissemination and communication works—Segment of the Torre del Fiscale Park, Claudio Aqueduct—Anio Novus)" (CIG ZC831BD87C—CUP F85F19000820001).

Data Availability Statement: The study did not report any data.

Acknowledgments: The authors thank the Appia Antica Archaeological Park Superintendence, Pro-tempore Director Arch. Simone Quilici for making it possible to conduct the survey and analysis activities on the Claudius Anio Novus aqueduct. Thanks to the archaeologist Francesca Romana Paolillo, the architects Raffaella Rocchetta and Aura Picchione and the restorer Sara Iovine. This work is part of the topographic survey and scan-to-BIM process named Scientific Responsible (DABC, Politecnico di Milano): R. Brumana, F. Banfi and M. Previtali. Collaborators: F. Roncoroni, C. Stanga.

Conflicts of Interest: The authors declare no conflict of interest.

References

1. Brogiolo, G.P. L'Archeologia dell'architettura in Italia nell'ultimo quinquennio (1997–2001). *Arqueol. De La Arquit.* **2002**, *1*, 19–26. [CrossRef]
2. Brogiolo, G.P.; Cagnana, A. Archeologia Dell'architettura. In *Metodi e Interpretazioni*; All'Insegna del Giglio: Florence, Italy, 2012.
3. Mannoni, T. *Caratteri Costruttivi Dell'edilizia Storica*; Sagep Editrice: Genova, Italy, 1994.
4. Doglioni, F. Stratigrafia e Restauro. In *Tra Conoscenza e Conservazione Dell'architettura*; Lint Editoriale Associati: Trieste, Italy, 1997.
5. Boato, A.; Pittaluga, D. Building Archaeology: A non-destructive archaeology. In Proceedings of the 15th World Conference on Nondestructive Testing, Roma, Italy, 15–21 October 2000; pp. 1–5. Available online: https://www.ndt.net/article/wcndtOO/papers/idn365/idn365.htm (accessed on 6 August 2023).
6. Boato, A. *L'archeologia in Architettura*; Marsilio Editori: Venezia, Italy, 2008.
7. Bartolomucci, C. I GIS per la conservazione dell'architettura storica. *MondoGIS* **2008**, *66/67*, 13–17.
8. Brumana, R.; Della Torre, S.; Oreni, D.; Previtali, M.; Cantini, L.; Barazzetti, L.; Franchi, A.; Banfi, F. HBIM challenge among the paradigm of complexity, tools and preservation: The Basilica di Collemaggio 8 years after the earthquake (L'Aquila). *Int. Arch. Photogramm. Remote Sens. Spat. Inf. Sci.* **2017**, *42*, 97–104. [CrossRef]
9. Valente, R.; Brumana, R.; Oreni, D.; Banfi, F.; Barazzetti, L.; Previtali, M. Object-oriented approach for 3d archaeological documentation. *Int. Arch. Photogramm. Remote Sens. Spat. Inf. Sci.* **2017**, *42*, 707–712. [CrossRef]
10. Demetrescu, E. Archaeological stratigraphy as a formal language for virtual reconstruction. *Theory Pract.* **2015**, *57*, 42–55.
11. Barazzetti, L.; Brumana, R.; Banfi, F.; Lostaffa, F.; Piraino, F.; Previtali, M.; Oreni, D.; Roncoroni, F.; Villa, L. BHIMM e Augmented Information: Il rilievo per la conoscenza e la valorizzazione di Castel Masegra. In *ASITA*; Federazione ASITA: Milano, Italy, 2015; pp. 35–45. ISBN 978-88-941232-2-7.
12. Brusaporci, S.; Ruggeri, G.; Maiezza, P.; Tata, A. AHBIM per l'analisi stratigrafica dell'architettura storica. *Restauro Archeol.* **2018**, *27*, 112–131.
13. Trizio, I.; Savini, F. Archaeology of buildings and HBIM methodology: Integrated tools for documentation and knowledge management of architectural heritage. In Proceedings of the IMEKO International Conference on Metrology for Archaeology and Cultural Heritage, MetroArchaeo, Rome, Italy, 22–24 October 2020; pp. 84–89.
14. Trizio, I.; Savini, F.; Giannangeli, A.; Boccabella, R.; Petrucci, G. The archaeological analysis of masonry for the restoration project in HBIM. *Int. Arch. Photogramm. Remote Sens. Spat. Inf. Sci.* **2019**, *42*, 715–722. [CrossRef]
15. Diara, F.; Rinaudo, F. Building archaeology documentation and analysis through open source HBIM solutions via nurbs modelling. *Int. Arch. Photogramm. Remote Sens. Spat. Inf. Sci.* **2020**, *43*, 1381–1388. [CrossRef]
16. Lo Turco, M.; Caputo, F.; Fusaro, G. From integrated survey to the parametric modeling of degradations. A feasible workflow. In Proceedings of the Digital Heritage, Progress in Cultural Heritage: Documentation, Preservation, and Protection: 6th International Conference, EuroMed 2016, Nicosia, Cyprus, 31 October–5 November 2016; Springer International Publishing: Berlin/Heidelberg, Germany, 2016; Proceedings, Part I 6; pp. 579–589.
17. Delpozzo, D.; Treccani, D.; Appolonia, L.; Adami, A.; Scala, B. Hbim and Thematic Mapping: Preliminary Results. *Int. Arch. Photogramm. Remote Sens. Spat. Inf. Sci.* **2022**, *46*, 199–206. [CrossRef]
18. Di Stefano, F.; Gorreja, A.; Malinverni, E.S.; Mariotti, C. Knowledge modeling for heritage conservation process: From survey to HBIM implementation. *Int. Arch. Photogramm. Remote Sens. Spat. Inf. Sci.* **2020**, *44*, 19–26. [CrossRef]
19. Lanciani, R. Topografia di Roma antica. In *I comMentari di Frontino Intorno le Acque e Gli Acquedotti*; Salviucci: Roma, Italy, 1880.
20. Radicioni, L. L'importanza delle infrastrutture idriche a Roma: Analisi tecnico-strutturale di un tratto di Aqua Claudia e Anio Novus e delle sue opere manutentive e di consolidamento antiche. In *Archaeology and Economy in the Ancient World. Proceedings of the 19th International Congress of Classical Archaeology, Cologne/Bonn 2018, Volume 55*; Bentz, M., Heinzelmann, M., Eds.; pp. 453–465. [CrossRef]

21. Motta, R. La decadenza degli acquedotti Claudio e Marcio ed il riuso delle loro strutture dalla fine del mondo antico al XIX secolo. In *Gli Acquedotti Claudio e Aniene Nuovo nell'area della Banca d'Italia in via Tuscolana*; Mancioli, D., Pisani Sartorio, G., Eds.; Istituto Poligrafico della Zecca dello Stato: Roma, Italy, 2001; pp. 81–90.
22. Banfi, F.; Roascio, S.; Paolillo, F.R.; Previtali, M.; Roncoroni, F.; Stanga, C. Diachronic and Synchronic Analysis for Knowledge Creation. *Energies* **2022**, *15*, 4598. [CrossRef]
23. Sharma, V. Advances in Drone Communications, State-of-the-Art and Architectures. *Drones* **2019**, *3*, 21. [CrossRef]
24. Campana, S. Drones in archaeology. State-of-the-art and future perspectives. *Archaeol. Prospect.* **2017**, *24*, 275–296. [CrossRef]
25. Battulwar, R.; Zare-Naghadehi, M.; Emami, E.; Sattarvand, J. A state-of-the-art review of automated extraction of rock mass discontinuity characteristics using three-dimensional surface models. *J. Rock Mech. Geotech. Eng.* **2021**, *13*, 920–936. [CrossRef]
26. Adamopoulos, E.; Rinaudo, F. UAS-based archaeological remote sensing: Review, meta-analysis and state-of-the-art. *Drones* **2020**, *4*, 46. [CrossRef]
27. Remondino, F.; Barazzetti, L.; Nex, F.; Scaioni, M.; Sarazzi, D. UAV photogrammetry for mapping and 3d modeling–current status and future perspectives. *Int. Arch. Photogramm. Remote Sens. Spat. Inf. Sci.* **2012**, *38*, 25–31. [CrossRef]
28. Parrinello, S.; Barba, S.; Dell'Amico, A.; di Filippo, A. *D-SITE*; Pavia University Press: Pavia, Italy, 2022. Available online: http://archivio.paviauniversitypress.it/oa/9788869521607.pdf (accessed on 6 August 2023).
29. Bemis, S.P.; Micklethwaite, S.; Turner, D.; James, M.R.; Akciz, S.; Thiele, S.T.; Bangash, H.A. Ground-based and UAV-Based photogrammetry: A multi-scale, high-resolution mapping tool for structural geology and paleoseismology. *J. Struct. Geol.* **2014**, *69*, 163–178. [CrossRef]
30. Yang, W.; Zhao, C.; Westoby, M.; Yao, T.; Wang, Y.; Pellicciotti, F.; Miles, E. Seasonal dynamics of a temperate Tibetan Glacier revealed by high-resolution UAV photogrammetry and in situ measurements. *Remote Sens.* **2020**, *12*, 2389. [CrossRef]
31. Cottini, A.; Becherini, P.; Rolando, V. A 3D model for architectural analysis, using aerial photogrammetry, for the digital documentation of the convent of Santa Maria da Ínsua, on the northern border between Portugal and Spain. In *Drones. System of Information on Cultural Heritage for a Spatial and Social Investigation*; Pavia University Press: Pavia, Italy, 2022; Volume 2, pp. 94–103.
32. Picchio, F.; Parrinello, S.; Barba, S. Drones and Drawings-methods of data acquisition, management, and representation. *Disegnarecon* **2022**, *15*, 1–7.
33. De Marco, R.; Parrinello, S. Management of mesh features in 3D reality-based polygonal models to support non-invasive structural diagnosis and emergency analysis in the context of earthquake heritage in Italy. *Int. Arch. Photogramm. Remote Sens. Spat. Inf. Sci.* **2021**, *46*, 173–180. [CrossRef]
34. Chiabrando, F.; Sammartano, G.; Spanò, A.; Semeraro, G. Multi-temporal images and 3D dense models for archaeological site monitoring in Hierapolis of Phrygia (TK). *Archeol. E Calc.* **2017**, *28*, 469–484.
35. Stek, T.D. Drones over Mediterranean landscapes. The potential of small UAV's (drones) for site detection and heritage management in archaeological survey projects: A case study from Le Pianelle in the Tappino Valley, Molise (Italy). *J. Cult. Herit.* **2016**, *22*, 1066–1071. [CrossRef]
36. Chiabrando, F.; Spanò, A.; Sammartano, G.; Teppati Losè, L. UAV oblique data and laser scanning in an excavated area. In Proceedings of the 8th International Congress on Archaeology, Computer Graphics, Cultural Heritage and Innovation, Valencia, Spain, 27 October 2016; Editorial Universitat Politècnica de València: Valencia, Spain; pp. 350–353.
37. Patrucco, G.; Giulio Tonolo, F.; Sammartano, G.; Spanò, A. Sfm-Based 3D Reconstruction of Heritage Assets Using UAV Thermal Images. *Int. Arch. Photogramm. Remote Sens. Spat. Inf. Sci.* **2022**, *43*, 399–406. [CrossRef]
38. Kaimaris, D.; Kandylas, A. Small multispectral UAV sensor and its image fusion capability in cultural heritage applications. *Heritage* **2020**, *3*, 57. [CrossRef]
39. Feroz, S.; Abu Dabous, S. Uav-based remote sensing applications for bridge condition assessment. *Remote Sens.* **2021**, *13*, 1809. [CrossRef]
40. Ham, Y.; Han, K.K.; Lin, J.J.; Golparvar-Fard, M. Visual monitoring of civil infrastructure systems via camera-equipped Unmanned Aerial Vehicles (UAVs): A review of related works. *Vis. Eng.* **2016**, *4*, 1. [CrossRef]
41. Shadiev, R.; Yi, S. A systematic review of UAV applications to education. *Interact. Learn. Environ.* **2022**, 1–30. [CrossRef]
42. Alkaabi, K.; Abuelgasim, A. Applications of unmanned aerial vehicle (UAV) technology for research and education in UAE. *Int. J. Soc. Sci. Arts Humanit.* **2017**, *5*, 4–11.
43. Martínez-Carricondo, P.; Carvajal-Ramírez, F.; Yero-Paneque, L.; Agüera-Vega, F. Combination of nadiral and oblique UAV photogrammetry and HBIM for the virtual reconstruction of cultural heritage. Case study of Cortijo del Fraile in Níjar, Almería (Spain). *Build. Res. Inf.* **2020**, *48*, 140–159. [CrossRef]
44. Fabris, M.; Fontana Granotto, P.; Monego, M. Expeditious low-cost SfM photogrammetry and a TLS survey for the structural analysis of Illasi Castle (Italy). *Drones* **2023**, *7*, 101. [CrossRef]
45. Klapa, P.; Gawronek, P. Synergy of Geospatial Data from TLS and UAV for Heritage Building Information Modeling (HBIM). *Remote Sens.* **2022**, *15*, 128. [CrossRef]
46. Liu, J.; Azhar, S.; Willkens, D.; Li, B. Static Terrestrial Laser Scanning (TLS) for Heritage Building Information Modeling (HBIM): A Systematic Review. In *Virtual Worlds*; MDPI: Basel, Switzerland, 2023; Volume 2, pp. 90–114.
47. Karachaliou, E.; Georgiou, E.; Psaltis, D.; Stylianidis, E. UAV for mapping historic buildings: From 3D modelling to BIM. *Int. Arch. Photogramm. Remote Sens. Spat. Inf. Sci.* **2019**, *42*, 397–402. [CrossRef]

48. Mesas-Carrascosa, F.J.; Notario García, M.D.; Meroño de Larriva, J.E.; García-Ferrer, A. An analysis of the influence of flight parameters in the generation of unmanned aerial vehicle (UAV) orthomosaicks to survey archaeological areas. *Sensors* **2016**, *16*, 1838. [CrossRef]
49. Themistocleous, K. Themistocleous, K. The use of UAVs for cultural heritage and archaeology. In *Remote Sensing for Archaeology and Cultural Landscapes, Best Practices and Perspectives Across Europe and the Middle East*; Springer: Cham, Switzerland, 2020; pp. 241–269.
50. Tache, A.V.; Sandu IC, A.; Popescu, O.C.; Petrişor, A.I. UAV Solutions for the Protection and Management of Cultural Heritage. Case Study: Halmyris Archaeological Site. *Int. J. Conserv. Sci.* **2018**, *9*, 795–804.
51. Berrett, B.E.; Vernon, C.A.; Beckstrand, H.; Pollei, M.; Markert, K.; Franke, K.W.; Hedengren, J.D. Large-scale reality modeling of a university campus using combined UAV and terrestrial photogrammetry for historical preservation and practical use. *Drones* **2021**, *5*, 136. [CrossRef]
52. Sefercik, U.G.; Kavzoglu, T.; Nazar, M.; Atalay, C.; Madak, M. UAV-based 3D virtual tour creation. In Proceedings of the The 6th International Conference on Smart City Applications, Safranbolu, Turkey, 27–29 October 2021; pp. 255–264.
53. Pavelka, K., Jr. Photogrammetry, Laser Scanning and Hbim for Construction Diagnostic. *ISPRS Ann. Photogramm. Remote Sens. Spat. Inf. Sci.* **2022**, *46*, 171–176. [CrossRef]
54. Sun, Z.; Zhang, Y. Using drones and 3D modeling to survey Tibetan architectural heritage: A case study with the multi-door stupa. *Sustainability* **2018**, *10*, 2259. [CrossRef]
55. Masiero, A.; Chiabrando, F.; Lingua, A.M.; Marino, B.G.; Fissore, F.; Guarnieri, A.; Vettore, A. 3D Modeling of Girifalco Fortress. *Int. Arch. Photogramm. Remote Sens. Spat. Inf. Sci.* **2019**, *XLII-2/W9*, 473–478.
56. Bianchini, C.; Inglese, C.; Ippolito, A. The role of BIM Building Information Modeling) for representation and managing of built and historic artifacts. *Disegnarecon* **2016**, *9*, 10–11.
57. Brusaporci, S. (Ed.) *Handbook of Research on Emerging Digital Tools for Architectural Surveying, Modeling, and Representation*; Igi Global: Hershey, PA, USA, 2015.
58. Giordano, A.; Borin, P. HBIM: Analisi critica tra didattica e ricerca. In *Brainstorming BIM. Il Modello Tra Rilievo e Costruzione*; Maggioli Editore: Romagna, Italia, 2016; Volume 1, pp. 44–53.
59. Moyano, J.; León, J.; Nieto-Julián, J.E.; Bruno, S. Semantic interpretation of architectural and archaeological geometries: Point cloud segmentation for HBIM parameterisation. *Autom. Constr.* **2021**, *130*, 103856. [CrossRef]
60. Bevilacqua, M.G.; Russo, M.; Giordano, A.; Spallone, R. 3D reconstruction, digital twinning, and virtual reality: Architectural heritage applications. In Proceedings of the 2022 IEEE Conference on Virtual Reality and 3D User Interfaces Abstracts and Workshops (VRW), Christchurch, New Zealand, 12–16 March 2022; pp. 92–96.
61. Jouan, P.; Hallot, P. Digital twin: Research framework to support preventive conservation policies. *ISPRS Int. J. Geo-Inf.* **2020**, *9*, 228. [CrossRef]
62. Turco, M.L.; Mattone, M.; Rinaudo, F. Metric survey and BIM technologies to record decay conditions. *Int. Arch. Photogramm. Remote Sens. Spat. Inf. Sci.* **2017**, *42*, 261–268. [CrossRef]
63. Banfi, F. BIM orientation: Grades of generation and information for different type of analysis and management process. *Int. Arch. Photogramm. Remote Sens. Spat. Inf. Sci.* **2017**, *42*, 57–64. [CrossRef]
64. Ioannides, M.; Magnenat-Thalmann, N.; Papagiannakis, G. *Mixed Reality and Gamification for Cultural Heritage*; Springer: Berlin/Heidelberg, Germany, 2017.
65. Giordano, A.; Russo, M.; Spallone, R. *Representation Challenges New Frontiers of AR and AI Research for Cultural Heritage and Innovative Design*; Franco Angeli: Milan, Italy, 2022.
66. Marco Vitrubio Polión, M. Vitruuiis Pollionis De Architectura Libri Decem. No Copyright—Other Known Legal Restrictions. Available online: https://www.europeana.eu/en/item/496/EVA_207_A308230 (accessed on 22 July 2023).
67. Perrault, C. *Les Dix Livres d'Architecture de Vitruve Corigés et Traduits en 1648 per Claude Perrault*; Jean Baptiste Coignard: Paris, France, 1676.

drones

Article

Machine Learning-Based Monitoring for Planning Climate-Resilient Conservation of Built Heritage

Lidia Fiorini [1,2,*], Alessandro Conti [1], Eugenio Pellis [1], Valentina Bonora [1], Andrea Masiero [3] and Grazia Tucci [1]

1 Department of Civil and Environmental Engineering, University of Florence, Via di Santa Marta, 3, 50121 Florence, Italy; alessandro.conti@unifi.it (A.C.); eugenio.pellis@unifi.it (E.P.); valentina.bonora@unifi.it (V.B.); grazia.tucci@unifi.it (G.T.)
2 La Sapienza University of Rome, Piazzale A. Moro, 5, 00185 Rome, Italy
3 Interdepartmental Research Center of Geomatics (CIRGEO), TESAF Department, University of Padua, 35020 Padua, Italy; andrea.masiero@unipd.it
* Correspondence: lidia.fiorini@unifi.it

Abstract: The increasing frequency and intensity of extreme weather events are accelerating the mechanisms of surface degradation of heritage buildings, and it is therefore appropriate to find automatic techniques to reduce the time and cost of monitoring and to support their planned conservation. A fully automated approach is presented here for the segmentation and classification of the architectural elements that make up one of the façades of Palazzo Pitti. The aim of this analysis is to provide tools for a more detailed assessment of the risk of detachment of parts of the pietraforte sandstone elements. Machine learning techniques were applied for the segmentation and classification of information from a DEM obtained via a photogrammetric drone survey. An unsupervised geometry-based classification of the segmented objects was performed using K-means for identifying the most vulnerable elements according to their shapes. The results were validated through comparing them with those obtained via manual segmentation and classification, as well as with studies carried out by experts in the field. The initial results, which can be integrated with non-geometric information, show the usefulness of drone surveys in the context of automatic monitoring of heritage buildings.

Keywords: Palazzo Pitti; sandstone; stone deterioration; built heritage management; planned conservation; photogrammetry; machine learning; segmentation; monitoring; cultural heritage

Citation: Fiorini, L.; Conti, A.; Pellis, E.; Bonora, V.; Masiero, A.; Tucci, G. Machine Learning-Based Monitoring for Planning Climate-Resilient Conservation of Built Heritage. *Drones* **2024**, *8*, 249. https://doi.org/10.3390/drones8060249

Academic Editors: Daniela Oreni, Fabrizio Banfi and Davide Mezzino

Received: 10 May 2024
Revised: 31 May 2024
Accepted: 3 June 2024
Published: 6 June 2024

1. Introduction

The current role attributed to cultural heritage as a founding value of society, coupled with the corresponding attention devoted to its protection, has given rise to a convergence of two trends. On the one hand, over the past century, there has been a growing awareness of the role played by cultural heritage in modern societies, which has in turn expanded the amount of assets to be protected. Conversely, the progressive widening of heritage recognition to ever broader categories of tangible and intangible assets necessitates the study of a wider range of processes for their documentation and protection [1].

On the other hand, awareness of the direct impact of human activities on cultural heritage has increased considerably. The very act of recognising the cultural value of a cultural asset can, in some cases, prove to be a catalyst for its destruction, as demonstrated by the impact of overtourism and deliberate destruction during conflicts and wars to dismantle the cultural identity of adversaries [2,3].

Therefore, alterations to environmental conditions caused by human activities can produce indirect, pervasive, and continuous damage [4]. Such results are determined by degradation mechanisms induced via pollutants, which have already been studied for a long time, and also by climate change effects, as revealed in recent studies [5].

Among the climate change-related factors, it is worth noticing that the increased intensity and frequency of extreme climatic events may accelerate the natural processes of alteration and degradation of the materials of artefacts that were originally designed to last in different environmental conditions. Furthermore, extreme climatic events are often highly correlated with an increase of landslide or flood risk, affecting not only cultural heritage but also large parts of the territory.

The creation of digital documentation of cultural heritage is a complex undertaking. The application of high-resolution geomatic sampling techniques has resulted in the generation of a vast quantity of spatial data in a relatively short period of time. The accuracy and quality of the data are sufficient and often higher than required in most applications. However, metric data represent only the initial material from which further processes, such as the production of CAD drawings, 3D models, GIS, BIM, and so forth, are derived. The generation of such outcomes and their subsequent interpretation in order to obtain useful information for the understanding and more effective management of cultural heritage still necessitates the intervention of specialists. Consequently, it is beneficial to identify processes that facilitate the management, processing, and interpretation of digital data, particularly in the context of large and complex buildings, archaeological areas, and cultural sites.

This study presents the initial findings of a research project investigating the potential of automated spatial data segmentation to identify the most vulnerable areas of the façades of the Palazzo Pitti in Florence, Italy. This is part of a larger project concerning the potential detachment of stone elements due to the characteristics of the material used for the construction of the building. Such an occurrence may have particularly grave consequences in a context with a high number of tourists, such as that of Palazzo Pitti. For this reason, the façades are periodically monitored through direct inspection by specialists.

In 2020, a complete survey of Palazzo Pitti (described in more detail in Section 3) was carried out using integrated geomatic techniques. The survey results were employed to generate computer-aided design (CAD) drawings like plans, elevations, and cross-sections. Subsequently, all the stone elements of the façades were manually segmented and classified, with the aim of creating a geographic information system (GIS) database to facilitate their management, monitoring, and conservation.

This paper presents a further research advancement, namely the development of a fully automatic approach for the segmentation and classification of these elements. The objective of this analysis is to facilitate the monitoring process through providing tools for a more detailed assessment of the threat of detachment of parts of the stone elements.

The study focusses on the south-west side of the main courtyard (known as "Cortile dell'Ammannati", Figures 1 and 2). However, the same method can be adapted to the entire range of rusticated façades of the Pitti Palace, even if they have slightly different finishes, as well as to other buildings with similar stonework.

Section 2 presents the role of digitalisation of cultural heritage in the broader implementation of a digital economy and its relation to European and Italian environmental policies. Section 3 presents the case study of Palazzo Pitti and the specific features of the degradation of its façades, which justifies a continuous surveillance programme. Section 4 describes the photogrammetric survey of its façades, carried out with a drone, and the elaboration of the documents and materials used as references in the subsequent sections.

Section 5 outlines the principal methodologies and the state of the art for segmenting the components of a façade. It also describes the system adopted, which is essentially based on the geometry of the components. In Section 6, the relationship between facade elements and associated risk factors is elucidated. The results obtained are presented in Section 7; Section 8 highlights strengths and weaknesses of the proposed research and outline possible future developments.

Figure 1. The Palazzo Pitti complex; The south-east facade of the courtyard is outlined in red.

Figure 2. A view of the "Cortile dell'Ammannati".

2. Digitalisation of Cultural Heritage and the Green Deal in Europe and Italy

The European Community has promoted innovation [6,7] in the cultural heritage sector for the documentation, preservation, and sharing of the European legacy. It has done this through directives, studies [8], guidelines [9], and supporting research projects and infrastructures such as Europeana [10], which is undoubtedly the widest and best known of these. The European Community's support for digitalisation can be traced back to the Lisbon Strategy [11,12], which was developed in 2000. The action plan highlighted the potential of digitalisation to promote European culture, while acknowledging the primary economic objective of "strengthening employment, economic reform and social cohesion within a knowledge-based economy".

At the beginning of 2020, the European Community initiated the Creative Europe [13,14] programme with the objective of providing support to cultural sectors, including those related to heritage, in order to facilitate the exploitation of the opportunities presented by the digital transition.

In addition, the Green Deal was presented to the European Parliament in 2019 [15]. This was a series of initiatives with the objectives of achieving climate neutrality by 2050 and of identifying prevention and adaptation strategies to be adopted in response to the challenges posed by climate change.

Although the Green Deal did not explicitly refer to cultural heritage, a document [16] produced in accordance with the Work Plan for Culture 2019–2022 [17] identifies threats to and critical issues for heritage arising from climate change and examines the potential contributions that cultural heritage can make to the Green Deal for Europe. The document presents an analysis of the current state of cultural heritage in Europe in relation to climate change and proposes a series of actions and good practices aimed at risk mitigation. Research and innovation play a central role, promoting an interdisciplinary approach and "more in-depth studies on the behaviour of cultural heritage materials (organic, inorganic and composite) in times of climate change, including studies based on modelling and simulation (including protection, conservation, green materials and consolidation), and studies on the compatibility of adaptation measures with cultural heritage guidelines and innovative materials monitoring (using sensors, three-dimensional documentation and AI/machine learning processing)".

Furthermore, the Commission Recommendation on a common European data space for cultural heritage [18] encourages member states to digitise their cultural heritage between 2025 and 2030 to improve its preservation, with particular reference to heritage at risk. It recalls how "the development of advanced digital technologies, such as 3D, artificial intelligence, machine learning, cloud computing, data technologies, virtual reality and augmented reality" has created great opportunities and that "cultural heritage assets digitised in 3D can be a source of relevant knowledge on climate-related impact, adaptation and resilience (e.g., 3D allows non-destructive analysis of assets, visualisation of damages and information for restoration, conservation, etc.)".

In July 2020, the European Council approved Next Generation EU [19], a fund with a value of EUR 750 billion to support member states in the aftermath of the COVID-19 pandemic. In Italy, the National Recovery and Resilience Plan (Piano Nazionale di Ripresa e Resilienza—PNRR) [20], a programme of reforms and investments over a 2021–2026 timeframe, was introduced in 2021 to manage Next Generation EU funds. The objectives were twofold: firstly, to address the economic and social damage caused by the pandemic; and secondly, over a longer time frame, to resolve the structural deficiencies of the Italian economy and to contribute to the digital and ecological transition. The programme, which underwent a partial modification in 2022 [21], is divided into 16 components grouped into six missions. Of these, the first, entitled "Digitalisation, Innovation, Competitiveness and Culture," and the fourth, "Education and Research," address the themes of cultural heritage in the context of the digital and ecological transition in greater depth.

In this context, the Ministry of Culture, through the National Digitalisation Plan (Piano Nazionale di Digitalizzazione—PND) [22], which was drafted by the Central Institute for the Digitalisation of Cultural Heritage Digital Library, intends to promote and organise the process of digital transformation in the period 2022–2026 in accordance with European directives. The process is oriented towards the preservation of collective heritage, represented by cultural assets, with the objective of making them more accessible to people and fostering the development of new professional actors and services. With these assumptions, the PND contributes to achieving some of the sustainable development goals (SDGs) of the United Nations 2030 Agenda [23], namely, quality education, decent work, economic growth, business innovation and infrastructure, and sustainable cities and communities.

The target groups are museums, archives, and other public and private cultural institutions, as well as professionals such as artists, scholars, stakeholders, etc.

The PND, which includes a number of appendices, is a highly detailed document and is divided into three sections. The first section, the vision, outlines the reference values and objectives. The second section, the strategy, lists the technologies and processes. The third section, the guidelines, provides the operational tools.

The vision is to move from digital conversion to digital transformation. This shift is not only a technological challenge but also a cultural one. It involves the creation of a digital ecosystem [24] based on the relationships between objects and between objects and people [25].

Cultural institutions should consider digitalisation as an integral aspect of their operations, rather than a standalone initiative to be implemented on an ad hoc basis. Rather than viewing digitalisation as a marginal, secondary aspect, cultural institutions should view it as an element closely connected to all the choices and actions they develop.

Among the purposes of digitalisation, the document [26] asserts that the digital copy of the original, if acquired correctly and for documentation purposes, assumes informative and cognitive value comparable to that of the asset itself, representing material characteristics and the relative state of preservation. The digital resource thus becomes an active tool of study and knowledge that enables simulations and, if consistently replicated over time, supports monitoring processes.

The CHANGES project [27], which is funded by PNRR resources within Mission 4 'Education and Research', comprises nine spokes. These are characterised by horizontal activities, including research, technology transfer, training, and the dissemination of results. The project is coordinated by eight state universities and one research organisation. Specifically, Spoke 7, entitled 'Protection and conservation of Cultural Heritage against climate change, natural, and anthropic risks', investigates the consequences for cultural heritage caused by the impact of climate change, which accelerates the effects of natural and anthropic risks. The investigation has been conducted with a multi-scalar approach, including territory, historical cities, architecture, and mobile cultural artefacts. This approach allows the consideration of different scenarios and the addressing of both general and specific issues. Spoke 7 selected artefacts of varying scales for the assessment of their state of conservation and exposure to risks. Among these, Palazzo Pitti was also chosen for assessing the potential for reusing the spatial data acquired during the survey campaign described in more detail below and on which this study is based. The considerations encompass the containment of the effects of degradation, which may be exacerbated by climate change, and the potential impact on the safety of visitors to a monument that is frequented by a large number of people.

In Italy, the protection of cultural heritage from the effects of climate change has been achieved through the formulation of strategies and implementation plans that provide guidance and good practices for monitoring and management in situations of risk from natural and man-made hazards.

Following the impetus provided by the 2013 European Adaptation Strategy [28], in 2015, Italy adopted the "National Climate Change Adaptation Strategy" [29], of which the operational instrument is the "National Climate Change Adaptation Plan" (Piano Nazionale di Adattamento al Cambiamento Climatico-PNACC) [30], approved on 21 December 2023.

Both documents have a section dedicated to 'Cultural Heritage and Landscape'. The design of protection, control, and damage-prevention strategies derives from the assessment of vulnerability and risks, the study of the heritage materials, with particular reference to built heritage, and the associated types of degradation in relation to environmental conditions. Among these factors, water plays a primary role. Intense and increasingly frequent rainfall causes significant damage to buildings, in particular through accelerating the deterioration of structures and wall surfaces. Considering the buildings in the Florentine area built with the typical "pietraforte" sandstone, the action of rainwater causes the dissolution of calcium carbonate present in the calcite veins and this mechanism can lead to the decohesion of the rock, with the complete opening of the veins and the possible detachment and falling of blocks, with possible personal injury and material loss.

The results of prediction models suggest that the dissolution of carbonate stone materials will be predominantly due to precipitation and an increased atmospheric CO_2 concentration in the near future [31].

The document, admitting its own shortcomings, outlines how the research on the impact of climate change on cultural heritage has thus far been relatively limited and notes the urgency of this issue. Furthermore, it underlines the dearth of observational data, emphasising the necessity of continuous monitoring in order to correlate the degradation of materials and systems characterising cultural heritage with climatic variables and their changes. It highlights how downscaling is necessary if one wants to predict the impact that changing climatic parameters have not only on historic centres and the landscape but also on areas more specifically related to individual artefacts.

The Extraordinary National Plan for Monitoring and Preservation of Immovable Cultural Heritage, adopted by the Italian Ministry for Cultural Heritage and Activities in 2018 [32], defines the criteria for the selection of artefacts to be monitored and the consequent conservation interventions, as well as the necessary control priorities. These are determined based on specific indices of territorial danger and the individual vulnerability of buildings, the instrumental control systems to be used, and methods for implementing safety, conservation, and protection measures. The Plan also claims that monitoring the state of cultural property in relation to risk situations has always been considered of utmost importance, which over the years, on several occasions, has resulted in attempts to implement suitable policies and tools. In order to ascertain the types and level of risk to which a heritage building is exposed, it is necessary to gain a comprehensive understanding of the artifact and to conduct regular and systematic assessments of the asset's condition and the site in which it is situated. It also states that a "holistic" approach to data identification is required, from the global to the particular and from the context to the artefact, using high-resolution diagnostics to assess the individual structure or individual element of the structure itself. To achieve this, it is necessary to integrate various technologies and techniques like remote sensing, drones, photogrammetry, and sensors to apply to the artefact, and periodic checks on the state of structural and material deterioration.

It should be noted that, as of the present date, the only operational (and only partially updated) tool available for planning interventions on cultural heritage according to the criterion of level of risk is the Cultural Heritage Risk Map [33], which has been elaborated by the Ministry of Culture and is managed by the General Directorate for Cultural Heritage Security.

The aforementioned Italian documents discuss these topics only in their general aspects. However, it is important to remember that raw data must be interpreted and evaluated. Processing is what turns data into information, and this requires a critical approach. Therefore, it is essential that documentation and monitoring are designed and carried out by qualified specialists, possibly with interdisciplinary skills. It is essential that collected data be accompanied by metadata and paradigms that allow for the correct use of the data, their quality and accuracy, their preservation, and the traceability of acquisition and processing steps [34].

The topic also aligns with the Italian National Research Plan 2021–2027 [35], which identifies 'Humanistic Culture, Creativity, Social Transformation, Society of Inclusion' among the major strategic research and innovation areas for development in Italy. "Digitalisation of protection, conservation and valorisation processes" is the title of the first articulation of this strategic research area. Therefore, a current research topic focuses on objectively quantifying the presence and effects of various forms of decay and then calculating damage indices to prioritize conservation and preservation works [36]. The present work is an example of the digital transition of the building analysis and monitoring process, initiating the long-sought transition to preventive conservation, maintenance, and monitoring [37] and, from a broader perspective, supporting the transition to digital transition in the field of documentation and protection of cultural heritage.

3. Palazzo Pitti and the Conservation of Pietraforte Sandstone

Palazzo Pitti is the largest and most magnificent mansion in Florence. The original part of the building was built in 1452 on the order of Luca Pitti. According to Giorgio Vasari, it may have been designed by Filippo Brunelleschi. In 1549, Eleonora of Toledo, the wife of Cosimo de' Medici (the future first Grand Duke of Tuscany), bought the whole property Even though the palace was already three storeys high, it is likely that the interior was still under construction [38].

After Cosimo ascended the throne, the building grew in importance and was used not only as a residence but also as the seat of the government. In the 19th century, it was briefly the seat of the Kingdom of Italy. The palace's current appearance is the result of a series of later extensions. The U-shaped courtyard designed by Bartolomeo Ammannati, which faces the hills behind the building, is one of the most important expansions. The courtyard was built between 1561 and 1577 and, like the rest of the building, features impressive rusticated stonework.

The façades of the courtyard were developed on three floors above ground level [39]. The architectural style employed by Ammannati involved the traditional superimposition of the three classical orders—Doric, Ionic, and Corinthian—with rustic yet regular ashlars (Figure 3).

(**a**) (**b**) (**c**)

Figure 3. The rusticated half-columns of the three classical orders in the courtyard of Palazzo Pitti. (**a**) the Doric order on the ground floor; (**b**) the Ionic order on the second floor; (**c**) the Corinthian order on the third floor.

On the ground floor, the rusticated blocks are domed and close together The column shafts and the wall were designed to accentuate the massiveness of the lower floor, which serves as the foundation of the building.

The upper floors exhibit an alternation of rustic and plain stonework. On the second floor, the rustic ashlars are squared, while on the third, they are domed. From an architectural perspective, this creates a strong and striking contrast between light and shadow. However, this type of stonework involves the presence of numerous protruding elements, from which some detachment of stone fragments can occur. This can lead to safety concerns and challenges in the conservation of the work [40–42].

The façades were constructed using a local kind of sandstone, called pietraforte, which was also used in many other Florentine buildings. The most important quarries for this kind of stone were located behind the Palazzo Pitti and right in the place where the courtyard was built. This sandstone is bluish-grey when freshly extracted, but over time it takes on a characteristic brown colour due to presence of iron oxide. It typically has complex stratification and calcite veins at different angles. Although the façade's stone texture is aesthetically pleasing, it compromises its consistency. This sandstone is susceptible to spalling and the detachment of large fragments due to mineral solubilisation and thermo-hygrometric cycles. As the rustication has overhanging elements, including large projections, it is important to regularly inspect the condition of the façades and promptly secure any elements at risk.

These problems are dependent on the intrinsic characteristics of the stone and have always been present, as evidenced by the numerous gaps in the façade and the metal anchors used in the past to hold the broken blocks together (Figure 4).

Figure 4. An example of a broken ashlar fixed with several types of metal ties over time.

However, pollution and climate change, altering the intensity and frequency of rainfall and the concentration of CO_2, have exacerbated these detachment problems, accelerating the degradation processes, and consequently increasing the building's preservation issues and the safety risks to people.

Motivated by the above considerations, a research agreement was established in 2020 between the administration of the "Gallerie degli Uffizi", which is responsible for the maintenance and preservation of Palazzo Pitti, and the University of Florence, to conduct a comprehensive 3D survey of the building, aiming at providing the tools for successively monitoring the building's health. This agreement involved a comprehensive survey of all interior and exterior spaces within the complex, which comprises over 1500 rooms and houses four museums and other cultural heritage institutions. The documentation encompassed both geometric and construction characteristics, as well as information on finishes, materials, and the state of conservation. It also included details on their management, use, and current and past designations. A three-dimensional information system was designed to support knowledge of the building and its future management and maintenance [43].

The survey was conducted utilising integrated geomatic techniques. This entailed the establishment of a control network employing topographic and geodetic tools (GNSS and a total station) and the utilisation of 3D scanning and structure-from-motion (SfM) photogrammetry for the architectural survey [44]. Laser scanners were primarily used for surveying the interior spaces, while the exterior ones, including all the roofs and façades, were acquired through the integration of data derived from scanning and photogrammetry.

4. UAV Photogrammetric Survey on the Courtyard of Palazzo Pitti

The detailed survey of the façades of the courtyard of Palazzo Pitti was carried out using UAV SfM photogrammetry, designing a suitable flight pattern in order to achieve the required result in the most efficient way [45–47]. Among the factors to be properly taken into account in the flight plan, the geometric and exposure characteristics played a pivotal role. The dimensions of the courtyard are approximately 40 m × 50 m, with façades approximately 35 m high on three sides, while the south-east side is about 12 m high. This implies that in July, when it was possible to carry out the survey, the lighting conditions were changing rapidly during the day. Consequently, the time periods during which each façade was in shadow were calculated, and the survey was conducted primarily in the early morning or late afternoon, also in order to avoid the presence of visitors close to the flight area.

On each of the three largest façades (the fourth is only one storey high), an average of one target per column or window was placed. The coordinates of these targets were measured with a total station. In addition, some easily recognisable natural points were measured on the cornice below the roof, where it was not possible to manually place the targets. This allowed for a redundant set of control measurements, to also take into account the possibility that some targets could be damaged or removed. The elaborations presented in the following sections focus on the south-west façade of the courtyard (Figure 5), whose photogrammetric reconstruction was obtained by exploiting 18 control points (CPs) and 11 check points (ChPs).

Figure 5. South-west façade of the courtyard of Palazzo Pitti with CPs (yellow dots) and ChPs (red dots).

The drone used for the survey was a DJI Phantom 4 Pro, which has the camera characteristics reported in Table 1.

Table 1. DJI Phantom 4 Pro camera specifications.

Specification	Dimension (in mm)	Dimension (in Pixels)
Focal length	8.8	
Sensor	13.2 × 8.8	5472 × 3648
Pixel size	0.00241	

The main products of the photogrammetric survey were planned to be a 3D model and an orthophoto, whose spatial resolution (ground sample distance, GSD) was required to be approximately 2 mm.

Images were taken orthogonal to the façade, at an average distance of 6 m, in order to achieve the required spatial resolution (Figure 6). The terraces on the second and third floors, as well as the roof, exhibit a projection of approximately 1.5 m from the façade. Consequently, the distance between the drone and the most prominent parts of the façade was approximately 4.5 m. In these circumstances, it was necessary to operate the drone manually, as the GNSS signal was irregular or completely absent within the courtyard and at a short distance from the walls.

Figure 6. A photo of the DJI Phantom Pro UAV in flight, surveying the "Cortile dell'Ammannati".

The camera network comprised a series of orthogonal photographs, with an 80% forward overlap and a 70% side overlap. This entailed that the strips were taken at approximately two-metre intervals. The acquisition of images was also extended beyond the eaves line to record the roof pitches. Additionally, images with an oblique axis were captured at the same distance to reinforce the camera network and to facilitate the detailed documentation of the side faces of the elements. It is important to note that the aforementioned terraces are equipped with cast-iron railings and protective nets, which would have prevented imaging the façade behind them. Consequently, the stripe configuration was implemented through pitching up and down over and under the terraces.

Each elevation terminates at a concave or convex angle to the adjacent ones. In order to ensure a safe distance from the walls, in the corners, photographs were taken along a vertical line with the axis oriented towards the intersection of the walls. These sets of images were subsequently employed in the production of the orthophotos of both the façades.

In consideration of the optimal exposure times to achieve the requisite images in the shade, in an effort to avoid motion blur, the flights were executed at a velocity of approximately 0.3–0.5 m/s. In total, approximately 3000 images were captured for the survey of the south-west façade. Although the drone was capable of saving images in the raw format (DNG), the image buffer had a limited capacity, which may have resulted in the loss of some frames, which motivated saving them in JPEG format.

Given the necessity of performing imaging in limited timeframes, flights were conducted over several days. This necessitated meticulous planning and comprehensive annotation of the metadata associated with each flight session. This was undertaken in order to prevent gaps or overlaps and to ensure the generation of consistent results.

Nowadays, several photogrammetric tools and software are available on the market. In the 2020 survey, all the photogrammetric processing tasks were performed with Agisoft

Metashape Pro software. The photogrammetric model of the south-west façade exhibited a root mean square error (RMSE) of 9.4 mm for the CPs and 8.1 mm for the ChPs, values compatible with the error level of the realized topographic control network. The outcomes included the creation of (i) a digital elevation model (DEM) with a resolution of 2 cm (Figure 7), upon which the studies presented here were based, and (ii) a 2 mm GSD orthophoto (Figure 8), which served as the cartographic basis for the 2D drawings. In particular, all ashlars and stone blocks were identified manually with great precision and this information has been used for the validation of the results of the current study (Figure 9).

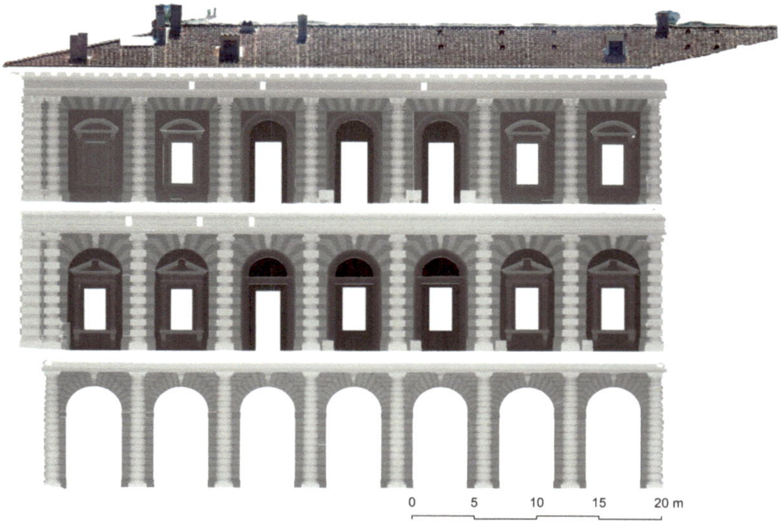

Figure 7. DEM of the south-west façade of the courtyard shown in Figure 1.

Figure 8. Orthophoto of the south-west façade of the courtyard shown in Figure 1.

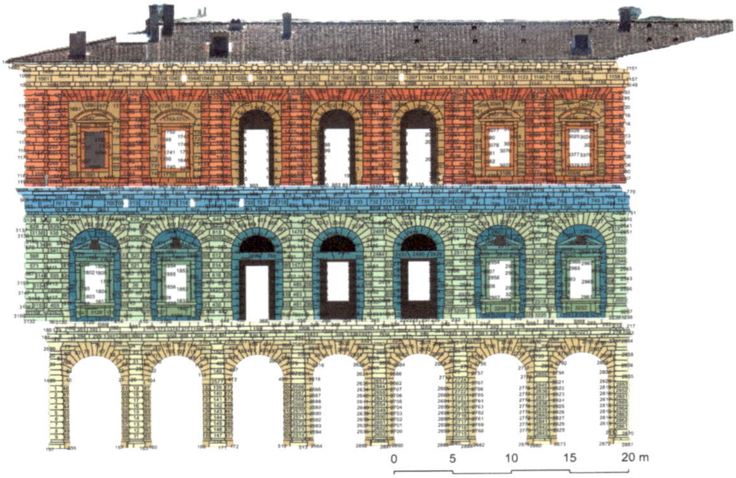

Figure 9. Manual segmentation and classification of the stone elements of the south-west façade of the courtyard shown in Figure 1. Each stone block is identified with a number, colours are related to the classes of the architectural components.

5. Segmentation of the Façade Elements

Segmentation entails partitioning a set of entities into several subsets, each of which is usually characterised by some kind of intra-set homogeneous characteristics. When considering spatial data related to a 3D model, segmentation aims at grouping the original data in different regions based on certain characteristic such as colour, intensity, typology, or semantic information. Since both photogrammetry and laser scanning, the two most commonly used 3D reconstruction techniques nowadays, produce point clouds as natural output, a number of techniques have been recently investigated in order to properly segment such data, when dealing with semantic segmentation of heritage buildings in particular [48–53]. Among the different proposed approaches, it is typically possible to distinguish between those directly dealing with the point cloud, which usually either deploy neural networks [54] or exploit geometric features [55], and those based on image segmentation instead. In the latter case, the image segmentation results should be properly reprojected on the original point cloud [56].

Although the approaches mentioned above aim to deal with the point clouds produced for quite generic heritage buildings, some simplifications may be introduced when considering certain specific cases, such as when dealing with only a building's façade, for instance. A dominant plane and its normal direction can usually be identified when considering a façade; hence, a 2D representation of the façade can be easily obtained via projecting the façade elements along the normal direction onto the dominant plane. The obtained representation, being either an orthophoto or a DEM, depending on the specific kind of operation that has been implemented, is clearly not completely equivalent to the original data, but the information loss, in this specific case, is often quite marginal. Several approaches have been tested for orthophoto segmentation, usually based on the use of proper convolutional neural networks [57,58], as well as DEM [59]; the latter approach usually aims at exploiting geometric information often including spatial derivatives, normals [60], or computations of level sets [61]. These approaches are often restricted to brick segmentation, often on masonry walls [57–59,62,63].

Motivated by the above observations, this work, quite similarly to [59,64], implemented geometry-based segmentation directly on the façade DEM (Figure 7). Since the main goal of this work is to monitor the detachment of portions of ashlars and the assessment of the riskiest elements, the purpose of the implemented segmentation procedure was principally that of automatically segmenting the ashlars. The DEM segmentation workflow was as follows:

- *Preliminary operations*. The DEM may include areas outside of the region of interest and/or areas not covered by ashlars. A simple depth threshold-based procedure was used in order to determine the DEM regions sufficiently close to the ideal wall surface to be considered ashlars. In our implementation, this corresponded to limiting the analysis to points within a 1.1 m depth interval;
- *Depth variation-based edge detection*. This step aimed at discriminating the potential object boundaries as those associated with depth variations in the DEM. A simple thresholding mechanism was used to determine the norm of the DEM gradient with such an aim, i.e., let $z = f(x, y)$ be the DEM depth value corresponding to the (x, y) coordinates and $\nabla f(x, y)$ be the value of the gradient in (x, y), which can be estimated, for instance, as follows:

$$\nabla f(x,y) = \begin{bmatrix} \frac{\partial f}{\partial x} & \frac{\partial f}{\partial y} \end{bmatrix} \approx \begin{bmatrix} \frac{f(x+1,y)-f(x-1,y)}{2} & \frac{f(x,y+1)-f(x,y-1)}{2} \end{bmatrix}, \tag{1}$$

where, for simplicity of notation, x and y are assumed to represent the location of the considered point in the DEM expressed in pixels.

Then, the following thresholding was applied to the norm of the gradient on all the (x, y) positions in the DEM:

$$|\nabla f(x,y)| > \overline{\Delta z}, \tag{2}$$

where $\overline{\Delta z}$ is a properly set depth threshold. An example of the thresholding results is shown in Figure 10. It is worth noticing that several more complex alternatives can be considered in spite of the described depth-variation-based edge detection approach, e.g., edge detection based on the Harris filter [65]. Nevertheless, in this work, other options have not yet been considered because of the simplicity of the proposed method and the reasonably consistent results obtained in the case study.

Figure 10. Example of depth-variation thresholding results.

- *Connected components computation*. Connected regions [66], of black pixels, i.e., not previously identified as potential edges, were determined in the binary image resulting from the previous step (e.g., Figure 10), considering 8 connectivity, i.e., black pixels were in the same region if either they had an edge or a corner in common.
- *Optimization of object boundaries*. Object boundaries derived from the depth-thresholding mechanism summarized in Equation (2) may be inaccurate, in particular when objects have a rounded shape, as visible in Figure 11a,b. This typically results in partially incomplete segmented ashlars (or, more in general, objects). Consequently, an optimization procedure was implemented in order to reduce the above-mentioned issue. With such an aim, first, the normal to the DEM surface in the neighbourhood area of each object border was computed, i.e., on the pixels identified as potential edges using Equation (2) (white pixels in Figure 10). Then, all the points of each border were moved to the corresponding closest local maxima of the angular difference between the normal to the surface and the z direction. Finally, a piecewise linear fit was computed on the derived object boundary in order to reject the outliers, if any. The result of this procedure for the region in Figure 11a is shown in Figure 11c.

(a) (b) (c)

Figure 11. Example of segmentation results: (**a**) UAV view of the examined area, (**b**) segmented objects after depth-thresholding and computation of the connected components, (**c**) segmented objects after boundary optimization.

6. Ashlar Classification and Risk Assessment

Risk assessment in buildings and infrastructures is a very important and popular topic, attracting the interest of researchers from different fields [67–69]. Nevertheless, the specific peculiarities of the façades of Palazzo Pitti require the development of ad hoc procedures in order to properly deal with their monitoring and maintenance.

The risk of detachment of portions of stone blocks is related to a number of factors, including the progress of lamination, the presence of discontinuities, cracks, and fractures, material loss and colouration, and erosion and exfoliation phenomena, as well as the geometric parameters described in detail below. In addition to material and decay analysis, non-destructive techniques (NDTs) have been applied for effective assessment of the façade's conservation: sclerometric measurements through the Schmidt hammer test to evaluate the stone surface strength and ultrasonic velocity testing to evaluate the consistency of masonry and allow the identification of internal defects like fractures, voids, and detachments [70]. Monitoring campaigns are periodically carried out on the various façade of the Palazzo Pitti, including the removal of detached parts of stone or the consolidation of blocks that need it; the last monitoring campaign was carried out in February 2022.

A first step in the risk assessment related to the detachment of ashlars has already been investigated [71] by an interdisciplinary group, including geologists, restorers, and geomatic experts. That work presented a method for managing both geometric data and results obtained through NDTs in an integrated manner. With regard to the geometric data, a virtual inspection was conducted manually on the same dataset used in the present study. A number of key aspects were considered, some of which were derived from the 3D survey output, while others were inferred by the virtual inspection of high-resolution, 3D-oriented images. Materials, forms of decay, and previous restoration works were also mapped.

On a test area, the co-occurrence of all of those key factors was considered and related to the shape model of the façade, and further analysis was limited to overhanging elements

The findings were corroborated through comparing the stone elements that were identified as presenting criticalities with those selected by expert geologists and restorers in a field investigation, thus validating the proposed approach as a valuable preliminary analysis prior to fieldwork, capable of reducing the time it takes and therefore its costs.

Despite the outcomes of such comparison proving the potential of the adopted approach, the main limitation was clearly the significant need for human intervention in the data processing and analysis. The present work aims to overcome the limitations of the manual approaches described in the aforementioned paper to ascertain the viability of deriving pertinent information to assess the risk of detachment of portions of stone elements through a more comprehensive and extensively automated analysis of the shape model, while the other previously identified factors influencing the criticalities of the blocks are, at present, overlooked. Introducing the discarded factors in the automated analysis will be considered in the future development of this work.

Analyses of a building conducted with a traditional approach require interpretation and, therefore, involve the work of specialized personnel according to study and "training" processes that can now be at least partially computerized. Courses in architectural history basically use methods analogous to machine learning: images are proposed, from different viewpoints, of buildings with some relationship to each other; comments on the images correspond to the training information. The architectonic order is a formalized code—yet, always declined in different ways—and its recognition in the elevation requires critically applying a knowledge model. The architectural order can then be broken down into its sub-elements, as well as the elements composing a general building, which can be identified and organised into more or less complex structures, depending on the objectives of the study, organising them, if necessary, into classes and subclasses [72–75]. In the masonry of Palazzo Pitti, there is not always a correspondence between the architectural decoration, apparently subdivided in three overlapping orders characterised by ashlars of different shapes and surface finish, and the façade's structure. In many cases, a single architectural element consists of different stone blocks; in others, the same block is carved to represent different architectural elements. Consequently, although at first glance the rustic order appears to consist of homogenous elements, on a closer inspection it is possible to distinguish the stone blocks from the ashlars (Figure 12). A single stone block used for masonry construction constitutes the minimum spatial reference for diagnostic and structural analysis.

In this study, on the other hand, the segmentation was aimed at identifying the individual ashlars, and subsequent analyses were concentrated on those elements. In accordance with this consideration, only a selection of the objects segmented as described in Section 5 was used for further analysis, in particular in order to (i) distinguish between different ashlar categories based on their shapes and (ii) compute a geometric-based ashlar detachment risk map. The rationale mostly motivating the procedure presented in this section is that, from a merely geometric point of view, the most protruding elements in the façade are those that are supposed to be associated with the highest level of (geomentry-only-based) risk. Hence, the above-mentioned restriction of object selection was implemented, considering only areas where protruding elements were present and, on the other hand, considering only objects within a certain area range, in order to select (almost) only ashlars.

Then, an unsupervised geometry-based classification of the segmented objects was performed using K-means [76]. It is worth noticing that segmentation and classification techniques can be used to support building information modelling (BIM) generation; e.g., semantic segmentation may be deployed, for instance, as a tool to support a scan-to-BIM workflow. Despite such potential of an automatic segmentation and classification procedure, the goal here was merely that of grouping the ashlars into categories according to their shapes and exploiting such information in the computation of the geometry-based risk assessment.

(a) (b)

Figure 12. Difference between ashlars and actual stone blocks. The apparently monolithic ashlars (**a**) are sometimes actually composed of several stone elements (**b**).

The use of unsupervised classification was motivated by the need of reducing as much as possible the human intervention in the analysis procedure, hence avoiding the need for creating an ad hoc training dataset. Nevertheless, supervised classification will also be considered in future investigations in order to check the potential performance of this kind of approach. Among the unsupervised approaches, K-means was chosen for its simplicity and limited computational complexity. Given a set of **n** vector data, where each data sample $\boldsymbol{x}_i$ is assumed to be a **d**-dimensional vector, K-means aims at partitioning the **n** samples into K clusters by minimising the functional V given by the sum of the intra-cluster squared distance with respect to the cluster mean $\boldsymbol{\mu}_j$:

$$V = \text{arg}min \sum_{j=1}^{K} \sum_{i \in C_j} \left| \boldsymbol{x}_i - \boldsymbol{\mu}_j \right|^2, \tag{3}$$

where C_j is the j-th cluster. It is well known that the outcome of the algorithm depends on the values $\{\boldsymbol{\mu}_j\}_{j=1,...,K}$ used to initialize it. Nevertheless, such dependence can be remarkably reduced via running several instances of the algorithm with different initialization conditions, as was carried out in this work.

In the application considered in this work, each vector $\boldsymbol{x}_i$ was associated with a different object, i.e., a different ashlar, and K-means was used to automatically group the ashlars into different categories based on their shape. K-means output clearly depends on the specific vectors used to describe the object's geometric characteristics. Taking into account the specific characteristics of the ashlars in the considered façade, the following geometric features were considered as input for the segmented objects for the K-means clustering algorithm:

- object area,
- width (along the x direction),
- height (along the y direction),
- depth variability, computed as the median absolute deviation (MAD) of the depths in the DEM area associated with the segmented object,
- variability of the horizontal angle of the surface normal with respect to the z axis, computed as the MAD of the horizontal angles for all the object pixels.

The values of the geometric features reported above were centred and normalized before being inputted in the K-means algorithm, in order to avoid an implicit different weighting between such factors. Nevertheless, future investigations will be dedicated to determining whether the use of different weighting criteria may lead to better classification results.

Then, a geometry-based risk index was computed for each of the segmented objects, based on how much it protruded. In accordance with the previously presented motivations, this index related to only the protruding volume of the considered ashlar. Let us consider the example in Figure 13a, where the protruding volume has to be computed for the blue ashlar; the red one is the one below it. In the computation implemented here, the upper surface of the red one is extended along the y direction up to the top border of the blue ashlar, hence defining the green solid shown in Figure 13b. The protruding volume is hereafter considered as the subtraction of the green parallelepiped volume from the blue ashlar one. The computation of such volumetric difference is achieved as follows:

- partitioning with a grid the projection of the blue ashlar surface on the x–y plane. A natural choice for such partitioning is clearly to consider grid cells equal to the DEM ones;
- for the grid cell of index (h,k), determining the corresponding parallelepiped of basis area A_{hk} and size z_{hk} along the z direction given by the difference of z depth between the blue and green boxes corresponding to such a position. Then, the volume for the parallelepiped is $V_{hk} = A_{hk} \times z_{hk}$, and the protruding volume is computed as $\sum_{h,k}(A_{hk} \times z_{hk})$.

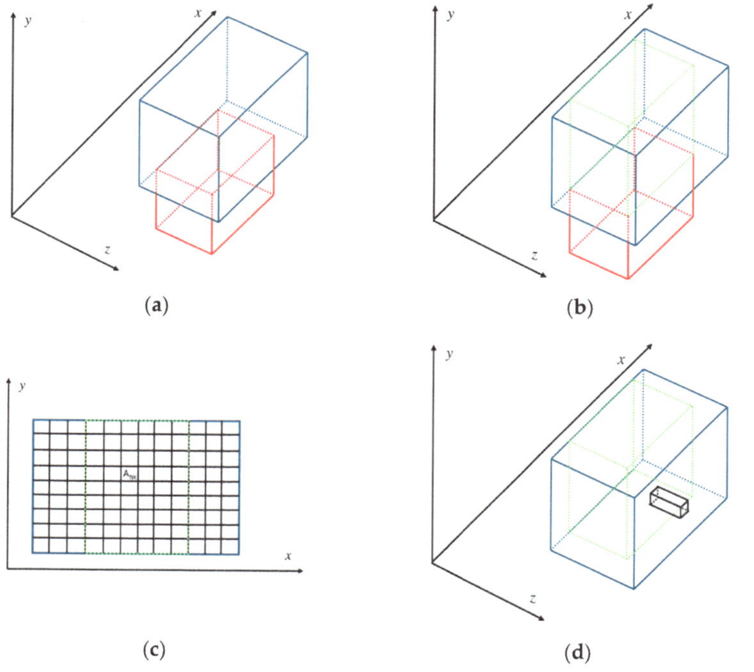

Figure 13. Computation of the protruding volume: (**a**) current ashlar (blue) and the one below (red); (**b**) extending the upper border surface of the red ashlar to the top border of the blue one; (**c**) partitioning grid of the blue ashlar's surface on the x–y plane; (**d**) parallelepiped defined on one of the grid cells in (**c**), with z size derived from the different z depth of the blue and green boxes on that cell.

The computation of the protruding volume presented above can easily be generalized to the irregular shapes of the ashlars. It is worth noting that when ashlars have a rounded shape, usually approximately circular on any section on the y = $\bar{k}$ plane, the cell representation may not provide a good fit to the real ashlar shape close to the two sides of the ashlar. The ashlar section being approximately circular, the circle radius r was computed as half of the ashlar width on the x direction. The area on the x–y plane of the ashlar slice in correspondence with y = $\bar{k}$ can be approximated as $\pi r^2/2$. Extending the volume computation to this case and its generalizations is also trivial.

Assuming a constant value for the stone density, i.e., 2.6 g/cm^3 [77], the protruding volume was converted to an approximate mass value. Hence, the following metrics were computed for each segmented object:

(a) the approximate mass m of the protruding volume;
(b) the linear density, that is $m/\Delta x$, where the mass m is defined as above, while Δx is the object width along the x direction;
(c) the weight of the protruding volume above each cell, i.e., for each cell in a segmented object, only the portion of the protruding mass above the cell was considered.

7. Results

The segmentation procedure presented in Section 5 was applied to the DEM of the south façade of the courtyard, obtaining the results shown in Figure 14.

Figure 14. Segmented objects according to the procedure described in Section 5.

In order to validate the obtained segmentation, a comparison was carried out with the manually identified stone elements shown in Figure 15, which were drawn with CAD tools on the orthophoto and were used here as a reference.

Since the main interest is in determining a geometrical vulnerability map for the protruding ashlars in the façade, the analysis of the segmentation results presented here and the subsequent vulnerability assessment reported here was limited to the area shown in Figure 16. Table 2 provides a numerical assessment of the implemented automatic segmentation procedure, reporting the following performance metrics: accuracy, F_1-score, the over segmentation index (OverSeg), median intersection over union (IoU), and the median absolute deviation of the IoU values. Accuracy and F_1-score are defined as follows:

$$\text{accuracy} = \frac{TP + TN}{TP + TN + FP + FN}, \tag{4}$$

$$F_1 = 2\frac{precision \times recall}{precision + recall}, \tag{5}$$

where TP are the true positives, TN the true negatives, FP the false positives, FN the false negatives, precision = $\frac{TP}{TP+FP}$, recall = $\frac{TP}{TP+FN}$. The OverSeg index, indicated as a percentage, is defined as the number of real stone elements that are segmented in more than one object divided by the total number of real stone elements automatically detected The IoU between a ground truth stone element and an automatically segmented object at least partially overlapped with it is equal to the overlapping area between the two divided by area of their union. It is worth noting that when a real stone element was segmented into more than one object, only the automatically segmented object with largest overlap with the real one was counted as TP, whereas the others were counted as FP. Figure 17 shows the IoU distribution limiting the analysis to only the correctly matched elements (true positives).

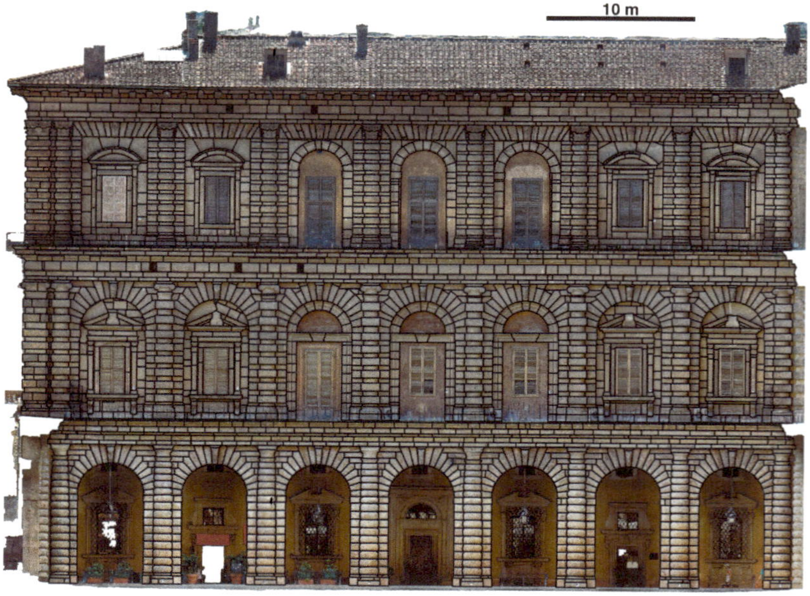

Figure 15. Manually segmented stone elements.

Table 2. Automatic segmentation vs. manually identified stone elements.

Accuracy	F_1	OverSeg [%]	Median IoU [%]	MAD IoU [%]
0.89	0.94	11.8	71.2	7.9

The accuracy and F_1 values in Table 2 can be considered sufficiently high to be quite satisfying. However, the median IoU was not as high as the accuracy and F_1. This value of the median IoU can be explained through the relatively large number of over-segmented stone elements (and a few of under-segmented ones), as shown in the OverSeg value in Table 2 and the left side of the IoU distribution in Figure 17. Figure 17 also shows that very few stone elements were identified with more than 90% IoU; this can probably be explained by differences mostly in the object borders.

Figure 16. DEM of the area selected for the analysis of the segmentation results and for the risk assessment.

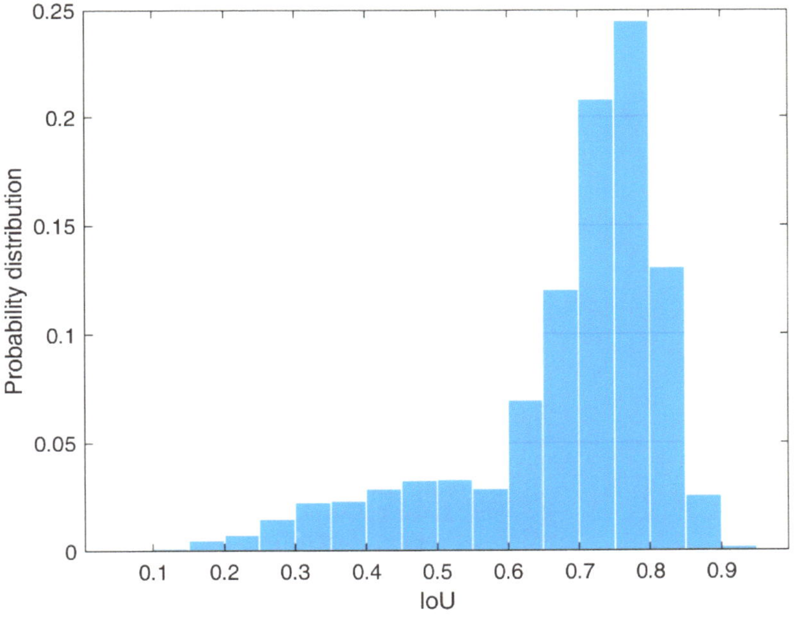

Figure 17. Distribution of the IoU values for the matched stone elements.

From a visual inspection of the façade, the presence of ashlars of different shapes is quite clear. For instance, in Figures 3 and 6, it is possible to distinguish at least the following categories: quite large rounded and squared stone elements on the half-columns, smaller squared elements on the sides of the half-columns, capitals, and less protruding elements on the arches and keystones. In accordance with this observation and with a few tests,

K-means algorithm was executed on the segmented objects, setting K = 5 and running 1000 instances of the algorithm with different initial conditions. The obtained classification is shown in Figure 18. Given the obtained classification results, the five identified classes can be quite well described as follows:

- Rounded large column elements (yellow);
- Squared large column elements (blue);
- Squared side column elements (green);
- Arch elements (orange);
- Rounded large column elements (light blue).

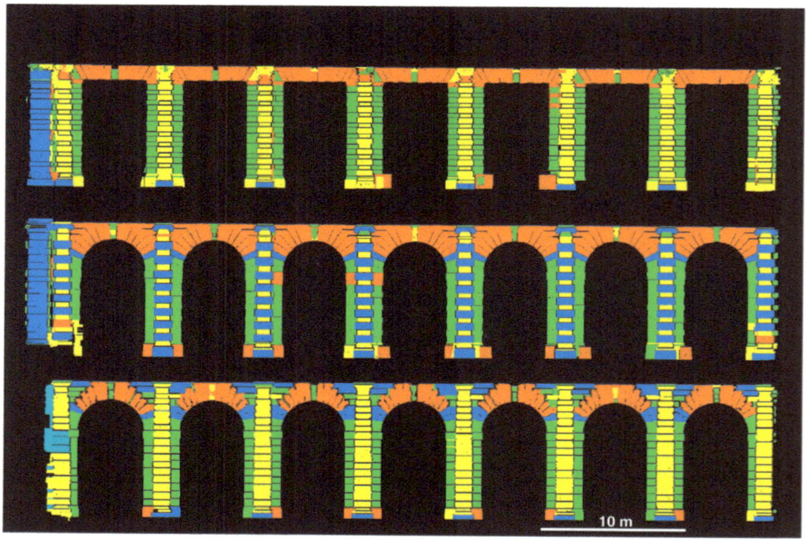

Figure 18. Unsupervised (K-means-based) classification of the segmented objects.

Regarding the light blue elements in Figure 18, it is clear that they are concentrated on the bottom left of the façade and that such a class is clearly a duplication of the first one. Nevertheless, the algorithm dedicated a different class to these elements because of some incorrect segmentations, leading to the formation of larger objects (under-segmentation issue) with shapes different from those described in the other classes. It is also worth noting that air conditioning systems installed on the façade (Figure 19) were also classified, mostly in the "arch elements" class. Although a few elements in Figure 18 were identified as wrongly classified (usually because of segmentation errors), the obtained classification appears overall to be quite reasonable.

In accordance with the indexes introduced at the end of Section 6, the following figures show the assessed mass of the protruding volume for each segmented object (Figure 20), the corresponding linear density, normalized to its maximum value within the test area (Figure 21), and the determined mass of the protruding volume over each point of the detected objects (Figure 22).

Figure 19. Air conditioning systems on the terrace of the second floor. Such elements can be wrongly classified.

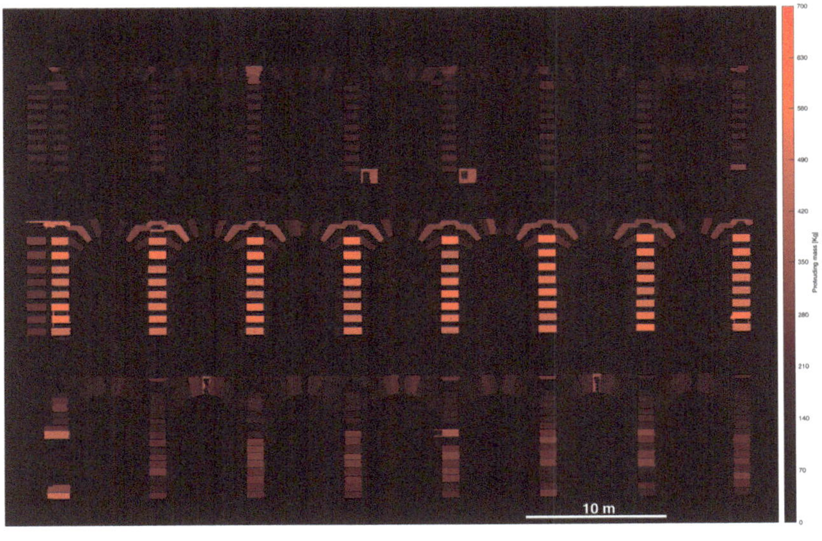

Figure 20. Approximate mass of the protruding volume for each segmented object, computed as described in Section 6.

Figure 20 allows identification of the stone elements associated with the largest protruding mass. Although this can give a general indication of risky locations, it is clearly largely correlated with the stone size. The normalized linear density, shown in Figure 21, allows partial reduction of the correlation with the stone size, while still highlighting the most protruding regions. Finally, Figure 22 shows that the overhanging mass index, normalized by its maximum value on the façade, may be useful to identify certain of the locations within the stones that may more subject to detachment.

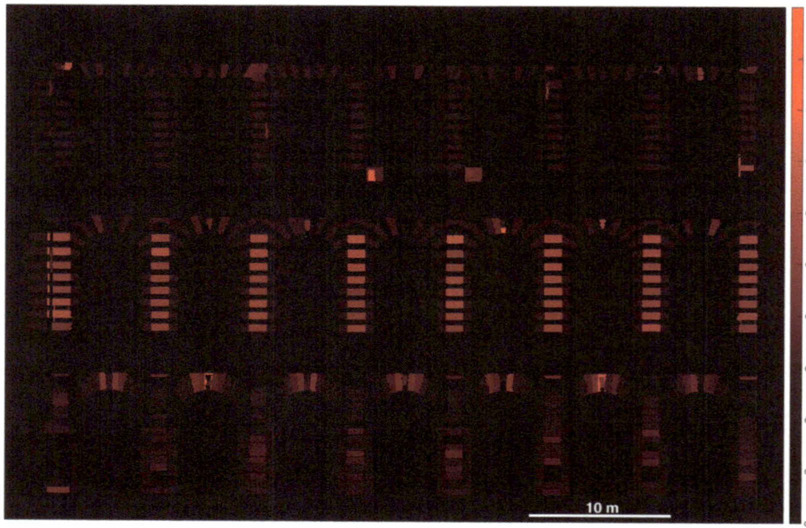

Figure 21. Normalized linear density of each segmented object, computed as described in Section 6.

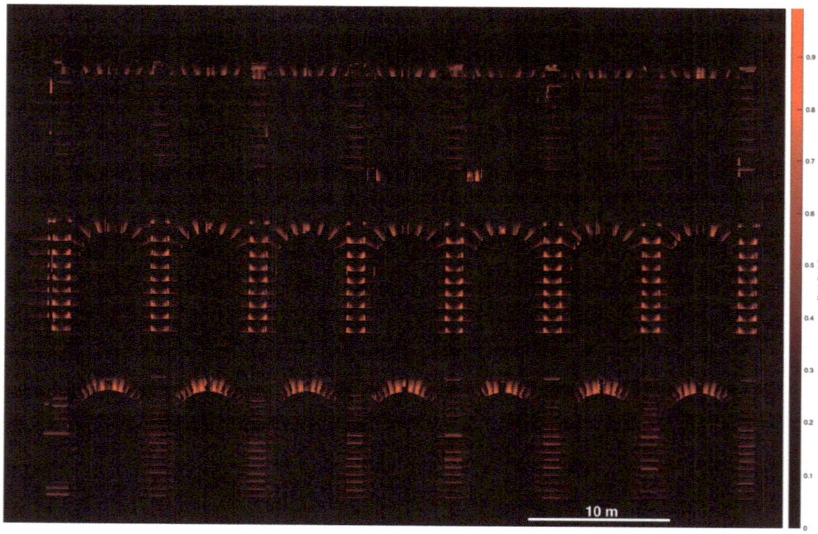

Figure 22. Normalized overhanging mass for each pixel of the segmented objects, computed as described in Section 6.

8. Discussion and Conclusions

8.1. Strengths and Weaknesses of the Proposed Research

The integration of knowledge into computer systems offers a promising solution to complex problems. Through considering geometric, structural, functional, and other aspects, this approach can significantly enhance the efficiency and effectiveness of conservation and preservation efforts. A pro-active approach towards monitoring has gained importance in recent times, and an objective approach based on measures and indicators has become the goal to pursue [78]. This study proposes the reuse of a 3D survey carried out using geomatic techniques and the development of methods for analysing the resulting 3D model to support the identification of possible problems relating to the conservation of the building and the safety of its visitors. It is appropriate to analyse the proposed approach's strengths and weaknesses by comparing it with the traditional mode of operation based on macroscopic analysis and close on-site inspection by experienced operators. This involves noting the number and type of calcite veins, open discontinuities, lamination, previous restoration interventions, and registering NDT measurements.

In particular, strengths must be considered from a long-term perspective, in both economic and technological terms. This encompasses, on the one hand, the economies of scale triggered via standardised processes repeated over time, transforming analysis from a one-off event to regular monitoring. On the other hand, technological developments, particularly those that may be supported by AI, may also be considered.

The potential of the work can be outlined as follows:

- cost effectiveness for built heritage owner–managers and limited time needed in the field, thus minimising the courtyard's inaccessibility to visitors and the costs of an aerial platform; the last monitoring campaign, carried out in February 2022, took two operators working for seven days on the platform, plus one operator to move the crane—during that time, the courtyard could not be visited by tourists;
- objectivity of the results produced, although this should not be confused with their reliability. Rather, reliability refers to the parameters analysed but not to the representativeness of the model adopted to describe the real levels of risk;
- once the analysis workflow has been defined, it can be completed in a very short computing time, thus contributing to the sustainability of repeating the process at shorter time intervals;
- the results produced as an outcome of each measurement campaign are directly comparable, if consistently realised in terms of resolution, georeferencing, etc., thus enabling effective monitoring;
- field campaigns carried out in the past resulted in paper-based reports with obvious problems of preservation and sharing, which are overcome by the digital transition in this approach, which allows effective multidisciplinary comparison, on-line cooperative work, remote consultation, and assessment by experts.

Current limits include the following:

- in this work, we have considered the effect of only the geometry of the examined objects, while cultural heritage conservation and restoration tasks concern the evaluation and modelling of a multitude of critical factors influencing structural or material deterioration. The more complex modelling approach proposed in [79] can be optimised by automating certain steps, as is proposed in this paper;
- automatic classification was applied to the rusticated construction elements that make up the columns, piers, and ashlars of the arches, and the infill above them. No criteria have yet been defined for the automatic evaluation of the overhang of the blocks of the stringcourse cornices;
- some stone elements are characterised by a complex three-dimensionality (as is the case for the capitals of the different orders, the decorations of the arch keystones, etc.), which has been neglected at this stage, having assumed the DEM as a significant shape model, thus limiting the analysis to 2.5 dimensions and depending on its resolution (4 cm/pixel).

8.2. Relation with Previous Works

Despite several works having already been proposed in the literature for identifying certain elements in building façades and for segmenting certain repetitive, quite regular, elements, such as bricks on masonry walls [57–63], the direct usage of such methods on quite complex façades of heritage buildings is often not so effective. This paper proposes a geometry-based segmentation method, quite similar to those in [59,64], combined with properly employed unsupervised classification that enabled distinguishing between different classes of ashlars. Despite some similarities with other segmentation methods, the obtained outcomes, i.e., the ashlar classification, have been used in a more general and original procedure, leading to the computation of three geometry-only-based metrics that can be employed in procedures to assess the detachment risk of portions of stones.

8.3. Future Research Outlook

The segmentation approach presented is based on the preliminary definition of geometric parameters characterising the shape and volume of stone elements. This research will continue investigating point cloud segmentation methods based on neural networks. Some of the critical elements for evaluating the strength of stone ashlars are also particularly minute and were not recorded in the point model, as was the case with calcite veins, for example; however, the high resolution of the images allowed them to be recognized in most cases, thus suggesting that segmentation systems based on oriented images should also be considered in the future, in order to reproject the results onto the 3D model at a later date. Indeed, further direction of the research development will have to concern the extension of critical factors from merely geometric to diagnostic ones, characterising the specific material considered and its state of preservation.

The implementation of automatic methods of analysis makes it possible to simplify monitoring activities, which are indispensable in this case study, given the high attendance by tourists. In particular, the repetition over time of survey campaigns using photogrammetry from UAVs will be able to provide information to be compared with previous states, allowing, for the first time, effective monitoring of the entire area under consideration.

8.4. Conclusions

Although considering a simplified approach with respect to the possible complexity of a multicriterial analysis, the proposed research has achieved significant results in terms of automating the segmentation and classification process, allowing the programming of monitoring not only of the envelope of the façade surface but also of its individual elements. For the time being, the factor assumed to prevail with respect to the danger of stone detachment is the geometric one: the greater the overhang (and thus the weight of the protruding portion) of the stone block with respect to the one below, the more damaging the effect of its fall could be.

In conclusion, the research presented here contributes to the long-term goal of supporting the management of a historic monumental building and allows the prioritisation of conservation and preservation responses following periodic monitoring campaigns conducted on the façades of Palazzo Pitti, identifying the stone elements presenting greater criticality.

Author Contributions: Conceptualization, L.F., A.C. and V.B.; methodology, L.F., A.C. and A.M.; software, A.M.; validation, L.F., A.C. and A.M.; formal analysis, L.F. and A.M.; investigation, L.F. and A.C.; resources, G.T. and V.B.; data curation, L.F.; writing—original draft preparation, L.F., A.C., A.M., E.P. and V.B.; writing—review and editing, A.C.; visualization, A.M. and V.B.; supervision, L.F. and A.M.; funding acquisition, G.T. All authors have read and agreed to the published version of the manuscript.

Funding: This project is supported by Spoke 7 Project PE 0000020 CHANGES-CUP B53C22004010006, NRP Mission 4 Component 2 Investment 1.3, funded by the European Union–Next Generation EU. L. Fiorini's research activity is conducted as part of the National PhD course in Heritage Science (La Sapienza University of Rome and University of Florence), A. Conti's research activity is conducted as part of the International Doctorate in Civil and Environmental Engineering (University of Florence).

Data Availability Statement: The raw data supporting the conclusions of this article will be made available by the authors on request.

Acknowledgments: The Geomatics for the Environment and Conservation of the Cultural Heritage Laboratory (GeCO Lab-UNIFI), directed by Grazia Tucci, was in charge of the 3D digitalisation project of Palazzo Pitti (2020–2022) in the frame of a research agreement with Le Gallerie degli Uffizi, directed at the time by Heike Schmidt. The authors gratefully thank Elena Pozzi, responsible for the Palace conservation, for the valuable insights and the fruitful dialogue. The authors gratefully thank architect Roberta Raffa for her attention and care in the realisation of 3D models and 2D drawings.

Conflicts of Interest: The authors declare no conflicts of interest. The funders had no role in the design of the study; in the collection, analyses, or interpretation of data; in the writing of the manuscript; or in the decision to publish the results.

References

1. Directorate-General for Research and Innovation (European Commission); Sonkoly, G.; Vahtikari, T. *Innovation in Cultural Heritage Research: For an Integrated European Research Policy*; Publications Office of the European Union: Luxembourg, 2018; ISBN 978-92-79-78019-6.
2. Vecco, M. A Definition of Cultural Heritage: From the Tangible to the Intangible. *J. Cult. Herit.* **2010**, *11*, 321–324. [CrossRef]
3. Rosén, F. The Dark Side of Cultural Heritage Protection. *Int. J. Cult. Prop.* **2020**, *27*, 495–510. [CrossRef]
4. Brimblecombe, P.; Grossi, C.M.; Harris, I. Climate Change Critical to Cultural Heritage. In *Survival and Sustainability: Environmental Concerns in the 21st Century*; Gökçekus, H., Türker, U., LaMoreaux, J.W., Eds.; Springer: Berlin/Heidelberg, Germany, 2011; pp. 195–205, ISBN 978-3-540-95991-5.
5. Sesana, E.; Gagnon, A.S.; Ciantelli, C.; Cassar, J.; Hughes, J.J. Climate Change Impacts on Cultural Heritage: A Literature Review. *WIREs Clim. Chang.* **2021**, *12*, e710. [CrossRef]
6. Di Giulio, R. Digital Transition Strategies for Architectural Heritage. In *ETHICS: Endorse Technologies for Heritage Innovation: Cross-Disciplinary Strategies*; Battisti, A., Baiani, S., Eds.; Springer: Cham, Switzerland, 2024; pp. 133–149, ISBN 978-3-031-50121-0.
7. Farella, E.M.; Rigon, S.; Remondino, F.; Stan, A.; Ioannidis, G.; Münster, S.; Medici, M.; Maietti, F.; Sánchez, A. Methods, Data and Tools for Facilitating a 3D Cultural Heritage Space. *Int. Arch. Photogramm. Remote Sens. Spat. Inf. Sci.* **2024**, *XLVIII-2-W4-2024*, 197–204. [CrossRef]
8. Directorate-General for Communications Networks, Content and Technology (European Commission). *Study on Quality in 3D Digitisation of Tangible Cultural Heritage: Mapping Parameters, Formats, Standards, Benchmarks, Methodologies, and Guidelines: Executive Summary*; Publications Office of the European Union: Luxembourg, 2022; ISBN 978-92-76-51316-2.
9. DHCE Expert Group Basic Principles and Tips for 3D Digitisation of Cultural Heritage | Shaping Europe's Digital Future. Available online: https://digital-strategy.ec.europa.eu/en/library/basic-principles-and-tips-3d-digitisation-cultural-heritage (accessed on 3 May 2024).
10. Fernie, K. 3D Content in Europeana: The Challenges of Providing Access. In *The 3 Dimensions of Digitalised Archaeology: State-of-the-Art, Data Management and Current Challenges in Archaeological 3D-Documentation*; Hostettler, M., Buhlke, A., Drummer, C., Emmenegger, L., Reich, J., Stäheli, C., Eds.; Springer International Publishing: Cham, Switzerland, 2024; pp. 167–177, ISBN 978-3-031-53032-6.
11. European Council Lisbon European Council 23-24.03.2000: Conclusions of the Presidency. Available online: https://www.europarl.europa.eu/summits/lis1_en.htm (accessed on 3 May 2024).
12. Hervás Soriano, F.; Mulatero, F. Knowledge Policy in the EU: From the Lisbon Strategy to Europe 2020. *J. Knowl. Econ.* **2010**, *1*, 289–302. [CrossRef]
13. About the Creative Europe Programme | Culture and Creativity. Available online: https://culture.ec.europa.eu/creative-europe/about-the-creative-europe-programme (accessed on 3 May 2024).
14. Pasikowska-Schnass, M. Creative Europe Programme 2021–2027. Available online: https://www.europarl.europa.eu/RegData/etudes/BRIE/2018/628229/EPRS_BRI(2018)628229_EN.pdf (accessed on 3 May 2024).
15. European Commission, Secretariat-General Communication from the Commission to the European Parliament, the European Council, the Council, the European Economic and Social Commitee and the Commitee of the Regions. The European Green Deal. Available online: https://eur-lex.europa.eu/legal-content/EN/ALL/?uri=COM:2019:640:FIN (accessed on 3 May 2024).
16. Directorate-General for Education, Youth, Sport and Culture (European Commission). *Strengthening Cultural Heritage Resilience for Climate Change: Where the European Green Deal Meets Cultural Heritage*; Publications Office of the European Union: Luxembourg, 2022; ISBN 978-92-76-48205-5.
17. Council Conclusions on the Work Plan for Culture 2019–2022. 2018; pp. 12–25. Available online: https://eur-lex.europa.eu/legal-content/EN/ALL/?uri=CELEX:52018XG1221(01) (accessed on 9 May 2024).
18. Commission Recommendation (EU) 2021/1970 of 10 November 2021 on a Common European Data Space for Cultural Heritage. 2021; pp. 5–16. Available online: http://data.europa.eu/eli/reco/2021/1970/oj (accessed on 9 May 2024).
19. Council Regulation (EU) 2020/2094 of 14 December 2020 Establishing a European Union Recovery Instrument to Support the Recovery in the Aftermath of the COVID-19 Crisis. 2020; pp. 23–27. Available online: http://data.europa.eu/eli/reg/2020/2094/oj (accessed on 9 May 2024).

20. Piano Nazionale Di Ripresa e Resilienza 2021. Available online: https://www.governo.it/sites/new.governo.it/files/PNRR_2021_0.pdf (accessed on 9 May 2024).
21. Piano Nazionale Di Ripresa e Resilienza 2022. Available online: https://www.governo.it/sites/governo.it/files/PNRR_0.pdf (accessed on 9 May 2024).
22. Istituto Centrale per la Digitalizzazione del Patrimonio Culturale—Digtal Library Piano Nazionale Di Digitalizzazione Del Patrimonio Culturale 2022–2023. Versione 1.1. Available online: https://digitallibrary.cultura.gov.it/wp-content/uploads/2023/10/PND_V1_1_2023-1.pdf (accessed on 9 May 2024).
23. United Nations Tranforming Our World: The 2032 Agenda for Sustainable Development A/RES/70/1. Available online: https://sdgs.un.org/sites/default/files/publications/21252030%20Agenda%20for%20Sustainable%20Development%20web.pdf (accessed on 9 May 2024).
24. Sotirova-Valkova, K. A Pilot Conceptualisation of the Data Space Ecosystems for Cultural Heritage. *Math. Educ. Math.* **2024**, *53*, 92–98. [CrossRef]
25. Rovelli, C. Meaning and Intentionality = Information + Evolution. In *Wandering towards a Goal: How Can Mindless Mathematical Laws Give Rise to Aims and Intention?* Aguirre, A., Foster, B., Merali, Z., Eds.; Springer International Publishing: Cham, Switzerland, 2018; pp. 17–27, ISBN 978-3-319-75726-1.
26. Istituto Centrale per la Digitalizzazione del Patrimonio Culturale—Digital Library Linee Guida per La Digitalizzazione Del Patrimonio Culturale Ve Rsione 1.0. Giugno 2022. Available online: https://docs.italia.it/italia/icdp/icdp-pnd-digitalizzazione-docs/it/v1.0-giugno-2022/index.html (accessed on 9 May 2024).
27. CHANGES—Cultural Heritage Active Innovation for Next-Gen Sustainable Society Extended Parnership. Available online: https://sites.google.com/uniroma1.it/changes (accessed on 9 May 2024).
28. Report from the Commission to the European Parliament and the Council on the Implementation of the EU Strategy on Adaptation to Climate Change. 2018; COM(2018) 738 Final. Available online: https://eur-lex.europa.eu/legal-content/EN/ALL/?uri=COM:2018:738:FIN (accessed on 9 May 2024).
29. Ministero dell'Ambiente e Della Sicurezza Energetica Strategia Nazionale Di Adattamento Ai Cambiamenti Climatici. Available online: https://www.mase.gov.it/sites/default/files/archivio/allegati/clima/documento_SNAC.pdf (accessed on 6 May 2024).
30. Ministero dell'Ambiente e Della Sicurezza Energetica Piano Nazionale Di Adattamento Ai Cambiamenti Climatici. Available online: https://www.mase.gov.it/notizie/clima-approvato-il-piano-nazionale-di-adattamento-ai-cambiamenti-climatici (accessed on 3 May 2024).
31. Bonazza, A.; Messina, P.; Sabbioni, C.; Grossi, C.M.; Brimblecombe, P. Mapping the Impact of Climate Change on Surface Recession of Carbonate Buildings in Europe. *Sci. Total Environ.* **2009**, *407*, 2039–2050. [CrossRef] [PubMed]
32. MiBACT Piano Straordinario Nazionale Di Monitoraggio e Conservazione Dei Beni Culturali Immobili. 18/11/2020 | 0000866-P. Available online: https://uss-sisma2016.cultura.gov.it/wp-content/uploads/2022/03/Piano-Straordinario-Nazionale-di-Monitoraggio-e-Conservazione-dei-Beni-Culturali-Immobili_18.11.2020.pdf (accessed on 9 May 2024).
33. Carta Del Rischio—MiC ICR. Available online: http://www.cartadelrischio.beniculturali.it/ (accessed on 9 May 2024).
34. Wetzel, S.; Mäs, S.; Bernard, L.; Vorobevskii, I.; Kronenberg, R. Spatial Data Infrastructure Components to Provide Regional Climate Information Services. *Clim. Serv.* **2024**, *34*, 100473. [CrossRef]
35. Programma Nazionale per La Ricerca | Ministero Dell'Università e Della Ricerca. Available online: https://www.mur.gov.it/it/aree-tematiche/ricerca/programmazione/programma-nazionale-la-ricerca (accessed on 9 May 2024).
36. Randazzo, L.; Collina, M.; Ricca, M.; Barbieri, L.; Bruno, F.; Arcudi, A.; La Russa, M.F. Damage Indices and Photogrammetry for Decay Assessment of Stone-Built Cultural Heritage: The Case Study of the San Domenico Church Main Entrance Portal (South Calabria, Italy). *Sustainability* **2020**, *12*, 5198. [CrossRef]
37. *Innovative Built Heritage Models: Edited Contributions to the International Conference on Innovative Built Heritage Models and Preventive Systems (CHANGES 2017), Leuven, Belgium, 6–8 February 2017*; van Balen, K.; Vandesande, A., Eds.; CRC Press: London, UK, 2018; ISBN 978-1-351-01479-3.
38. Vasari, G. *The Lives of the Artists*; Conaway Bondanella, J., Ed.; OUP: Oxford, UK, 1991; ISBN 978-0-19-160548-2.
39. Ferretti, E.; Baldini, L.; Le Due, R. *Ammannati e Vasari per la città dei Medici*; Polistampa: Firenze, Italy, 2011; ISBN 978-88-564-0174-5.
40. Ferretti, E. Palazzo Pitti 1550–1560: Precisazioni e Nuove Acquisizioni Sui Lavori Di Eleonora Di Toledo. *Opus Incert.* **2006**, *I*, 45–56.
41. Belluzzi, A. Gli Interventi Di Bartolomeo Ammannati a Palazzo Pitti. *Opus Incert.* **2006**, *I*, 56–73.
42. Belli, G. *Paramenti Bugnati e Architettura nella Firenze del Quattrocento*; Firenze University Press: Firenze, Italy, 2019; ISBN 978-88-6453-905-8.
43. Bonora, V.; Meucci, A.; Conti, A.; Fiorini, L.; Tucci, G. Knowledge Representation of Built Heritage Mapping an Ad Hoc Data Model in OGC Standards: The Case Study of Pitti Palace in Florence, Italy. *Int. Arch. Photogramm. Remote Sens. Spat. Inf. Sci.* **2023**, *XLVIII-M-2–2023*, 281–288. [CrossRef]
44. Bonora, V.; Maseroli, R.; Mugnai, F.; Tucci, G. GNNS Control Network Supporting Large Historical Building Architectural Survey. *Int. Arch. Photogramm. Remote Sens. Spat. Inf. Sci.* **2021**, *XLVI-M-1–2021*, 87–91. [CrossRef]

45. Murtiyoso, A.; Grussenmeyer, P.; Koehl, M.; Freville, T. Acquisition and Processing Experiences of Close Range UAV Images for the 3D Modeling of Heritage Buildings. In *Digital Heritage, Progress in Cultural Heritage: Documentation, Preservation, and Protection, Proceedings of the 6th International Conference, EuroMed 2016, Nicosia, Cyprus, 31 October–5 November 2016*; Ioannides, M., Fink, E., Moropoulou, A., Hagedorn-Saupe, M., Fresa, A., Liestøl, G., Rajcic, V., Grussenmeyer, P., Eds.; Springer International Publishing: Cham, Switzerland, 2016; pp. 420–431.
46. Palanirajan, H.K.; Alsadik, B.; Nex, F.; Oude Elberink, S. Efficient Flight Planning for Building Façade 3D Reconstruction. *Int. Arch. Photogramm. Remote Sens. Spat. Inf. Sci.* **2019**, *XLII-2-W13*, 495–502. [CrossRef]
47. Martínez-Carricondo, P.; Carvajal-Ramírez, F.; Yero-Paneque, L.; Agüera-Vega, F. Combination of Nadiral and Oblique UAV Photogrammetry and HBIM for the Virtual Reconstruction of Cultural Heritage. Case Study of Cortijo Del Fraile in Níjar, Almería (Spain). *Build. Res. Inf.* **2020**, *48*, 140–159. [CrossRef]
48. Matrone, F.; Grilli, E.; Martini, M.; Paolanti, M.; Pierdicca, R.; Remondino, F. Comparing Machine and Deep Learning Methods for Large 3D Heritage Semantic Segmentation. *ISPRS Int. J. Geo-Inf.* **2020**, *9*, 535. [CrossRef]
49. Pellis, E.; Murtiyoso, A.; Masiero, A.; Tucci, G.; Betti, M.; Grussenmeyer, P. An Image-Based Deep Learning Workflow for 3D Heritage Point Cloud Semantic Segmentation. *Int. Arch. Photogramm. Remote Sens. Spat. Inf. Sci.* **2022**, *XLVI-2-W1-2022*, 429–434. [CrossRef]
50. Wysocki, O.; Grilli, E.; Hoegner, L.; Stilla, U. Combining Visibility Analisys and Deep Learning for Refinement of Semantic 3D Building Models by Conflict Classification. *ISPRS Ann. Photogramm. Remote Sens. Spat. Inf. Sci.* **2022**, *X-4-W2-2022*, 289–296. [CrossRef]
51. Croce, V.; Caroti, G.; Piemonte, A.; De Luca, L.; Véron, P. H-BIM and Artificial Intelligence: Classification of Architectural Heritage for Semi-Automatic Scan-to-BIM Reconstruction. *Sensors* **2023**, *23*, 2497. [CrossRef] [PubMed]
52. Weinmann, M.; Jutzi, B.; Hinz, S.; Mallet, C. Semantic Point Cloud Interpretation Based on Optimal Neighborhoods, Relevant Features and Efficient Classifiers. *ISPRS J. Photogramm. Remote Sens.* **2015**, *105*, 286–304. [CrossRef]
53. Weinmann, M.; Jutzi, B.; Mallet, C.; Weinmann, M. Geometric Features and Their Relevance for 3D Point Cloud Classification. *ISPRS Ann. Photogramm. Remote Sens. Spat. Inf. Sci.* **2017**, *IV-1-W1*, 157–164. [CrossRef]
54. Qi, C.R.; Yi, L.; Su, H.; Guibas, L.J. PointNet++: Deep Hierarchical Feature Learning on Point Sets in a Metric Space. In Proceedings of the 31st International Conference on Neural Information Processing Systems, Long Beach, CA, USA, 4–9 December 2017; Curran Associates Inc.: Red Hook, NY, USA, 2017; pp. 5105–5114.
55. Shen, Y.; Lindenbergh, R.; Wang, J.G.; Ferreira, V. Extracting Individual Bricks from a Laser Scan Point Cloud of an Unorganized Pile of Bricks. *Remote Sens.* **2018**, *10*, 1709. [CrossRef]
56. Pellis, E.; Murtiyoso, A.; Masiero, A.; Tucci, G.; Betti, M.; Grussenmeyer, P. 2D to 3D Label Propagation for the Semantic Segmentation of Heritage Buildings Point Clouds. *Int. Arch. Photogramm. Remote Sens. Spat. Inf. Sci.* **2022**, *XLIII-B2-2022*, 861–867. [CrossRef]
57. Loverdos, D.; Sarhosis, V. Automatic Image-Based Brick Segmentation and Crack Detection of Masonry Walls Using Machine Learning. *Autom. Constr.* **2022**, *140*, 104389. [CrossRef]
58. Agrafiotis, P.; Talaveros, G.; Georgopoulos, A. Orthoimage-to-2D Architectural Drawing with Conditional Adversarial Networks. *ISPRS Ann. Photogramm. Remote Sens. Spat. Inf. Sci.* **2023**, *X-M-1-2023*, 11–18. [CrossRef]
59. Valero, E.; Bosché, F.; Forster, A. Automatic Segmentation of 3D Point Clouds of Rubble Masonry Walls, and Its Application to Building Surveying, Repair and Maintenance. *Autom. Constr.* **2018**, *96*, 29–39. [CrossRef]
60. Makka, A.; Pateraki, M.; Betsas, T.; Georgopoulos, A. 3D Edge Detection Based on Normal Vectors. *Int. Arch. Photogramm. Remote Sens. Spat. Inf. Sci.* **2024**, *XLVIII-2-W4-2024*, 295–300. [CrossRef]
61. Cremers, D.; Rousson, M.; Deriche, R. A Review of Statistical Approaches to Level Set Segmentation: Integrating Color, Texture, Motion and Shape. *Int. J. Comput. Vis.* **2007**, *72*, 195–215. [CrossRef]
62. Ibrahim, Y.; Nagy, B.; Benedek, C. CNN-Based Watershed Marker Extraction for Brick Segmentation in Masonry Walls. In *Proceedings of the Image Analysis and Recognition*; Karray, F., Campilho, A., Yu, A., Eds.; Springer International Publishing: Cham, Switzerland, 2019; pp. 332–344.
63. Kajatin, R.; Nalpantidis, L. Image Segmentation of Bricks in Masonry Wall Using a Fusion of Machine Learning Algorithms. In *Pattern Recognition. ICPR International Workshops and Challenges, Proceedings of the ICPR 2021, Virtual Event, 10–15 January 2021*; Del Bimbo, A., Cucchiara, R., Sclaroff, S., Farinella, G.M., Mei, T., Bertini, M., Escalante, H.J., Vezzani, R., Eds.; Springer International Publishing: Cham, Switzerland, 2021; pp. 446–461.
64. Masiero, A.; Costantino, D. TLS for Detecting Small Damages on a Building Façade. *Int. Arch. Photogramm. Remote Sens. Spat. Inf. Sci.* **2019**, *XLII-2-W11*, 831–836. [CrossRef]
65. Harris, C.; Stephens, M. A Combined Corner and Edge Detector. In Proceedings of the Alvey Vision Conference 1988, Manchester, UK, 31 August–2 September 1988; Alvey Vision Club: Manchester, UK, 1988; pp. 23.1–23.6.
66. Clark, J.; Holton, D.A. *A First Look at Graph Theory*; World Scientific: Singapore, 1991; ISBN 978-981-02-0489-1.
67. D'Ayala, D.F.; Copping, A.G.; Wang, H. A Conceptual Model for Multihazard Assessment of the Vulnerability of Historic Buildings. In *Structural Analysis of Historical Constructions: Possibilities of Numerical and Experimental Techniques, Proceedings of the Fifth International Conference, New Delhi, India, 6–8 November 2006*; Lourenco, P.B., Roca, P., Modena, C., Agrawal, S., Eds.; Macmillian: New Delhi, India, 2006; Volume 1, pp. 121–140, ISBN 978-1-4039-3155-9.

68. Ellingwood, B.R. Risk-Informed Condition Assessment of Civil Infrastructure: State of Practice and Research Issues. *Struct. Infrastruct. Eng.* **2005**, *1*, 7–18. [CrossRef]
69. Kumar, N.; Poonia, V.; Gupta, B.B.; Goyal, M.K. A Novel Framework for Risk Assessment and Resilience of Critical Infrastructure towards Climate Change. *Technol. Forecast. Soc. Chang.* **2021**, *165*, 120532. [CrossRef]
70. Salvatici, T.; Calandra, S.; Centauro, I.; Pecchioni, E.; Intrieri, E.; Garzonio, C.A. Monitoring and Evaluation of Sandstone Decay Adopting Non-Destructive Techniques: On-Site Application on Building Stones. *Heritage* **2020**, *3*, 1287–1301. [CrossRef]
71. Bonora, V.; Centauro, I.; Fiorini, L.; Conti, A.; Salvatici, T.; Calandra, S.; Raffa, R.; Intrieri, E.; Garzonio, C.A.; Tucci, G. Virtual Inspection Based on 3D Survey Supporting Risks Detachment Analisys in Pietraforte Stone Built Heritage. *Int. Arch. Photogramm. Remote Sens. Spat. Inf. Sci.* **2023**, *XLVIII-M-2–2023*, 273–280. [CrossRef]
72. Malinverni, E.S.; Mariano, F.; Di Stefano, F.; Petetta, L.; Onori, F. Modelling in HBIM to Document Materials Decay by a Thematic Mapping to Manage the Cultural Heritage: The Case of "Chiesa Della Pietà" in Fermo. *Int. Arch. Photogramm. Remote Sens. Spat. Inf. Sci.* **2019**, *XLII-2-W11*, 777–784. [CrossRef]
73. Korro Bañuelos, J.; Rodríguez Miranda, Á.; Valle-Melón, J.M.; Zornoza-Indart, A.; Castellano-Román, M.; Angulo-Fornos, R.; Pinto-Puerto, F.; Acosta Ibáñez, P.; Ferreira-Lopes, P. The Role of Information Management for the Sustainable Conservation of Cultural Heritage. *Sustainability* **2021**, *13*, 4325. [CrossRef]
74. Costamagna, E.; Spanò, A. Semantic Models for Architectural Heritage Documentation. In *Proceedings of the Progress in Cultural Heritage Preservation*; Ioannides, M., Fritsch, D., Leissner, J., Davies, R., Remondino, F., Caffo, R., Eds.; Springer: Berlin/Heidelberg, Germany, 2012; pp. 241–250.
75. Brumana, R.; Oreni, D.; Barazzetti, L.; Cuca, B.; Previtali, M.; Banfi, F. Survey and Scan to BIM Model for the Knowledge of Built Heritage and the Management of Conservation Activities. In *Digital Transformation of the Design, Construction and Management Processes of the Built Environment*; Daniotti, B., Gianinetto, M., Della Torre, S., Eds.; Springer International Publishing: Cham, Switzerland, 2020; pp. 391–400, ISBN 978-3-030-33570-0.
76. Lloyd, S. Least Squares Quantization in PCM. *IEEE Trans. Inf. Theory* **1982**, *28*, 129–137. [CrossRef]
77. Coli, M.; Donigaglia, T.; Cristofaro, M.T.; Tanganelli, M.; Viti, S. Assessments on the Material Properties of the Pietraforte Stone of Florence (Italy) in Conservation, Restoration and Construction. *Case Stud. Constr. Mater.* **2022**, *16*, e00986. [CrossRef]
78. Stovel, H. Origins and Influence of the Nara Document on Authenticity. *APT Bull.* **2008**, *39*, 9–17.
79. Centauro, I.; Vitale, J.G.; Calandra, S.; Salvatici, T.; Natali, C.; Coppola, M.; Intrieri, E.; Garzonio, C.A. A Multidisciplinary Methodology for Technological Knowledge, Characterization and Diagnostics: Sandstone Facades in Florentine Architectural Heritage. *Appl. Sci.* **2022**, *12*, 4266. [CrossRef]

drones

MDPI

Article

Three-Dimensional Documentation and Reconversion of Architectural Heritage by UAV and HBIM: A Study of Santo Stefano Church in Italy

Guiye Lin [1], Guokai Li [1,*], Andrea Giordano [1], Kun Sang [2], Luigi Stendardo [3] and Xiaochun Yang [4]

[1] Department of Civil, Environmental and Architectural Engineering, University of Padua, 35131 Padova, Italy; guiye.lin@phd.unipd.it (G.L.); andrea.giordano@unipd.it (A.G.)
[2] School of Humanities and Communication, Xiamen University Malaysia, Sepang 43900, Malaysia; kun.sang@xmu.edu.my
[3] Department of Civil, Architectural and Environmental Engineering, University of Naples Federico II, 80138 Naples, Italy; luigi.stendardo@unina.it
[4] School of Architecture & Urban Planning, State Key Laboratory of Subtropical Building and Urban Science, University of Shenzhen, Shenzhen 518000, China; yangxc@szu.edu.cn
* Correspondence: guokai.li@studenti.unipd.it

Abstract: Historic buildings hold significant cultural value and their repair and protection require diverse approaches. With the advent of 3D digitalization, drones have gained significance in heritage studies. This research focuses on applying digital methods for restoring architectural heritage. It utilizes non-contact measurement technology, specifically unmanned aerial vehicles (UAVs), for data collection, creating 3D point cloud models using heritage building information modeling (HBIM), and employing virtual reality (VR) for architectural heritage restoration. Employing the "close + surround" oblique photography technique combined with image matching, computer vision, and other technologies, a detailed and comprehensive 3D model of the real scene can be constructed. It provides crucial data support for subsequent protection research and transformation efforts. Using the case of the Santo Stefano Church in Volterra, Italy, an idealized reconstructed 3D model database was established after data collection to preserve essential resources such as the original spatial data and relationships of architectural sites. Through the analysis of relevant historical data and the implementation of VR, the idealized and original appearance of the case was authentically restored. As a result, in the virtual simulation space, the building's style was realistically displayed with an immersive experience. This approach not only safeguards cultural heritage but also enhances the city's image and promotes tourism resources, catering to the diverse needs of tourists.

Keywords: documentation; cultural heritage; HBIM; drone photogrammetry; UAV; VR

Citation: Lin, G.; Li, G.; Giordano, A.; Sang, K.; Stendardo, L.; Yang, X. Three-Dimensional Documentation and Reconversion of Architectural Heritage by UAV and HBIM: A Study of Santo Stefano Church in Italy. *Drones* **2024**, *8*, 250. https://doi.org/10.3390/drones8060250

Academic Editors: Fabrizio Banfi, Daniela Oreni and Davide Mezzino

Received: 15 April 2024
Revised: 31 May 2024
Accepted: 3 June 2024
Published: 6 June 2024

1. Introduction

Architectural heritage encompasses three key categories: urban heritage, landscape heritage, and architectural heritage. Each of them are imbued with cultural, material, and social significance [1,2]. Not only does architectural heritage embody a local area's unique cultural essence, but it also serves as a vital conduit for preserving beliefs, documenting local characteristics, and chronicling societal norms and customs, thereby playing a pivotal role in the ongoing evolution of human society [3,4]. In recent years, architectural heritage, being a finite resource, has encountered numerous challenges. These include degradation stemming from social urbanization, deterioration and collapse induced by natural disasters, and inappropriate alteration of materials and color schemes due to anthropogenic development and conservation efforts [5]. Since the release of the UN's Sustainable Development Goals (SDGs) in 2015, the world has been explicitly called upon to make efforts to protect cultural and natural heritage [6]. Furthermore, the UN's New Urban Agenda 2030 clearly emphasized the importance of tangible and intangible cultural and landscape heritage; it also proposed that research on the application of digital technology to heritage conservation be encouraged [7]. It has been demonstrated that amid the era of rapid technological advancement, the integration of digital tools such as 3D scanning, modeling, VR, 3D point clouds, and other digital resources holds increasingly significant importance in the preservation and documentation of cultural heritage [8,9]. Furthermore, these technologies enable the recreation of historical sites that were previously unable to be fully represented [6]. Presently, this stands as the most effective method for showcasing lost artifacts without impacting the physical integrity of architectural sites, particularly through virtual mediums like digital displays [10]. An architectural site is an important type of artifact, and its value is expressed through the site itself. Reconstruction of destroyed buildings on the site is not allowed, and virtual methods can fulfill this requirement [11]. To accurately build the virtual restoration model, it is important to obtain accurate site data, which will affect the accuracy of the restoration model data [12,13].

HBIM, as a new paradigm for digital design and management [14], was proposed by Murphy in 2009 for the first time. The purpose of his research was to outline, in detail, the procedure of remote data capture using laser scanning and the subsequent processing required in order to identify a new methodology for creating full engineering drawings (orthographic and 3D models) from laser scan and image survey data for historic structures [15]. Coşgun et al. [16] defined HBIM as a parametric object prototype repository based on historical architectural data. With the deepening of the research and updates in the technology, Barrile et al. [17] stated that HBIM is a 3D computer model based on digital technology with the means to embed heritage asset data throughout the preservation lifecycle.

The traditional surveying method for architectural heritage typically involves data recording in various formats, such as manual measurements, textual descriptions, hand-drawn sketches, photographs, videos, and audio recordings. However, these methods lack efficiency, are labor-intensive, and yield more varied data records, which hinder subsequent data application and analysis [18]. During the process of establishing architectural heritage archives, surveying and mapping are the core components. Surveying and mapping are the basic premise of establishing archives [19]. Through the observation, measurement, and systematic documentation of architectural heritage, the acquisition of specific data and related information emerges as the cornerstone of heritage protection [20], establishing preconditions for subsequent protective measures. Without the data support obtained through surveying and mapping, research, evaluation, planning, design, and implementation of protection initiatives may be compromised. Thus, it is imperative to prioritize surveying and mapping to protect architectural heritage.

With the rapid development of UAV technology, it has been widely used in military affairs, agriculture, architectural heritage measurements, natural resource surveys, urban and rural planning, and other fields [21]. In the field of architectural heritage protection, when investigating and researching architectural heritage, it is often difficult to show the whole picture using manual surveying and mapping alone because of the complex surrounding environment of the building, and some detailed structures need to be photographed and measured by setting up a frame [22,23]. The surveyed and mapped buildings may be damaged due to the surveyors' behavior. On the other hand, since the ancient buildings have been exposed to the natural environment for a long time and most of the main structures have problems, there will be certain safety hazards for the surveying and mapping personnel [24]. UAVs can perform close-range photogrammetry of ancient buildings, obtain detailed information in all directions, and make up for the limitations of ground photography modeling in terms of perspective, creating refined 3D models. They can accurately record the material information and damage of a building and are suitable for the recording of buildings or components of various scales. They can also be used for tilt analysis, like the results of 3D laser scanning.

Although the model generated by drone modeling is less accurate compared to 3D laser scanning, drones are low-cost, simple to operate, and can record the shape, color, and other information of the measured object in all directions. The application of drones can help conservators understand and present the characteristics and values of historical buildings more comprehensively and accurately while also providing the public with a more intuitive, vivid, and rich heritage visiting experience.

2. Literature Review

HBIM, as a specific subcategory of building information modeling (BIM), has received a lot of attention in the field of archaeology and architectural heritage since its implementation [25], importantly because of its significant breakthrough in modeling accuracy. The HBIM models developed by Brumana et al. [26] demonstrate a diverse spectrum of precision levels and geometric intricacies and are precisely tailored to the specific functional demands of each target. Moreover, Antón et al. [27] have further enhanced HBIM precision by leveraging remote sensing technologies. Additionally, other techniques such as laser scanning [28,29], computer science, GIS, 3D modeling, and 3D printing have been employed to meet the intricate modeling requirements of various architectural component shapes and complex surfaces [30–33]. In addition, HBIM, as a quintessential parametric design tool, facilitates element modifications through parameter adjustments, thereby enabling seamless modification and information interaction [34]. The HBIM modeling workflow typically integrates descriptive and reality-based data to document architectural heritage, as illustrated in Figure 1. This framework divides HBIM implementation into seven steps. The first step is initial data collection and processing, followed by point cloud generation, building block creation, and automation, culminating in the formation of the final HBIM structure.

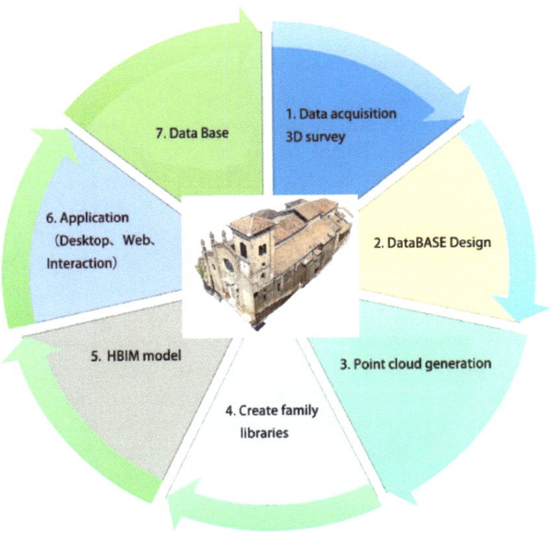

Figure 1. Framework for HBIM.

2.1. Methodology of Modeling

Three-dimensional computer graphics was the first technology used to model heritage properties, often dedicated to virtual reconstruction [35]; combining it with BIM techniques gave birth to HBIM. Wang et al. proposed a new approach to HBIM that allows for virtual reconstruction based on the collection and processing of laser/image measurement data [36]. Virtual reconstruction usually refers to a virtual replica of a heritage object (disappeared, damaged, or present), where the data sources may come from digitized historical materials documenting the heritage, reality-based photographs, and measurable drawings representing surfaces. Thus, virtual reconstruction may resurrect the past, virtualize the present, and predict the future. Yang et al. [37] have discussed how to solve the problem of 3D modeling of historic buildings and apply it to HBIM documentation and analysis.

As a 3D information modeling environment, BIM software increasingly supports 3D point clouds representing solid surfaces, which is called the scan-to-BIM process [38]. In the last decade, reality capture technologies that generate 3D point cloud data have become increasingly sophisticated and available, enabling more accurate and cost-effective point cloud data acquisition. This means that the points in three-dimensional space defined by X, Y, and Z can accurately characterize the exterior surfaces of architectural objects [36]. Three-dimensional point cloud data can, in turn, be obtained by means of laser scanning, 3D computer graphics, and videos, among others. This semi-automated modeling method has been studied several times to obtain a certain accuracy [39,40]. The accuracy of the model was enhanced by combining other methods such as using NURBS surfaces and curves to solve more complex building component shapes [41]. However, this semi-automated modeling still requires some manual effort to improve its efficiency, and the process is time-consuming, tedious, and laborious as well as somewhat subjective. Therefore, many scholars' studies have focused on developing a system/process that takes the point cloud of a facility as input and combines it with other techniques to automatically generate a fully annotated as-built BIM of the facility as output. For example, Pađen et al. [42] proposed a workflow for automatic reconstruction of simulated 3D city models by combining 2D geographic datasets and elevation data based on aerial point clouds. Moyano et al. [43] explored the insertion of point cloud data into a BIM environment, classifying the data and applying digital twin probabilities for automatic modeling of HBIM. The use of artificial intelligence and deep learning to improve modeling accuracy and automated modeling will be a research trend in the future [44,45].

2.2. HBIM and Drone Technology Modeling

The emerging field of drone technology as a digitization methodology offers new perspectives on digitizing and preserving architectural heritage [39]. A study by Zaker et al. [46] emphasized the importance of drones in the documentation of architectural heritage. Drones can carry various sensors, such as LiDAR and cameras, which are used to acquire high-resolution 3D data from buildings [47]. Drones offer significant advantages over traditional surveying methods. Firstly, they can measure large areas in a shorter period, thus increasing efficiency. Second, UAVs can perform non-invasive measurements, reducing damage to buildings [48]. This is particularly important for old and vulnerable built heritage. Liu et al. [49] highlighted the complementary nature of drone data and HBIM modeling. High-quality 3D data acquired by drones can be directly integrated into HBIM to create a more comprehensive record of built heritage [50]. This integration provides not only the geometric data of the building but can also include its exterior, structural, and material characteristics. This provides more accurate information for the maintenance and restoration of built heritage, helping to reduce unnecessary damage and costs.

UAVs can also meet different measurement angle requirements and greatly enrich modeling data. Özdemir and Remondino [51] used aerial RGB orthophotography to measure architectural objects and combined it with GIS to obtain detailed object information. Moreover, in this paper, we adopt the "close + surround" tilt photography technique, using a UAV for data collection, create a 3D point cloud model from UAV data, and construct a detailed and comprehensive 3D model of the actual scene using VR technology, image matching, and other techniques for architectural heritage restoration.

3. Materials and Methods

3.1. Study Area

Volterra city in Italy was established more than 3000 years ago and contains historic sites dating back to the fourth century BC. It has been tentatively recognized as part of the World Heritage List since 2006. Being one of the oldest cities in the world, it is also home to the world's oldest standing Etruscan arch, as seen in Figure 2. The authors chose the Santo Stefano Church, located on the edge of Volterra city, as the case study. It was built in its present form in 1784 on an ancient medieval church dating back to the 11th century. However, it is heavily damaged; only the lower part of the façade and part of the side walls remain from the original phase. The church has three access portals, of which the central one is the largest. The jambs are made up of red stone pillars and support an architrave. They are surmounted by capitals decorated with different decorative motifs, as is typical of Romanesque churches. The architraves are also decorated. Today, only the portal on the right, which has a floral motif, has been preserved, as shown in Figure 3. The façade's high degree of decorative accuracy, which is indicated by the use of numerous materials to create a chromatic effect of strong impact and by the architectural articulation, undoubtedly informs us of the wealth and prestige enjoyed by the church during the Middle Ages.

Figure 2. The map of Volterra city and the location of Santo Stefano Church.

Figure 3. Santo Stefano Church in Volterra, Italy.

Many architectural heritage sites have fallen into disuse due to many economic, technological, social, and cultural factors. A lack of documentation is an essential factor that makes restoring architectural heritage difficult. Architectural restoration often requires detailed documentation and records to understand the history of the building, its structure, its design intent, and its original materials. The present study case, which can be traced back to the 12th century, is of great significance for the study of architectural heritage in the Romanesque style. This case's point cloud data can provide ideas and reference points for less costly restoration work. At the same time, this low-cost way of restoring architectural heritage helps to preserve and pass on the architectural and cultural heritage of a community, country, or region; these buildings usually carry traces of history, and through restoration, the buildings act as witnesses to history. Restoration of architectural and cultural heritage contributes to the preservation of the common cultural heritage of humanity, the promotion of cultural diversity, and the sustainable development of society

3.2. Workflow

The structure and appearance of Santo Stefano Church have changed greatly after hundreds of years of wind and rain. Without drawings and historical documents, it is even more difficult to trace its history. In addition, architectural heritage itself has a certain function of cultural heritage and artistic edification, and both carving details and color details need to be accurately restored. Traditional surveying and mapping methods lack precision, require a lot of manpower and material resources, and are more likely to damage the building itself during the process. Therefore, this research focuses on using digital technology to protect and revitalize the Santo Stefano Church in the city of Volterra.

Drones are mainly used as aerial photography tools for obtaining the orthographic projection of building roofs and orthographic projection photos of buildings and surrounding environments. They are an auxiliary tool for building information collection in areas with low visibility and accessibility and for spatial analysis. The church was originally a single-family Romanesque building. After demolition and damage, the surrounding urban residences have undergone many expansions and repairs, and now they are closely connected with the church. Therefore, photos taken using the orthographic projection capabilities of the drone can not only identify the details of the main body of the building remains but also enable the relationship between the entire building and the surrounding environment to be studied.

In this study, we chose to use Mavic 3 Pro UAV equipment to conduct low-altitude photogrammetry for the research case. The drone has a 24 mm Hasselblad camera, a high-precision anti-shake gimbal, and a 20-megapixel image sensor. The photo formats that can be taken are JPEG and RAW. The maximum flight speed is 20 m/s, and the battery can last for about 30 min. In this photogrammetry project, 200 UAV remote sensing images were obtained, with a heading overlap rate of 80% and a side overlap rate of 70%. The UAV flight height was set to be two meters above the building. The three main steps are data acquisition, information management, and application. They will be implemented through a suite of work packages fulfilling the following workflow (Figure 4).

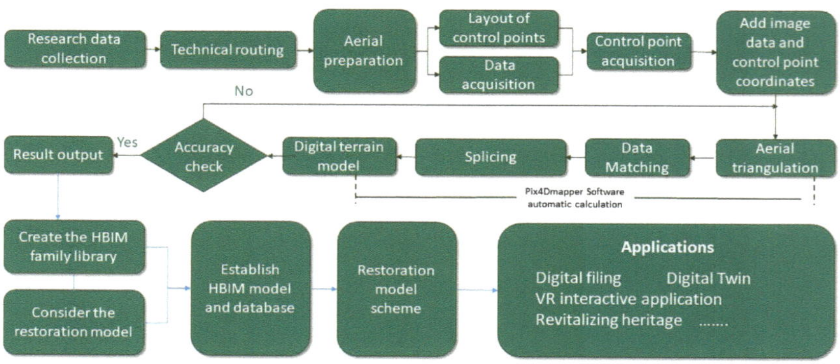

Figure 4. Workflow of the project.

In the first step, control point measurement is integral to the UAV low-altitude photogrammetry technology. A GPS is used to measure the control points in the research area.

To ensure the accuracy of the data, the control points are evenly distributed throughout the research area. Therefore, five control points are arranged at the four corners of the surveying and mapping object, targets need to be placed, and the color of the targets must be obvious to ensure that the center of the target can be clearly seen during the information collection process. To measure the target coordinates, it is better to use a prism-free electronic total station to ensure accuracy. RTK (real-time kinematic) measurement technology may produce large errors due to small-scale measurement, and the accuracy could be better, so it is not recommended for use. For investigating the weather environment in the aerial survey area, researchers should consider whether there are any obstacles that may affect the safe flight of drones in the surrounding area and whether there are any interference sources that may affect information transmission. They should also choose a cloudy day with better light to avoid overexposure of the lens and distortion of the picture caused by direct sunlight, significantly increasing the workload and difficulty of image processing, and select the operating point to provide a reference for UAV aerial survey planning. This is then followed by formulating the aerial survey plan, determining the UAV's take-off and landing point, and setting the route based on the UAV parameters. After setting the route, the test flight should be carried out first, and the aerial survey should be officially started after the relevant parameters are calibrated. During the aerial survey, the researcher should pay attention to the drone's power and flight attitude to ensure that it can return safely. A stepped route can solve the problem of different heights in the shooting area. When shooting, the height difference of each area changes as little as possible. For this project, the vertical distance between the UAV and the church façade is two meters, and it flies vertically along each façade from bottom to top. Figure 5 shows a working process by which the Santo Stefano Church data was scanned by UAV technology.

Figure 5. UAV collecting church data.

After capturing images, it is essential to review them promptly for missed shots. Subsequently, the images are uploaded onto a computer for processing (Photoshop, v.24.7.1). It utilizes batch processing to standardize parameters (saturation and brightness) while eliminating flawed images. Next, the data are transferred to the specialized software Pix4Dmapper 4.6.1 for comprehensive UAV data and aerial image processing. It efficiently processes hundreds of images, minimizing manual intervention. The software automatically identifies and reads the data after importing the JPG aerial photos and corresponding coordinates of the control point into Pix4Dmapper. After setting parameters for feature point extraction, product specifications, and contour line parameters, Pix4Dmapper generates orthophoto maps, point cloud data, digital surface models (DSM), and digital terrain models (DTM) with high precision and accuracy. Finally, the point cloud data obtained from the aerial photography data are processed as follows (Figure 6).

Figure 6. The point cloud of the Santo Stefano Church.

4. Results and Discussion

HBIM is a technical means of heritage building information management. The standard parametric component library is built through a cross-platform program that maps point cloud and photographic modeling data to parametric components based on existing historical building data. Compared with traditional two-dimensional information management methods such as text, pictures, and drawings, HBIM can express the original appearance of buildings more intuitively and accurately and create three-dimensional records of non-geometric data, improving data management and providing information on the current performance of architectural heritage.

4.1. Acquisition and Recovery of Individual Object Data

For this article's research case, the Santo Stefano Church, the authors analyzed the point cloud data model obtained by the drone. The model can not only analyze and evaluate the status of each part of the remaining building but can also assist in investigating the structural problems of the current building, including crooked wall flashing, tilting, foundation subsidence, etc. Using HBIM technology, we were able to establish a parametric component library, build a building information model, input survey data, realize the visual management of Santo Stefano Church's survey data, and use the output to draw the general plan, elevation, and structure of the building, a large sample diagram of nodes, and other archival information technology drawings. Figure 7 shows the creation of a family of HBIM models of the Santo Church from point cloud data. However, due to the lack of historical documents, this project's restoration model is an idealized model based on the partially preserved wreckage.

The family library created based on point cloud data contains standard historical architectural elements, such as windows, doors, columns, etc., so that experts and scholars can reuse these elements during the conservation and restoration process. Figure 8 shows a built-up combined member based on the columns and beams of Figure 7. From repairing individual components to systematic restoration of progressively assembled elements, specialized personnel undertake manual repair processes, ensuring precision and accuracy. This meticulous approach contributes significantly to enhancing the HBIM database but also aids in preserving the style and coherence of historic structures. The model's comprehensive family library is a repository for detailed information concerning historic building elements. This encompasses material specifications, historical context, maintenance records, and more.

Such extensive documentation enables architects, engineers, and preservation experts to gain deeper insights into the unique requirements of historic buildings throughout the design and maintenance phases. At the same time, the family library of HBIM models can be used to create virtual simulations of various aspects of historical buildings, including structure, lighting, ventilation, etc. This helps evaluate the effectiveness of retrofit design options while maximizing the preservation of the historic building's original appearance. In general, the model family library created based on point cloud data plays a crucial role in HBIM, helping to protect and maintain these precious cultural heritages by standardizing, sharing, and managing the information.

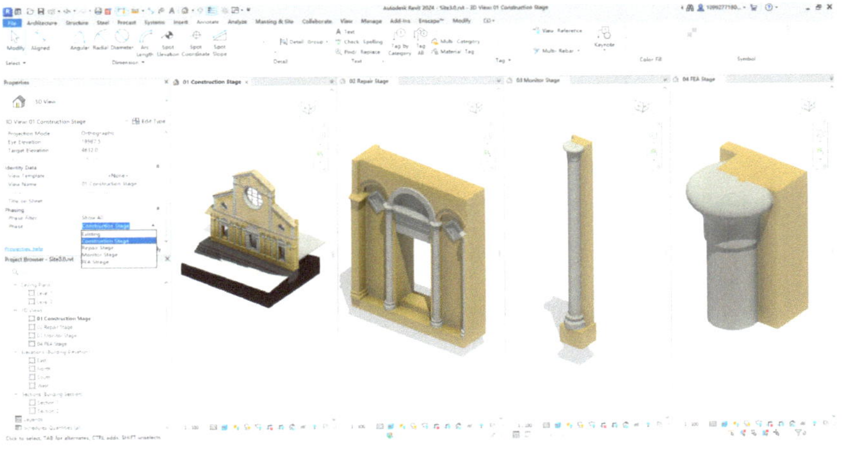

Figure 7. Repair process for individual components.

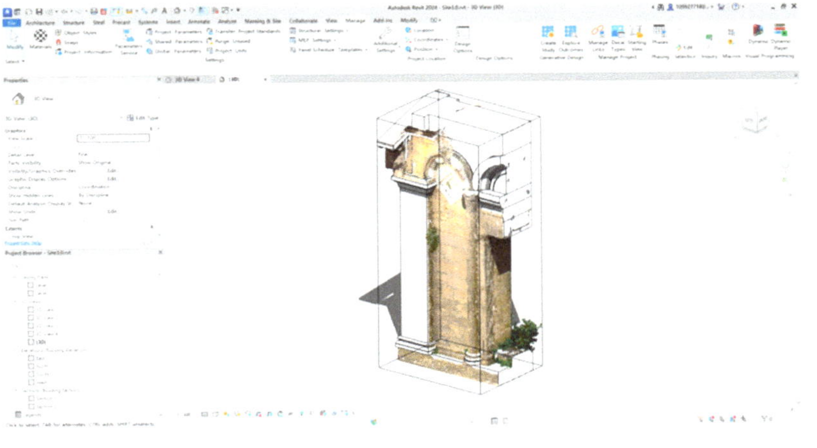

Figure 8. *Cont.*

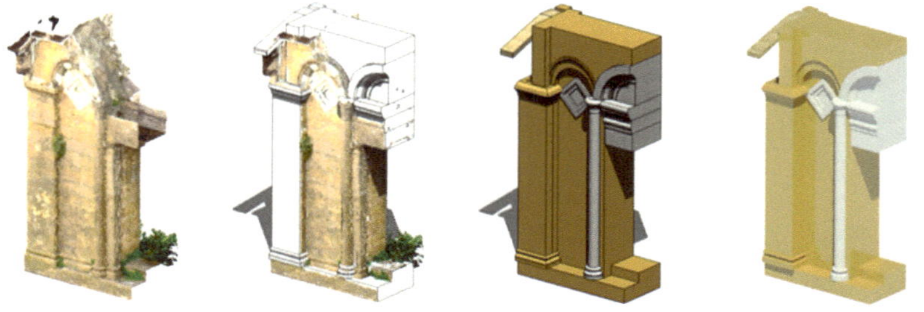

Figure 8. HBIM family library of the Santo Stefano Church.

4.2. Partial Restoration of the Church

Expert technicians meticulously engage in the manual repair processes, methodically addressing individual components before systematically restoring progressively assembled elements, guaranteeing utmost precision and accuracy. This painstaking methodology significantly enriches the HBIM database and plays an instrumental role in safeguarding the integrity and authenticity of historic structures. Within the HBIM framework, an exhaustive family library stands as a veritable treasure trove of detailed insights into historic building elements (Figure 9). This repository encompasses a wealth of information ranging from intricate material specifications and nuanced historical contexts to meticulous maintenance records and beyond. Such comprehensive documentation is invaluable, empowering architects, engineers, and preservation experts to delve deeply into the intricate nuances and unique demands of historic edifices at every stage—from initial design conceptualization to ongoing maintenance endeavors. Through this holistic understanding, practitioners can navigate the complexities of preserving and revitalizing these architectural treasures with heightened sensitivity and expertise.

Without historical materials and archeology, surveying and mapping data and current models should be fully utilized to carry out reasonable and hierarchical restoration design based on the building's traces and other architectural styles of the same period. The idea of restoration is to divide it into two parts, the well-preserved part and the poorly preserved part, according to the overall preservation status, and to restore the better-preserved part before the poorer-preserved part. Individual pieces of the brick decoration of the church of Santo Stefano, found in various locations and often in multiple pieces, are being recorded in 3D using photogrammetry. They are being virtually restored to their original location in the 3D model of the structure (Figure 10). Due to the lack of historical documents in this case, apart from the remains that can be restored to an identical HBIM model, such as the data on the upper part of the façade, we can only estimate the original appearance of the entire church based on the style of the remaining parts and by using churches of similar style from the same period as reference cases for restoration.

Figure 9. HBIM model of the Santo Stefano Church.

Figure 10. Restoration of church fragments.

4.3. Application of VR Tools

VR offers a promising solution by integrating 3D UAV data with HBIM to provide users with immersive experiences. Users can explore indoor and outdoor architectural heritage sites within virtual simulation spaces, including virtual reconstructions of incomplete areas, fostering a deeply immersive experience. HBIM realizes the integration of all-round information on historical buildings. Therefore, the combination of HBIM and VR can provide a visual reference for the restoration and protection of architectural heritage. Unity 2019.4 is a game engine software that effectively realizes this link using Revit 2023.1 software. Unity 3D supports multiple virtual reality devices: HTC Vive, Oculus Rift, and PlayStation VR, and provides rich interactive control options. It also supports technologies such as head tracking, spatial sound effects, and high-quality graphics rendering to improve the realism and immersion of the virtual environment while focusing on performance optimization to ensure smooth operation. The workflow of interactive virtual architectural heritage based on Unity mainly includes the following steps: realization of interactive functions, design of dynamic navigation system, application of intelligent interactive technology, use of an AI-driven virtual tour guide, and eye tracking. During the VR roaming process, relevant building information can be queried in real time, edited online, and the real-time effects displayed in a visual way, as shown in Figure 11a,b.

This approach enhances the preservation of cultural relics and diversifies the dissemination of the city's image and tourism resources, catering to a wide range of user interests. The virtual restoration of architectural heritage enhances the city's image and boosts tourism resources through online promotion. Additionally, the significant human–computer interaction capabilities facilitate research in tourism planning and drive the advancement of virtual tourism.

(a)

Figure 11. *Cont.*

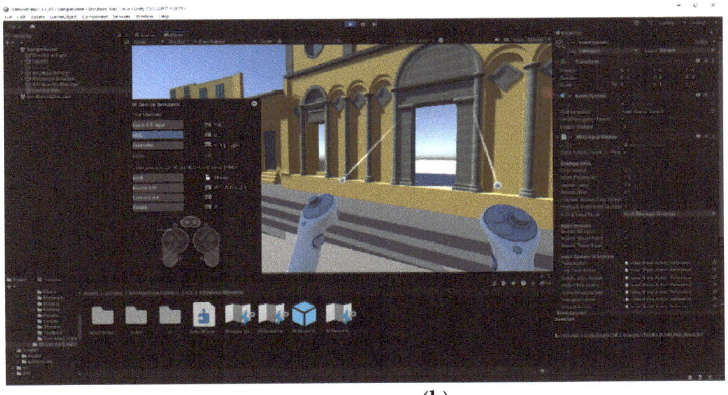

(b)

Figure 11. (a) Point cloud data and restored HBIM model. (b) The restored model on the VR device platform.

Furthermore, the integration and standardization of diverse cultural relic resources through computer networks enable virtual technology to depict cultural relics comprehensively and realistically, transcending geographical constraints and facilitating resource sharing. This strategy enables cultural relics to reach broader audiences and fosters a deeper appreciation for heritage preservation efforts.

5. Conclusions

Architectural heritage is a crucial part of cultural heritage, and its preservation and recording are essential for transmitting historical, cultural, and artistic values. This paper focuses on the use of drone close-range photogrammetry technology to collect basic information about some architectural ruins that cannot be repaired and restored. In the case of Santo Stefano Church, drones and HBIM are used to improve the efficiency of recording and conservation of architectural heritage, thereby reducing labor and time costs. Through data collection, data processing, HBIM model creation, virtual restoration of the original plan, and VR application, the church's original façade effect was restored. This technique not only provides digital protection data for architectural heritage but also meets various needs, such as tourism resource promotion, heritage cultural education, etc. The results show that drones are powerful tools for acquiring high-resolution three-dimensional data of buildings in a way that is more efficient than traditional surveying methods and avoids damaging contacts. As computer science continues to develop, related digital technologies will become more available for heritage studies. In the foreseeable future, by combining technologies such as 3D printing, big data, AR, artificial intelligence, robotics, and computer vision, the digital protection means and methods of architectural cultural heritage will become more systematic, which will surely promote the informatization of architectural cultural heritage and provide more innovative methods for architectural heritage protection.

Author Contributions: Conceptualization, G.L. (Guiye Lin) and X.Y.; methodology, G.L. (Guiye Lin), G.L. (Guokai Li), and A.G.; software, G.L. (Guiye Lin) and L.S.; validation, L.S.; investigation, G.L. (Guiye Lin) and X.Y.; resources, A.G.; writing—original draft preparation, G.L. (Guiye Lin), G.L. (Guokai Li) and K.S.; writing—review and editing, G.L. (Guiye Lin), G.L. (Guokai Li) and K.S.; visualization, G.L. (Guiye Lin); supervision, G.L. (Guokai Li) and X.Y. All authors have read and agreed to the published version of the manuscript.

Funding: This research was supported by University of Padua, Italy; Shenzhen University 2035 Program for Excellent Research (NO: 2022B005); Guangzhou-Hongkong-Macau Joint Laboratory for Smart Cities Open Research Fund and State Key Laboratory of Subtropical Building and Urban

Science, Shenzhen, China. We would like to express appreciation to colleagues for their constructive suggestions and comments. Additionally, we thank the anonymous reviewers and members of the editorial team for their constructive comments and contributions.

Data Availability Statement: Data supporting the findings of this study are available from the author G. Lin, upon reasonable request.

Conflicts of Interest: The authors declare no conflicts of interest.

References

1. Martínez-Carricondo, P.; Carvajal-Ramírez, F.; Yero-Paneque, L.; Agüera-Vega, F. Combination of Nadiral and Oblique UAV Photogrammetry and HBIM for the Virtual Reconstruction of Cultural Heritage. Case Study of Cortijo Del Fraile in Níjar, Almería (Spain). *Build. Res. Inf.* **2020**, *48*, 140–159. [CrossRef]
2. Tan, S.B.; Ti, E.S.W. What Is the Value of Built Heritage Conservation? Assessing Spillover Effects of Conserving Historic Sites in Singapore. *Land Use Policy* **2020**, *91*, 104393. [CrossRef]
3. Li, J.; Krishnamurthy, S.; Pereira Roders, A.; van Wesemael, P. Community Participation in Cultural Heritage Management: A Systematic Literature Review Comparing Chinese and International Practices. *Cities* **2020**, *96*, 102476. [CrossRef]
4. Xu, L.; Sang, K.; Li, G.; Lin, G.; Luo, Q.; Giordano, A. Heritage Evaluation and Analysis Based on Entropy Weight Method: The Study of Wengji Ancient Village in China. *J. Hous. Built Environ.* **2023**, *38*, 1843–1868. [CrossRef]
5. Coombes, M.A.; Viles, H.A. Integrating Nature-Based Solutions and the Conservation of Urban Built Heritage: Challenges, Opportunities, and Prospects. *Urban For. Urban Green.* **2021**, *63*, 127192. [CrossRef]
6. Della Torre, S. Italian Perspective on the Planned Preventive Conservation of Architectural Heritage. *Front. Archit. Res.* **2021**, *10*, 108–116. [CrossRef]
7. Cotella, V. From 3D Point Clouds to HBIM: Application of Artificial Intelligence in Cultural Heritage. *Autom. Constr.* **2023**, *152*, 104936. [CrossRef]
8. Xie, Y.; Tian, J.; Zhu, X.X. Linking Points with Labels in 3D: A Review of Point Cloud Semantic Segmentation. *IEEE Geosci. Remote Sens. Mag.* **2020**, *8*, 38–59. [CrossRef]
9. Sztwiertnia, D.; Ochałek, A.; Tama, A.; Lewińska, P. HBIM (Heritage Building Information Modell) of the Wang Stave Church in Karpacz–Case Study. *Int. J. Archit. Herit.* **2021**, *15*, 713–727. [CrossRef]
10. Li, Y.; Zhao, L.; Huang, J.; Law, A. Research Frameworks, Methodologies, and Assessment Methods Concerning the Adaptive Reuse of Architectural Heritage: A Review. *Built Herit.* **2021**, *5*, 6. [CrossRef]
11. Azmin, A.K.; Kassim, M.H.; Abdullah, F.; Sanusi, A.N.Z. Architectural Heritage Restoration of Rumah Datuk Setia via Mobile Augmented Reality Restoration. *Plan. Malays.* **2017**, *15*. [CrossRef]
12. Moyano, J.; Justo-Estebaranz, Á.; Nieto-Julián, J.E.; Barrera, A.O.; Fernández-Alconchel, M. Evaluation of Records Using Terrestrial Laser Scanner in Architectural Heritage for Information Modeling in HBIM Construction: The Case Study of the La Anunciación Church (Seville). *J. Build. Eng.* **2022**, *62*, 105190. [CrossRef]
13. Sun, Z.; Zhang, Y. Using Drones and 3D Modeling to Survey Tibetan Architectural Heritage: A Case Study with the Multi-Door Stupa. *Sustainability* **2018**, *10*, 2259. [CrossRef]
14. Albano, R. Investigation on Roof Segmentation for 3D Building Reconstruction from Aerial LIDAR Point Clouds. *Appl. Sci.* **2019**, *9*, 4674. [CrossRef]
15. Al-kheder, S.; Al-shawabkeh, Y.; Haala, N. Developing a Documentation System for Desert Palaces in Jordan Using 3D Laser Scanning and Digital Photogrammetry. *J. Archaeol. Sci.* **2009**, *36*, 537–546. [CrossRef]
16. Coşgun, N.T.; Çügen, H.F.; Selçuk, S.A. A Bibliometric Analysis on Heritage Building Information Modeling (HBIM) Tools. *ATA Plan. Ve Tasarım Derg.* **2021**, *5*, 61–80. [CrossRef]
17. Barrile, V.; Bernardo, E.; Bilotta, G. An Experimental HBIM Processing: Innovative Tool for 3D Model Reconstruction of Morpho-Typological Phases for the Cultural Heritage. *Remote Sens.* **2022**, *14*, 1288. [CrossRef]
18. Brumana, R.; Oreni, D.; Barazzetti, L.; Cuca, B.; Previtali, M.; Banfi, F. Survey and Scan to BIM Model for the Knowledge of Built Heritage and the Management of Conservation Activities. In *Digital Transformation of the Design, Construction and Management Processes of the Built Environment*; Daniotti, B., Gianinetto, M., Della Torre, S., Eds.; Research for Development; Springer International Publishing: Cham, Switzerland, 2020; pp. 391–400, ISBN 978-3-030-33570-0.
19. Lin, G.; Giordano, A.; Sang, K. From Site Survey to HBIM Model for the Documentation of Historic Buildings: The Case Study of Hexinwu Village in China. *Conserv. Sci. Cult. Herit.* **2020**, *20*, 111–123. [CrossRef]
20. Zhang, Z.; Zou, Y. Research Hotspots and Trends in HBIM: A Review Based on CiteSpace Analysis. *Humanit. Soc. Sci. Commun.* **2022**, *9*, 394. [CrossRef]
21. Germanese, D.; Pascali, M.A.; Berton, A.; Leone, G.; Moroni, D.; Jalil, B.; Tampucci, M.; Benassi, A. Architectural Heritage: 3D Documentation and Structural Monitoring Using UAV; 2019. Available online: https://openportal.isti.cnr.it/data/2019/403343/2019_%20403343.published.pdf (accessed on 25 March 2024).
22. Karataş, L.; Alptekin, A.; Yakar, M. Detection and Documentation of Stone Material Deterioration in Historical Masonry Structures Using UAV Photogrammetry: A Case Study of Mersin Aba Mausoleum. *Adv. UAV* **2022**, *2*, 51–64.

23. Moyano, J.; Gil-Arizón, I.; Nieto-Julián, J.E.; Marín-García, D. Analysis and Management of Structural Deformations through Parametric Models and HBIM Workflow in Architectural Heritage. *J. Build. Eng.* **2022**, *45*, 103274. [CrossRef]
24. Nieto-Julián, J.E.; Farratell, J.; Bouzas Cavada, M.; Moyano, J. Collaborative Workflow in an HBIM Project for the Restoration and Conservation of Cultural Heritage. *Int. J. Arch. Herit.* **2023**, *17*, 1813–1832. [CrossRef]
25. Hamed, W.; El Antably, A. Knowledge-Based HBIM for Conservation: The Case of Yahya al-Shabih Mausoleum. *Digit. Appl. Archaeol. Cult. Herit.* **2023**, *30*, e00278. [CrossRef]
26. Brumana, R.; Banfi, F.; Cantini, L.; Previtali, M.; Della Torre, S. Hbim Level of Detail-Geometry-Accuracy and Survey Analysis for Architectural Preservation. *ISPRS Int. Arch. Photogramm. Remote Sens. Spat. Inf. Sci.* **2019**, *XLII-2/W11*, 293–299. [CrossRef]
27. Antón, D.; Pineda, P.; Medjdoub, B.; Iranzo, A. As-Built 3D Heritage City Modelling to Support Numerical Structural Analysis: Application to the Assessment of an Archaeological Remain. *Remote Sens.* **2019**, *11*, 1276. [CrossRef]
28. Gámiz-Gordo, A.; Antón, D.; Barrero-Ortega, P. Urban Environment of Disappeared Heritage: Graphic Analysis of Puerta Real in Seville. *Heritage* **2023**, *6*, 5469–5499. [CrossRef]
29. Rocha, G.; Mateus, L.; Fernández, J.; Ferreira, V. A Scan-to-BIM Methodology Applied to Heritage Buildings. *Heritage* **2020**, *3*, 47–67. [CrossRef]
30. Moyano, J.; Pili, A.; Nieto-Julián, J.E.; Della Torre, S.; Bruno, S. Semantic Interoperability for Cultural Heritage Conservation: Workflow from Ontologies to a Tool for Managing and Sharing Data. *J. Build. Eng.* **2023**, *80*, 107965. [CrossRef]
31. Quattrini, R.; Malinverni, E.S.; Clini, P.; Nespeca, R.; Orlietti, E. From TLS to HBIM. High Quality Semantically-Aware 3D Modeling of Complex Architecture. *Int. Arch. Photogramm. Remote Sens. Spat. Inf. Sci.* **2015**, *XL-5-W4*, 367–374. [CrossRef]
32. Yang, X.; Grussenmeyer, P.; Koehl, M.; Macher, H.; Murtiyoso, A.; Landes, T. Review of Built Heritage Modelling: Integration of HBIM and Other Information Techniques. *J. Cult. Herit.* **2020**, *46*, 350–360. [CrossRef]
33. Sammartano, G.; Previtali, M.; Banfi, F. Parametric Generation in Hbim Workflows for Slam-Based Data: Discussing Expectations on Suitability and Accuracy. In Proceedings of the 9th International Congress & 3rd GEORES—GEOmatics and pREServation, Valencia, Spain, 26–28 April 2021.
34. Sánchez-Aparicio, L.J.; del Blanco-García, F.L.; Mencías-Carrizosa, D.; Villanueva-Llauradó, P.; Aira-Zunzunegui, J.R.; Sanz-Arauz, D.; Pierdicca, R.; Pinilla-Melo, J.; Garcia-Gago, J. Detection of Damage in Heritage Constructions Based on 3D Point Clouds. A Systematic Review. *J. Build. Eng.* **2023**, *77*, 107440. [CrossRef]
35. Murphy, M.; McGovern, E.; Pavia, S. Historic Building Information Modelling–Adding Intelligence to Laser and Image Based Surveys of European Classical Architecture. *ISPRS J. Photogramm. Remote Sens.* **2013**, *76*, 89–102. [CrossRef]
36. Wang, Q.; Kim, M.-K. Applications of 3D Point Cloud Data in the Construction Industry: A Fifteen-Year Review from 2004 to 2018. *Adv. Eng. Inform.* **2019**, *39*, 306–319. [CrossRef]
37. Yang, X.; Koehl, M.; Grussenmeyer, P.; Macher, H. Complementarity of Historic Building Information Modelling and Geographic Information Systems. *ISPRS Int. Arch. Photogramm. Remote Sens. Spat. Inf. Sci.* **2016**, *XLI-B5*, 437–443. [CrossRef]
38. Zeynalpoor Asl, M.; Samadzadegan, F.; Dadras Javan, F.; Talebian, M.H. 3D Modeling of Architectural Heritage Using UAV Photogrammetry. Case Study: Deir-e Gachin Caravansary. *J. Fine Arts Archit. Urban Plan.* **2022**, *26*, 61–73. [CrossRef]
39. Moyano, J.; León, J.; Nieto-Julián, J.; Bruno, S. Semantic Interpretation of Architectural and Archaeological Geometries: Point Cloud Segmentation for HBIM Parameterisation. *Autom. Constr.* **2021**, *130*, 103856. [CrossRef]
40. Lin, G.; Giordano, A.; Sang, K.; Stendardo, L.; Yang, X. Application of Territorial Laser Scanning in 3D Modeling of Traditional Village: A Case Study of Fenghuang Village in China. *ISPRS Int. J. Geo-Inf.* **2021**, *10*, 770. [CrossRef]
41. Li, X.-X.; Deng, E.; Wang, Y.-W.; Ni, Y.-Q. 3D Laser Scanning for Predicting the Alignment of Large-Span Segmental Precast Assembled Concrete Cable-Stayed Bridges. *Autom. Constr.* **2023**, *155*, 105056. [CrossRef]
42. Pađen, I.; García-Sánchez, C.; Ledoux, H. Towards Automatic Reconstruction of 3D City Models Tailored for Urban Flow Simulations. *Front. Built Environ.* **2022**, *8*, 899332. [CrossRef]
43. Moyano, J.; Carreño, E.; Nieto-Julián, J.E.; Gil-Arizón, I.; Bruno, S. Systematic Approach to Generate Historical Building Information Modelling (HBIM) in Architectural Restoration Project. *Autom. Constr.* **2022**, *143*, 104551. [CrossRef]
44. Tang, S.; Li, X.; Zheng, X.; Wu, B.; Wang, W.; Zhang, Y. BIM Generation from 3D Point Clouds by Combining 3D Deep Learning and Improved Morphological Approach. *Autom. Constr.* **2022**, *141*, 104422. [CrossRef]
45. Croce, V.; Caroti, G.; Piemonte, A.; De Luca, L.; Véron, P. H-BIM and Artificial Intelligence: Classification of Architectural Heritage for Semi-Automatic Scan-to-BIM Reconstruction. *Sensors* **2023**, *23*, 2497. [CrossRef] [PubMed]
46. Zaker, R.; Eghra, A.; Pahlavan, P. Documentation and HBIM of Industrial Heritage Using Drone Images: Petroleum Reservoir of Mashhad. *Int. Arch. Photogramm. Remote Sens. Spat. Inf. Sci.* **2021**, *XLVI-M-1–2021*, 917–923. [CrossRef]
47. Martínez-Carricondo, P.; Carvajal-Ramírez, F.; Yero-Paneque, L.; Agüera-Vega, F. Combination of HBIM and UAV Photogrammetry for Modelling and Documentation of Forgotten Heritage. Case Study: Isabel II Dam in Níjar (Almería, Spain). *Herit. Sci.* **2021**, *9*, 95. [CrossRef]
48. Santos, D.; Sousa, H.S.; Cabaleiro, M.; Branco, J.M. HBIM Application in Historic Timber Structures: A Systematic Review. *Int. J. Arch. Herit.* **2023**, *17*, 1331–1347. [CrossRef]
49. Liu, J.; Willkens, D.S.; Foreman, G. An introduction to technological tools and process of Heritage Building Information Modeling (HBIM). *EGE Rev. Expresión Gráfica Edif.* **2022**, *16*, 50–65.

50. Pepe, M.; Alfio, V.S.; Costantino, D. UAV Platforms and the SfM-MVS Approach in the 3D Surveys and Modelling: A Review in the Cultural Heritage Field. *Appl. Sci.* **2022**, *12*, 12886. [CrossRef]
51. Özdemir, E.; Remondino, F. Segmentation Of 3d Photogrammetric Point Cloud For 3d Building Modeling. *Int. Arch. Photogramm. Remote Sens. Spat. Inf. Sci.* **2018**, *XLII-4-W10*, 135–142. [CrossRef]

drones

MDPI

Article

Digital Recording of Historical Defensive Structures in Mountainous Areas Using Drones: Considerations and Comparisons

Luigi Barazzetti [1,*], Mattia Previtali [1], Lorenzo Cantini [1] and Annunziata Maria Oteri [2]

[1] Department of Architecture, Built Environment and Construction Engineering (ABC), Politecnico di Milano, Via Ponzio 31, 20133 Milan, Italy; mattia.previtali@polimi.it (M.P.); lorenzo.cantini@polimi.it (L.C.)
[2] Department of Architecture and Urban Studies (DASTU), Politecnico di Milano, Via Bonardi 3, 20133 Milan, Italy; annunziatamaria.oteri@polimi.it
* Correspondence: luigi.barazzetti@polimi.it

Abstract: Digital recording of historic buildings and sites in mountainous areas could be challenging. The paper considers and discusses the case of historical defensive structures in the Italian Alps, designed and built to be not accessible. Drone images and photogrammetric techniques for 3D modeling play a fundamental role in the digital documentation of fortified constructions with non-contact techniques. This manuscript describes the use of drones for reconstructing the external surfaces of some fortified structures using traditional photogrammetric/SfM solutions and novel methods based on NeRFs. The case of direct orientation based on PPK and traditional GCPs placed on the ground is also discussed, considering the difficulties in placing and measuring control points in such environments.

Keywords: 3D modeling; drone; fortifications; mountain; NeRFs; ruins; photogrammetry; PPK; SfM

Citation: Barazzetti, L.; Previtali, M.; Cantini, L.; Oteri, A.M. Digital Recording of Historical Defensive Structures in Mountainous Areas Using Drones: Considerations and Comparisons. *Drones* **2023**, *7*, 512. https://doi.org/10.3390/drones7080512

Academic Editor: Giordano Teza

Received: 15 July 2023
Revised: 31 July 2023
Accepted: 1 August 2023
Published: 3 August 2023

1. Introduction

Nowadays, 3D modeling from drone images has reached significant maturity and a high automation level. Digital photogrammetry coupled with computer vision methods (e.g., Structure from Motion/SfM) can automate most phases of the reconstruction workflow [1]. In addition, flight planning software also allow users to automate the data acquisition phase, which can be remotely planned before reaching the site [2].

This paper aims to show some of the latest advancements in drone image processing for 3D modeling applications, focusing on using Post-Processed Kinematic (PPK) for direct georeferencing [3], reducing the need for ground control points (GCPs) measured on-site. The manuscript then discusses Neural Radiance Fields (NeRF) [4] compared with the more traditional extraction of point clouds and meshes with dense image-matching techniques.

The particular case of fortified heritage constructions in mountainous areas is discussed. Such defensive structures were designed to be not easily accessible [5], making drones a powerful solution to capture their irregular geometry. Here, digital recording for restoration and conservation can pose several challenges, including vegetation on ruins and around the construction. Drones can become an essential tool for the opportunity to reach parts that are not accessible, providing images and videos not only limited to visual documentation.

After orienting images with photogrammetric techniques, SfM and NeRFs were used to reconstruct the geometry of three fortified constructions in the Italian Alps. A preview of the achieved results with NeRFs is shown in Figure 1, which depicts some frames extracted from a rendered video. Input data for NeRFs training are photogrammetrically oriented images acquired with a Mavic 3E drone supported by PPK for georeferencing. A simple qualitative inspection shows the excellent quality of the vegetated background achieved with NeRFs.

Overall, the main topics considered in this manuscript can be summarized as follows

- The use of drones with PPK for digital recording heritage ruins with limited accessibility and comparison with more traditional surveying strategy (laser scanning and photogrammetry with GCPs);
- Comparison between traditional photogrammetry/SfM and new methods based on NeRFs generated with Nerfstudio [6].

Figure 1. Frames of a rendered video created using NeRFs and a set of oriented images. The fortified construction is Mancapane Castle. The frames were generated with Nerfstudio using a set of images oriented with Metashape.

1.1. Digital Recording of Ruins for Restoration and Conservation

The versatility of advanced digital recording tools proves to be invaluable in the field of conservation. Restoring and conserving historical buildings necessitates a precise understanding of their geometrical features. Digital recording techniques, such as laser scanning and digital photogrammetry, offer a very high level of detail, providing geometrically accurate information. This data can be utilized to analyze the historic building based on its geometrical proportions and to facilitate further investigations, including assessing the material composition and state of conservation.

The successful preparation of various conservation actions relies on creating a comprehensive recording dossier, instrumental for analyzing the proportions and architectural heritage. This includes verticality deviations or other deformations, understanding the construction logic, identifying materials, and assessing their conservation conditions.

Digital recording techniques offer the advantage of flexible 2D and 3D representations utilized throughout the conservation process, from preliminary studies and intervention identification to devising enhancement strategies to ensure ongoing use and accessibility of the buildings.

This study conducted three digital recording campaigns using drones on architectural ruins, namely, Castel Grumello and Castel Mancapane in Valtellina, Italy, and the Rocca di Vogogna in Val d'Ossola, Italy.

Following the guidelines of the Italian Code of Cultural Heritage and upholding a long-standing tradition in preservation, conducting meticulous surveys is crucial for the correct development of conservation design. The three cases discussed here involve archaeological remains—medieval fortifications still preserving some volumes of their original configuration, including square towers and defensive circuits. These tangible physical remnants bear testimony to values that must be preserved.

Upon initial observation, these defensive buildings conjure a distinctive spatial organization of the territory, known as the fortifications period, bridging the transition from the administrative structure of the Roman Empire to the fragmented power landscape of the Middle Ages. Furthermore, the material remnants preserve the vestiges of ancient construction traditions, employing stone components and organizing massive masonry structures.

Unraveling a comprehensive understanding of all the elements encompassing these archaeological sites demands a meticulous analysis of the buildings, ranging from their geometrical proportions to the characteristics of their structural elements and components. Advanced surveying techniques provide the crucial information necessary for further conservation strategy analyses. This includes studying the building's various phases of transformation, mapping materials, and their deterioration, evaluating mechanical damages, and more.

Digital techniques able to record complex heritage sites are a valid support for preserving ruined buildings, which are particularly fragile in terms of materials. By creating digital records, experts can document the condition of the existing structures and identify deteriorated areas. Digital tools also facilitate advanced data analysis, such as studying material composition, structural integrity, and environmental factors, aiding in interpreting the transformations of the building over time and, hence, the potential risks they could face in the future. Moreover, they assist in assessing historical buildings' structural stability.

Structural modeling software can simulate different conditions, helping to identify potential vulnerabilities and decisions on conservation and safety measures. Moreover, digital tools eventually simulate various restoration scenarios, interpreting the previously defined conservation strategies and allowing experts to assess their potential impact (e.g., the perception by the public) before implementing physical interventions.

1.2. Challenges in Heritage Recording in Mountainous Areas

Creating a detailed and accurate digital record of historical fortifications requires a combination of digital recording methods and tools. Photogrammetry and laser scanning are helpful solutions that capture the surfaces without requiring direct contact with the object. However, this paper does not explicitly focus on using different digital tools and their integrated use for surveying complex geometry. The reader is referred to several technical papers describing real case studies [7–18].

As mentioned in the previous section, this paper aims to evaluate and discuss the use of drones (and photogrammetry with images captured by drones) for the particular case of fortified constructions in mountainous areas. Photogrammetric processing using SfM methods is highly automated today thanks to different commercial and open-source solutions that can generate accurate point clouds and textured meshes. Fortified heritage often includes complex structures (castles, towers, forts, among others) with irregular geometry. An accurate and comprehensive digital record requires techniques capable of revealing the actual shape with all irregularities. Point clouds produced by drone images captured with suitable GSD (ground sampling distance) can show small details and geometric anomalies. Planning a complete survey requires knowledge of project requirements and images with a sufficiently short distance to reach the desired GSD while keeping a suitable geometry of the image block for photogrammetric processing.

In addition, drones equipped with GNSS receivers can also provide direct georeferencing using RTK or PPK, reaching centimeter-level precision. For fortifications built on mountains, placing and measuring control points with traditional surveying equipment (e.g., a GNSS antenna on top of a pole) could be challenging. Points with a homogeneous distribution on the site are also necessary. The opportunity to measure camera poses with centimeter-level precision coupled with a photogrammetric block with suitable geometry and sufficient overlap can be exploited to georeference the project without using external control points.

Drones are particularly suitable for capturing elevations and producing orthophotos, which are used for condition mapping (materials, construction technologies, degradation) and designing interventions. External elevations could also be rather challenging to reach with traditional terrestrial methods. At the same time, accurate and detailed site plans can be created using high-resolution orthophotos produced in the cartographic plane or a local Cartesian system. The user must be aware that products in a cartographic system could be affected by cartographic effects, making angles, distances, and areas different from the "real" values, which can be captured with laser scanning.

Although photogrammetry with drones can be helpful in these types of surveying activities, several limitations exist. For instance, parts covered with thick vegetation cannot be reconstructed. Vegetation in abandoned ruins is often present. A reconstruction from images would provide a model of the vegetated areas, which is also often incomplete for the intrinsic difficulties in modeling such irregular areas.

The case studies in this work represent three medieval fortified structures, partially collapsed, realized in two different contexts in the Lombardy and Piedmont regions in Italy. They were built on mountain cliffs and have an irregular shape in the plan's organization for adapting the different parts of the buildings to the local topography.

2. Case Study Description

Three case studies are considered in this paper:

- Castel Grumello;
- Castel Mancapane;
- Rocca di Vogogna.

Detailed information about the case studies (Figure 2) are presented in the following paragraphs and summarized in Table 1.

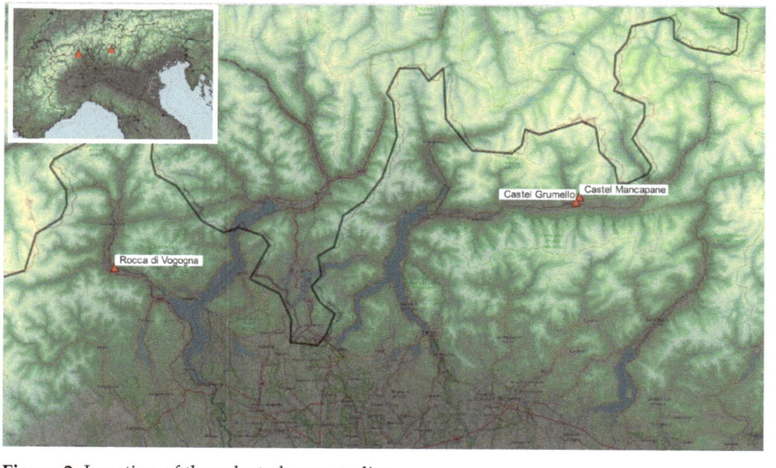

Figure 2. Location of the selected case studies.

Table 1. Summary of presented cases study.

Case Study	Location	Approximate Elevation [m]
Castel Grumello	Montagna di Valtellina (Sondrio, Italy)	490
Castel Mancapane	Montagna di Valtellina (Italy)	890
Rocca di Vogogna	Vogogna (Verbania, Italy)	320

2.1. Castel Grumello

Castel Grumello is a medieval fortress in Montagna di Valtellina, Italy. The ruins of the castle, which are surrounded by the famous vineyards of Valtellina, strongly characterize the site where built heritage and landscape are merged into a fascinating combination of architecture and nature.

The fortress is a rare example of a twin castle used for residential and military purposes (Figure 3). It was part of a more complex system of fortifications that controlled Valtellina in the Middle Age. Together with Mancapane Castle (see the following sections) and other fortresses along the valley, it symbolized the economic and political powers of the families that controlled the territory.

Figure 3. The first case study proposed in the paper: Castel Grumello.

The fortress's history shows how the destiny of medieval castles is strictly related to the landlords' fortune (or misfortune). The castle is mentioned in a document currently preserved in the Como national archive, written between 1329 and 1333. Historians agree on the possible construction of the castle in the 14th century by the De Piro, a powerful feudatory family in conflict with the Capitanei. The contrasts between the two families (reflecting the conflict between the Guelfi and the Ghibellini at the local level) are at the base of the ruin of Castel Grumello, which was then dismantled in 1526 by Grigioni, who conquered the valley. After a period of abandonment, the destiny of the castle was connected to the productive history of the Valtellina vineyards (probably cultivated since the 16th century), as it was used as a quarry to reorganize the cultivation of the vineyards through an enlargement of the terraces. Only in 1990, the FAI (Fondo per l'Ambiente Italiano) bought the ruins, and the fortress was restored thanks to a vast conservation program designed by the architect, S. Tirinzoni [19].

The castle is characterized by two main constructions along the crest of the hill. The west building was probably used for residential purposes, while the east was for military purposes. The different uses were mainly documented during the archaeological excavations at the beginning of this century. The restoration by Tirinzoni aimed at preserving the authenticity of the ruins. For this reason, the interventions fully respected the original materials and preserved the transformations over time that document the site's complex history.

The masonry underwent a cleaning process that did not remove the patina chromatic variations, and oxidation happened over time. The cornices were carefully dismantled stone by stone and reassembled with the same blocks and new lime mortar. Reconstruction was carried out only if strictly necessary for consolidation and preservation. Mortar joint repointing was conducted using selected and differentiated lime and aggregates tailored to match the existing chromatic variations and textures in different areas.

The main goal of the restoration was to preserve the perception of the site by the local community, who traditionally use the site in their spare time to relax and enjoy the fantastic view from the ruins. For this reason, the interventions were aimed at improving the accessibility of the site by adding stairways, services, and barriers. Thanks to the conservation strategy and the beautiful landscape with vineyards, local communities and other tourists continuously visit the castle.

2.2. *Castel Mancapane*

In the mountainous context surrounding Sondrio, not far from Castel Grumello, another fortified structure with a square tower can be reached following a path on the mountains (Figure 1). Because of a lack of documentation, the origin of the fortress is unknown, but thanks to an accurate evaluation of the masonry pattern, the building was dated back to the 13th century [20]. It became the property of the De' Capitanei family, who tried to establish their hegemony in the valley since the 14th century.

The structure comprises irregular stones with roughly rectangular blocks at the corners. As observed in the case of Castel Grumello, also this castle was abandoned during the period of domination by Grigioni, when the fortress system of the valley was dismantled. The ruins of the complex are composed of a simple rectangular wall, realized along the slope of the mountain, with a square tower set in the middle of the north wall.

The complex is more isolated than the Grumello Castle. It is surrounded by dense vegetation. Stefano Tirinzoni, who designed the conservation plan also for this archaeological site, removed some plants from the internal courtyard and the outer sides of the walls to guarantee the accessibility of the castle [21].

2.3. *Rocca di Vogogna*

The Rocca di Vogogna (Figure 4) was a defensive structure of a comprehensive military system composed of strongholds diffused along the Ossola Valley.

The origin of the building is commonly posted in the medieval period. Still, a systematic studio [22] of the fortress and the stratigraphy of its structures show that the archaeological remains can be divided into different phases, revealing that several parts were built after the 15th century.

Nowadays, the ruins of the fortress are surrounded by the vegetation of the Val Grande natural park, a protected area formed by the forests on the mountains in a large natural basin. The building dominates the historical center of Vogogna from the south side of Orsetto Mountain, facing the Simplon road. The ruins appear behind the medieval buildings characterizing the town, an important administrative center that, in 1348, became the capital of the lower Ossola Valley. The municipal palace was realized by Giovanni Visconti, the future Duke of Milan, who reinforced the fortifications of the town, connecting the city walls to a castle (known as the Castello Visconteo of Vogogna) located on the top of the city center and facing the Rocca of Vogogna on the mountain.

Figure 4. Overview of the third case study: Rocca di Vogogna.

The relationship between the Castle of Vogogna, the medieval city center, the valley's natural environment with the river Toce, the mountain hamlet, and the archaeological remains of the fortress required a complementary analysis carried out by using historical research and the recording of the complex. The southwest side of the building is composed of different bodies realized along the slope of the mountain, creating a irregular hypogeal distribution of the spaces. An existing record of the plan of the various levels was not available.

According to Babbini [22], who provides a stratigraphic analysis of the complex starting from a detailed survey of the main masonry patterns characterizing various areas of the fortress, the building can be divided into an ancient core, conserving the remains of a central square tower (which dates back to the 12th century), connected to a yard enclosed into defensive walls and sub-circular minor towers at the edges. This part of the fortress is connected to other spaces displayed on different levels. They appear as rooms with large openings facing the valley.

The complex results from several additions, including the works in the 15th century when Vitaliano Borromeo bought the Feud of Vogogna and introduced significant renewals to the town's fortifications [23]. The fortress was deeply damaged after an attack in 1514 organized by Domodossola [23], the capital of the north district of Ossola Valley. The complex was therefore abandoned.

With the rearrangement of the territorial administration after the unification of Italy, Vogogna lost its administrative role and became an ordinary town. Its defensive structures were transformed, changing the military use into rural ones. The fortress was misused with the changing economic assets, moving from rural to industrial. The surviving structures, with the buttresses, the patrol walkways, and the bombards, are proof of the military origin of the building. In the same way, the ruins showing the addition of new spaces used as storage, and the subdivision of pre-existing large rooms into minor ones, require additional studies for interpreting the complex stratification of uses characterizing the complex and its transformations over time.

The difficult accessibility of the external walls, directly placed on the sides of the mountain, and the dangerous paths connecting the different levels of the structures require a detailed survey of these spaces, aimed at overcoming the lack of graphic documentation about the fortress, actually documented by two tables published by Airoldi [23].

3. Image Acquisition and Orientation

Acquisition and processing of the images for the proposed case studies were carried out using different approaches depending on the characteristics of the buildings and the availability of specific drones. The three approaches (Table 2) can be summarized as follows:

- Castel Grumello: complete direct orientation combining photogrammetric processing and PPK-automatic flight;
- Castel Mancapane: partial direct orientation integrating PPK in photogrammetric processing, only for a specific set of images for the presence of thick vegetation and lack of fixed-flight conducted manually;
- Rocca di Vogogna: georeferencing using a set of GCPs measured with RTK-manual flight (the used drone did not support RTK or PPK).

More details are provided in the following subsections.

Table 2. Acquisition an orientation strategy for the presented case studies.

Case Study	Image Acquisition	Drone Used	Orientation Method	Expected Deliverables	Software Used	NeRF Test
Castel Grumello	Automated flight plan (normal and oblique images)	Mavic 3 Enterprise	Bundle adjustment constrained by camera positions (PPK)	3D model-site orthophoto-GIS	REDtoolbox-Agisoft Metashape	No
Castel Mancapane	Manual flight (normal and 360° around the castle)	Mavic 3 Enterprise	Bundle adjustment constrained by camera positions (PPK)	3D model	REDtoolbox-Agisoft Metashape	Yes
Rocca di Vogogna	Manual flight (normal and oblique images)	DJI Air 2S	Bundle adjustment constrained GCPs	3D model-site orthophoto-elevation orthophotos	Agisoft Metashape	Yes

3.1. Castel Grumello

Image acquisition was carried out with a Mavic 3 Enterprise, using an automated flight plan with normal and oblique images (40°). The acquisition took about 30 min, obtaining 1224 images at an average elevation (above ground) of 30 m organized in different strips.

A reference GNSS antenna (master) was also placed on-site, recording raw data in static mode for about 45 min (RINEX). The coordinates of the master were then calculated in static post-processing using 1-second RINEX from a permanent GNSS station and Leica Geo Office. Data were provided by the SPIN3 GNSS service (https://www.spingnss.it/ (accessed on 1 August 2023), a network of 33 permanent GNSS receivers covering three Italian regions (Piedmont, Lombardy, Valle D'Aosta).

PPK was used to calculate camera poses using the raw data acquired with the drone. The software used was REDtoolbox (https://www.redcatch.at/redtoolbox (accessed on 1 August 2023)), obtaining a fixed solution for all images. The flight was conducted with complete sky visibility, without occlusions, and in a fully automatic way.

Images were imported in Agisoft Metashape. The calibration parameters were loaded from a previous (specific) calibration project and were assumed to have constant values during bundle adjustment. Georeferencing was obtained with a rigid seven-parameter transformation (scale, rotation, translation), obtaining RMSE values on camera poses of 16 mm for East, 30 mm for North, and 22 mm for altitude. Figure 5 shows the results after image orientation. More details about generating additional deliverables are provided in the next section.

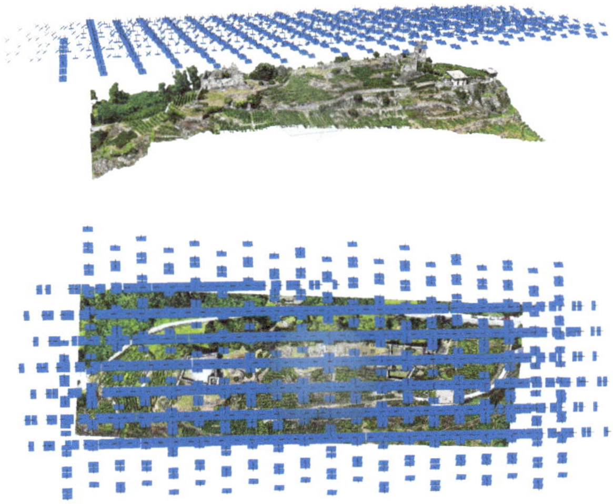

Figure 5. Image orientation results for the Castel Grumello dataset.

3.2. Castel Mancapane

The acquisition was conducted with images acquired with a Mavic 3 Enterprise with an RTK module. Rinex data were stored and post-processed using PPK. The master GNSS station is a part of the permanent network SPIN3 GNSS (https://www.spingnss.it/ (accessed on 1 August 2023)). The closest permanent receiver is located in Sondrio, a few kilometers from Castel Mancapane.

As the castle is surrounded by vegetation, a manual flight composed of several vertical strips was carried out. A total number of 549 images was acquired in about 30 min, also featuring a convergent geometry around the castle. Before running photogrammetric processing, PPK was used to calculate precise camera poses, which were used to georeference the project. However, a precise (fix) camera pose was achieved only for 347 images (63.2%), whereas 202 images (36.8%) had a float solution. Such a result was expected because vegetation strongly reduced sky visibility, especially for images near the ground. A trigger map and a height map after PPK processing are shown in Figure 6. As can be seen, the fix was lost and obtained again after a few vertical strips. The software used was still REDtoolbox.

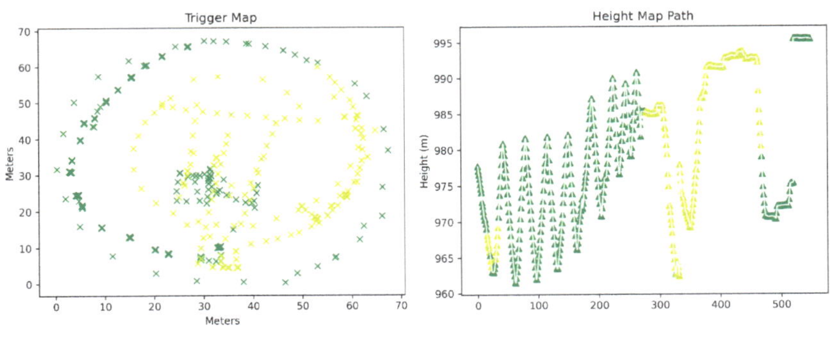

Figure 6. Trigger and height maps showing fixed (green) and float (yellow) solutions.

Image orientation was carried out using only camera poses with a fixed solution. Images with a float solution were added to the project and included in image matching and bundle adjustment so that they were oriented together with the other images in a combined orientation procedure. In other words, coordinates from the float solution were excluded from processing, and georeferencing was carried out with a seven-parameter transformation calculated only using images with a fixed solution.

The software used for processing was Agisoft Metashape. The camera was also calibrated beforehand, and the calibration parameters (interior orientation and distortion coefficients) were imported and assumed as fixed values in processing. The RMSE values on camera poses after matching and bundle adjustment were about 9 mm for East, 7 mm for North, and 12 mm for altitude. An image showing camera poses around the castle is shown in Figure 7.

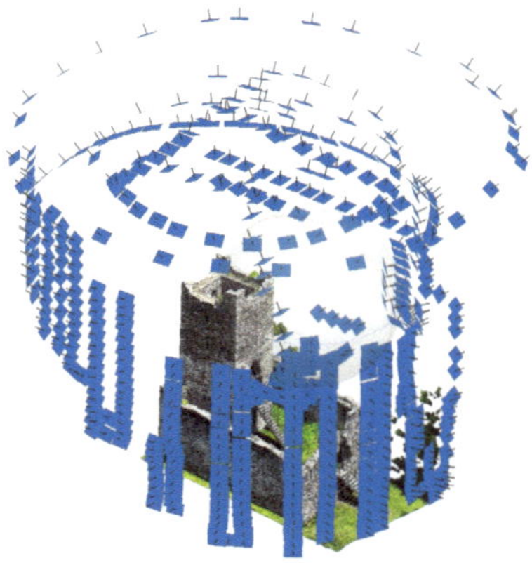

Figure 7. Image orientation results for the Castel Mancapane dataset.

The achieved metric accuracy using PPK is sufficient for producing deliverables at a metric scale of 1:50. From this point of view, the results are satisfying, considering the complexity of accessing the place, especially with light equipment: only the drone was taken to the castle. More information about this point is also compared with a traditional laser scanning digital survey.

3.3. Rocca di Vogogna

Images of the Rocca di Vogogna case study were acquired using a DJI Air 2S. Image acquisition was carried out using a manual strategy. The choice of a manual flight was due to the atmospheric conditions. A strong wind characterized the area during the acquisition, and the drone pilot manually operated it to maintain control of drone movement during wind gusts. Apart from the manual flight, the acquisition was carried out using a schema combining normal images (aimed at creating an orthophoto of the fortress and nearby structures) and a set of oblique (45°) and convergent images to reconstruct the vertical surfaces.

Indeed, the reconstruction of vertical structures and their texture, like the remains of the fortress tower and the external fortification walls, are of fundamental importance for any stratigraphic analysis. However, this is not possible starting from ground acquisition due to the hill's steep slope on which the fortification walls are built. In more detail, 373 normal images were acquired at a flight altitude of approximately 25 m above the ground organized in different strips, and 390 oblique and convergent images were captured covering the entire fortress area. A total of 763 images were acquired in approximately 1 h.

The project's georeferencing was carried out using a set of GCPs placed on the ground and measured with RTK. Indeed, as previously mentioned, the drone used in this project, the DJI Air 2S, does not support RTK or PPK. In particular, 16 points have been materialized by using checkerboard targets (dimension 30 × 30 cm) in the area of the fortress. The measurement of targets was performed using a receiver Emlid Reach RS2 receiving corrections from service provided by the network of permanent stations developed by SPIN3 GNSS. The precision of the measured target positions is about ±2 cm.

The images were oriented using the software Agisoft Metashape (Figure 8). Due to the good geometry of the image acquisition, no fixed calibration was imported, and the calibration parameters were estimated during bundle adjustment. For the georeferencing of the project, 10 points were defined as GCPs and 6 were used as Check Points (CPs) for the evaluation of the orientation (Figure 9).

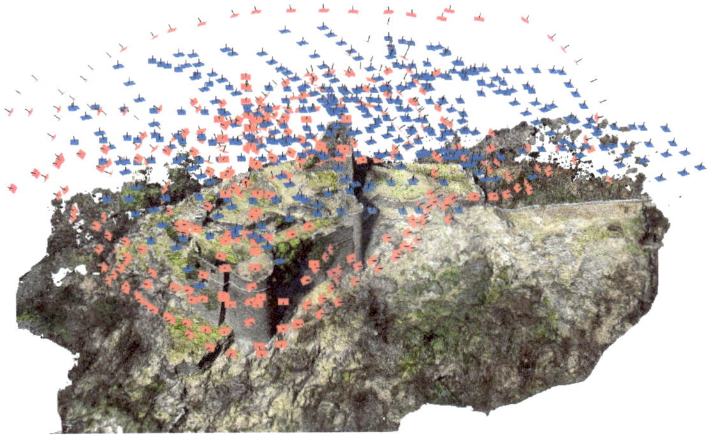

Figure 8. Image orientation results for the Vogogna dataset: normal images (blue) and oblique images (red).

The RMSE on the GCPs target position after bundle adjustment is 1.0 cm for East, 1.1 cm for North, and 0.8 cm for altitude. Similar results can be observed with CPs. RMSE is 0.9 cm for East, 1.4 cm for North, and 0.8 cm for altitude.

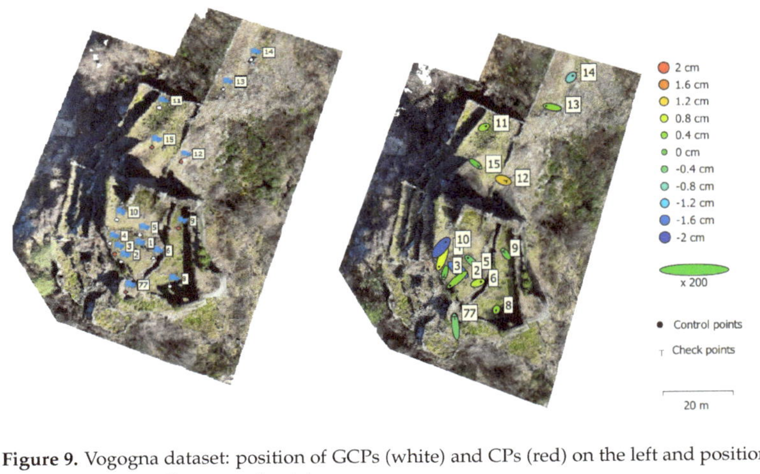

Figure 9. Vogogna dataset: position of GCPs (white) and CPs (red) on the left and position residuals represent as error ellipses on the right.

4. Production of Deliverables with Digital Photogrammetry: Some Considerations

Different deliverables can be produced after orienting the images with different approaches (described in the previous section). We discuss the case of a traditional acquisition with laser scanning (often integrated with terrestrial and drone images) compared to the approach proposed in this paper, i.e., only drone images.

4.1. Castel Grumello

A textured 3D model was generated from the set of oriented images supported by PPK. A digital orthophoto (top view in the UTM cartographic plane) was then produced and imported into an existing GIS system developed to host different datasets related to Castel Grumello. Some images of the obtained textured 3D model are shown in Figure 10.

Figure 10. Castel Grumello: the 3D model obtained with the new drone acquisition.

The GIS was created using ArcGIS Pro and ArcGIS Online, starting from several photogrammetric and laser scanning datasets acquired in previous years. Creating a complete GIS required a combined approach for the survey, which was georeferenced using some ground control points placed on the ground and measured with a GNSS in RTK. The new acquisition presented in this paper is instead directly integrated into the GIS thanks to the camera poses calculated using PPK.

A drone orthophoto was also available in the GIS. The previous digital recording was carried out with another drone (Mavic Air 2S) about two years before. As mentioned, some GCPs were necessary for the lack of PPK for the drone previously used. Another difference between the two drone surveys is the different geometry for acquisition. The old capture had only normal images, and vertical surfaces were integrated with laser scanning and terrestrial photogrammetry. The new drone flight also has oblique images, which provide a good reconstruction of walls, although without the same level of detail as a terrestrial capture. Of course, internal spaces could not be captured only with the drone.

An image of the GIS of Castel Grumello is shown in Figure 11. Creating vector files and the association of information requires (long) manual work. Digital orthophotos and georeferenced point clouds are valid geometric support for producing this kind of deliverables, in which automation in the restitution phase is still minimal and require manual work conducted by specialists.

Figure 11. The GIS system for Castel Grumello.

4.2. Castel Mancapane

The set of oriented images with the (partial) direct orientation provided by PPK was used to generate a textured mesh of the castle. The software is Agisoft Metashape, which produced a textured mesh of about 5 million faces. As can be seen, the software was able to provide also a good model of the vegetation around the castle. A large bounding box, including trees around the castle, was necessary for the texture-mapping phase. The extraction of a mesh limited to the volume occupied by the castle resulted in a wrongly textured model, in which the vegetation was also projected on the external walls. This is not only a visual artifact; it could lead to a wrong interpretation of the conditions of the external surfaces, which would present biological colonization during the condition mapping phase, an essential step of the restoration process.

Some pictures of the final 3D model are shown in Figure 12. The reconstruction seems complete and includes the defensive wall (both external and internal surfaces) and the external surfaces of the tower, including the top level.

Figure 12. Castel Mancapane: the 3D model obtained with the new drone acquisition.

As mentioned in the previous section, comparing the results of such an acquisition with a laser scanning survey carried out in 2020 using a Faro Focus HDR is also interesting. Although the time required to capture the castle was less than a day, it was necessary to bring a total station (and a tripod) and a GNSS antenna (on a 2 m pole) to georeference the laser scanning survey. Static GNSS was used because of the dense vegetation in the area.

Using a drone with PPK (as in the latest acquisition) is a more comfortable choice, notwithstanding the quality of the laser scanning points is higher (in terms of data density and metric accuracy). In addition, the laser scanner also allowed for capturing the tower's interior, whereas the latest drone acquisition could only capture the external surfaces.

Using both techniques is still the best approach to obtain the complete acquisition. Still, the use of a laser scanner can also be limited to internal parts and those external elements in which a higher level of detail can be necessary to generate an accurate 3D model (for instance, the staircases added during the restoration intervention). At the same time, the laser scanning acquisition for the walls close to vegetation could only capture the lower parts of the walls.

Another consideration is related to the reconstruction of small details and the need for specialists involved in the conservation process. The staircase added during the restoration could be an example (Figure 13). An automatic reconstruction obtained from images does not separate the different elements, like walls, the metal arch supporting the ruins, and the metal-wooden staircase. Drone images also have a GSD insufficient to provide a sufficiently good geometry.

Figure 13. Textured mesh from the drone, laser scanning point clouds, and 3D model obtained with manual modeling operations using the laser point cloud.

The laser scanning point clouds have a better resolution instead. They were used in manual modeling enormity to create a 3D model which separates the different constructive elements. Point clouds and textured meshes greatly support heritage documentation, restoration, and conservation. However, other deliverables are often required when other specialists need to design interventions, and direct outputs from photogrammetry and laser scanning should be considered as tools for producing additional deliverables.

4.3. *Rocca di Vogogna*

The orientation results generated a textured 3D model of the Rocca di Vogogna. Bentley ContextCapture was used for this project. The choice of Bentley ContextCapture instead of Agistoft Metashape relates to the completeness of the mesh obtainable with the software. External and internal orientation results obtained with Agistoft Metashape were exported and used as input in Bentley ContextCapture. The obtained textured model is presented in Figure 14.

Once the textured 3D model is obtained, two main deliverables and results are expected:

- Creation of an overall orthophoto of the Rocca di Vogoga site;
- Creation of orthophotos of the different elevations facing steep slope hills for stratigraphic analysis.

Figure 14. Two different views of the 3D textured mesh model of the Vogogna Fortress.

The 3D model generated in Bentley ContextCapture was imported back into Agistoft Metashape to generate those two typologies of deliverables. Indeed, Agistoft Metashape allows for a more straightforward definition of the projection planes for orthophoto definition and has some editing functions.

The Rocca di Vogogna's overall orthophoto was generated using only normal images. This choice is motivated by the need for a uniform ground sampling distance of 1.0 cm and the reduction of illumination and shadow differences among the images used to generate the deliverable. The final orthophoto is shown in Figure 15.

Figure 15. Orthophoto of the Vogogna Fortress and its nearby area.

The second deliverable typology consists of orthophotos of the fortress's different elevations (Figure 16). As previously mentioned, the fortress is built on the top of a steep slope hill, preventing an effective acquisition of ground data for stratigraphic analysis of the different masonry structures constituting the walls. For this task, a drone survey was fundamental to reconstruct the different fortress units' textures properly. Orthophotos were defined for the different elevations of the structure: external main fortifications walls and remains of the fortress tower. Only oblique and convergent images were utilized to generate the elevation orthophotos. The final orthophotos have a GSD of 0.5 cm.

Figure 16. Two different orthophotos of the fortress elevations: South (**top**) and West (**bottom**) elevation.

5. NeRFs for Processing Drone Images

5.1. Preliminary Considerations

Reconstruction of 3D scenes starting from images is a research topic in which scholars have been developing new methods since the beginning of the 20th century. The approach receiving more attention so far is the one based on the geometric consideration of image geometry and image-matching technique for radiometric consistency [1,24–26]. New approaches are emerging in the field based on implicit representations. In those methods, a reconstruction model can be trained from a set of 2D images, and the trained model can learn implicit 3D shapes and textures [27]. In particular, Neural Radiance Field (NeRF) models are gaining popularity recently. NeRF models use volume rendering with implicit neural scene representation by using Multi-Layer Perceptrons (MLPs) to learn the geometry and lighting of a 3D scene starting from 2D images. The first paper introducing the concept of NeRF was published by Mildenhall et al. [4] in 2020, and to date (June 2023) it has received more than 3000 citations (source: Google Scholar), proving the increasing interest in this topic. Providing a literature review of this topic is out of the scope of this paper, and the reader is referred to recent review papers [28–30]. In this paper, we introduce the central concepts of NeRF, as well as the required input data, and we propose some considerations on NeRF in general and their application to Cultural Heritage applications. NeRF models are gaining attention due to their advantages compared to other similar approaches:

- NeRF models can be trained using only multi-view images of a scene. Unlike other neural representations, NeRF models require only images and poses to learn a scene;
- NeRF models are photo-realistic;
- NeRF models require low memory and perform computations faster.

In its basic form, a NeRF model represents three-dimensional scenes as a radiance field approximated by a neural network. The radiance field describes color and volume density for every point and every viewing direction in the scene. NeRF models are self-supervised. This means that they do not need extra data (like 3D or depth information) for the 3D reconstruction: the input of a NeRF neural network are camera poses (i.e., spatial location and viewing directions) of a set of 2D images of a scene. The output is the volume density of the scene and the view emitted radiance (RGB color) in any direction and for each scene location. The function mapping camera poses to a radiance field (color and volume density) are approximated by one or more Multi-Layer Perceptrons (MLPs). The prediction of the volume density (i.e., the scene's content) is constrained as independent concerning the viewing direction, while the color is set as dependent on both view direction and scene position. In its standard formulation, NeRF models are designed in two stages. The first stage takes as input the camera poses, passes them to a Multi-Layer Perceptron, and outputs volume density and a high-dimensional feature vector. In the second stage, the feature vector is concatenated with the viewing direction, using them as input for a further MLP, which outputs colors. Then, starting from a trained NeRF, new synthetic views can be created using this workflow:

- For each pixel to be created in the new synthetic image, camera rays and some sampling points are generated into the scene;
- Local color and density are computed for each sampling point by using the view direction, the sampling location, and the trained NeRF;
- The computed colors and densities are the input to produce the synthetic image using volume rendering.

NeRFs are an emerging technology that may provide novel outcomes in the heritage documentation field. A review of possible applications in digital recording and a comparison between NeRFs and traditional photogrammetric approaches was recently proposed in [31–34]. Previous work was also carried out by [35]. The author recognized the advantages in the case of complex objects with occlusions or irregular shapes, such as vegetated areas (one of the main problems faced in this work). The results also seem promising for homogenous surfaces or transparent/shining objects, which are difficult to model with photogrammetric methods. The metric accuracy of the achieved results is still questionable, especially when compared to photogrammetric or laser scanning point clouds. However, recent neural surface reconstruction methods seem very powerful [36].

Images and videos acquired from drones can also be processed using NeRFs, and previous work was carried out by [37]. The growing interest in such novel technology is demonstrated by the number of examples generated with Luma AI, accessible at https://lumalabs.ai/featured (accessed on 1 August 2023). We tried to process the high-resolution datasets oriented with photogrammetric methods using Nerfstudio, although advancements and updates in NeRFs are expected in the short term.

5.2. Castel Mancapane

5.2.1. Processing with Nerfstudio

The first test with NeRFs was carried out with the images oriented in Metashape and Nerfstudio (https://docs.nerf.studio/en/latest/ (accessed on 1 August 2023)). Exterior, interior, and calibration parameters were exported from the software and processed using Nerfacto. Processing was very rapid (a few minutes, faster than the generation of the point clouds and mesh with Metashape), even considering the significant number of images in the project (549). Default parameters were used in the processing. An image of the viewer after training is shown in Figure 17, whereas some frames of the rendered video were illustrated at the beginning of the manuscript (Figure 1).

Figure 17. Castel Mancapane processed using Nerfstudio.

Rendering a video is very straightforward, and the quality of the results was excellent. In this sense, NeRFs provided excellent results for Castel Mancapane. Creating a camera path is also relatively simple, making NeRFs an alternative to traditional approaches for rendering based on point clouds or meshes.

Nerfstudio also allows for exporting the geometry. The point cloud was then exported, setting the number of points to 20 million and using the default setting provided by Nerfstudio. A visual comparison with the point cloud generated by Metashape is shown in Figure 18, which comprises 90 million points. The results achieved with Metashape are more complete and less noisy. However, processing was carried out using the default set of options, and the authors know that results could be improved. NeRFs are also relatively new compared to traditional photogrammetric processing, and improvements are indeed expected with future research work.

Figure 18. Point clouds produced with Nerfstudio (**left**) and Metashape (**right**) using the same set of images.

5.2.2. A Few Considerations on (Some) Other NeRFs Implementations

The same dataset was processed using Luma AI (https://lumalabs.ai/ (accessed on 1 August 2023) and Instant-ngp (https://github.com/NVlabs/instant-ngp (accessed on 1 August 2023)). Luma AI provides a fully automated workflow through a website in which images can be loaded as a single zip file. As the maximum size is 5 gigabytes, the images were resized by 50%. Instant-ngp can be downloaded from the provided website, and we used COLMAP for image orientation. Additionally, in this case, compressed images were used in the workflow, knowing that image resizing affects the final results.

Rendering a video was straightforward with both solutions. The visual quality was excellent, especially for areas with vegetation, which is always difficult to model from a set of images and photogrammetric techniques. The background was also visible in the videos, including elements far from the castle. This interesting result could significantly improve traditional rendering from point clouds or meshes. A couple of sample frames from both videos are shown in Figure 19.

Figure 19. Two frames from the rendered videos using Lumia AI (**top**) and Instant-ngp (**bottom**).

5.3. Rocca di Vogogna

A second test with NeRF was carried out for the Rocca di Vogogna dataset. This test aimed to compare results obtained with traditional image-matching techniques, those obtained with NeRF, and a common benchmark represented by a terrestrial laser scanning point cloud. A complete survey of the area for the Rocca di Vogogna is available. It was carried out using a Faro Focus X130 and a Leica BLK360.

For this case study, the comparison is between Metashape and Nerfstudio. The same images were used for both reconstructions: 390 oblique and convergent images. The set of normal images was discarded since it determined some issues with the NeRF training. Default parameters were used in the processing workflows with both software. Once the training in Nerfstudio was completed, the point cloud was exported, setting the target points to 20 million. The point cloud obtained with Metashape has about 35 million points and was then subsampled to 20 million for further comparison.

As previously noticed, also in the case of the Rocca di Vogogna dataset, a visual comparison shows a NeRF point cloud that is more noisy and less detailed than the one obtained with Metashape (Figure 20).

Figure 20. Point clouds produced with Nerfstudio (**left**) and Metashape (**right**) using the same set of images for the Vogogna dataset.

During the comparison with the terrestrial laser scanning data, the absolute distance between the point cloud is considered the test metric, and the laser scanning point cloud is considered the reference one. The computation of the distance between point clouds was performed by using the software CloudCompare (https://www.danielgm.net/cc/ (accessed on 1 August 2023)). Results of the comparison for image matching and NeRF are reported in Figure 21. The color scale is the same in both figures.

No systematic error or discrepancy can be visualized in comparing both point clouds. However, some more significant problems with the NeRFs solution can be observed in correspondence of vegetated areas, which are not detectable visually.

As shown in Figure 21, the discrepancies between image matching and laser scanning data are in the order of 2–3 cm. In particular, 50% of the discrepancies are below 3.6 cm.

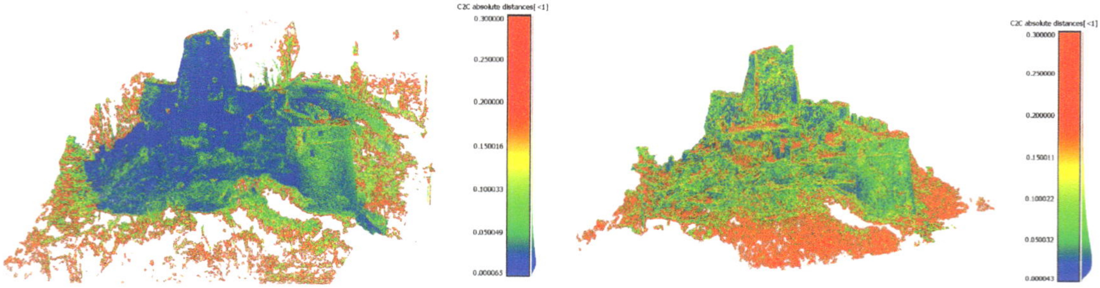

Figure 21. Comparison with TLS data: absolute distance. False color colorized point cloud (**left**) for image matching—Metashape (**left**) and NeRF—Nerfstudio (**right**).

On the other hand (Figure 21-right), discrepancies for the NeRF reconstruction are in the order of magnitude of 4–5 cm, and 50% of the discrepancies are below 4.9 cm. In addition, the distribution of discrepancies for the image matching shows a lower dispersion of this dataset than the NeRF results, as it can be observed both in the density and the cumulative distribution of discrepancies for the two datasets (Figure 22). This is probably related to the higher noise in the NeRF solution.

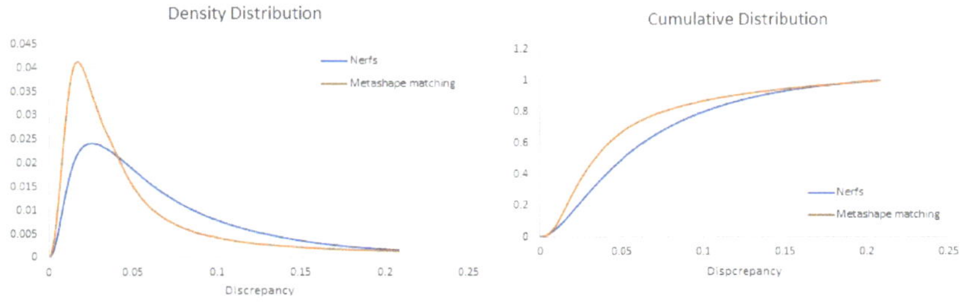

Figure 22. Density (**left**) and cumulative (**right**) distributions of the discrepancies for image matching with Metashape (orange) and the Nerfstudio solution (blue).

6. Discussion

We discussed a workflow using PPK for direct georeferencing after bundle block adjustment for two case studies (Castel Grumello and Castel Mancapane). The advantage of PPK compared to other solutions based on Ground Control Points for the specific case of castles and fortresses is related to the location of such monuments. Indeed, these structures are frequently built on the tops of steep hills, often surrounded by wooded areas. This circumstance makes it challenging, and, at times, not feasible, to identify natural points and set targets to be used as Ground Control Points.

A solution based on the measurement of camera position can be the only alternative. Both RTK and PPK can be used for camera pose measurement. The RTK connection allows for better waypoint navigation in the case of autonomous flight, which can be useful for challenging environments. However, RTK requires an on-field connection between the master and the rover (in this case, the drone) that may be difficult to obtain in mountain areas. In the PPK model, an on-field connection is unnecessary, and data can be post-processed. The baseline between the master and drone should be carefully planned to avoid a long baseline which may end up in low-accuracy corrections and/or may determine the impossibility of determining a fixed solution. For the same reasons, it is important to set up the cut-off angles and the noise-to-signal ratio in the master antenna. In addition, as presented for the Castel Mancapane case study, the solution may remain a float solution in areas with high vegetation, making those positions unusable for georeferencing after bundle block adjustment.

A second topic explored in this paper is the possibility of using NeRFs (specifically the Nerfstudio implementation) as an alternative to traditional image-matching algorithms for geometry reconstruction. While NeRF approaches present a new paradigm in computer graphics, their utilization for the metric reconstruction of large and complex sites is not fully explored. In addition, a reduction of image size (with a factor of 2, 4, and 8) may reduce the final metric quality of the results, especially for the definition of details. In this paper, a first test was conducted comparing the metric accuracy of NeRFs and a laser scanning dataset considered ground truth.

The metric adopted for the comparison is the unsigned distance between the laser scanning and the NeRF point cloud. This metric can provide some insights into the overall quality of the NeRF solution. The aim was to understand if such results can be used for further processing (e.g., production of architectural drawings) needed to define a conservation strategy. The NeRF point cloud has a good overall consistency with the laser scanning point cloud but also shows large noise and metric discrepancies compared to traditional photogrammetric approaches. However, in the future, significant improvements in NeRF models are expected.

Metrics other than unsigned distanced are also worth testing and evaluating. For example, the definition of geometrical features like point normal and eigenvalues are of primary importance for automated classification algorithms. Their consistency in NeRF-based point clouds is fundamental for evaluating their applicability in conservation projects requiring accurate metric documentation. This topic will be studied in future works.

7. Conclusions

The paper presented some projects using recording techniques based on digital photogrammetry with drones. PPK was a powerful solution to overcome traditional limitations for heritage ruins in mountainous areas. Camera poses provided by PPK allow for a reduction (or a total substitution) of GCPs on the ground, although having some GCPs is always helpful, as well as some check points to verify metric accuracy.

One of the main advantages of drone-based photogrammetry is the opportunity to extract dense point clouds and meshes of parts that are difficult to reach with terrestrial surveying methods. The case of heritage ruins considered in this paper (especially fortified constructions on mountains designed and built to be not accessible) is a clear example where drones help capture external elevations, the remains of tall structures, and the local topography.

Advanced digital surveying techniques can drive the definition of more reliable 3D geometrical models of the built heritage, which is a fundamental step for elaborating the analyses characterizing the conservation process. The digital model provides the common 2D graphic layout for describing the ruined building from the geometrical, technological, and conservation points of view. Moreover, according to the level of complexity of the site, the management of the conservation plan can be structured through HBIM models, providing an interoperable tool for those subjects involved in the preservation process, such as archaeologists, architects, engineers, and stakeholders.

Creating a detailed digital record cannot be carried out only using drones. Laser scanning, terrestrial photogrammetry, and their integrated use with GNSS and total stations are often necessary for the complete acquisition of the ruins. In addition, point clouds and meshes can rarely be considered the final deliverables for restoration and conservation, which usually require additional outputs such as orthophotos, technical drawings (CAD with plans, sections, elevations), GIS, or 3D models from modeling environments allowing further processing for designing interventions. Most of such deliverables still need manual work by specialists, which can use the point clouds/meshes as a reference and produce additional deliverables. Interpreting the different constructive elements (and their separation in the modeling phase), materials, and condition assessment is still conducted manually for complex projects.

The manuscript also considered the case of novel approaches based on NeRFs. Two datasets were also processed with Nerfstudio starting from the initial orientation results achieved with Metashape. Rendering provided very satisfactory results and excellent outputs for complex scenes with vegetation. The extraction of 3D geometry (point clouds in the proposed examples) did not reach the same quality (in terms of metric accuracy and point density) as traditional photogrammetric outputs. However, the rapid development of this relatively new technology is surely expected, and the application of NeRFs in heritage documentation could provide novel alternative processing methods.

Author Contributions: Conceptualization, methodology, formal analysis, and writing—all authors; software, investigation and validation, L.B. and M.P. All authors have read and agreed to the published version of the manuscript.

Funding: This research received no external funding.

Data Availability Statement: Data sharing is not applicable to this article.

Acknowledgments: The authors would like to thank the different institutions that provided access to the three sites: FAI (Fondo per l'Ambiente Italiano), Comune di Montagna in Valtellina, and Comune di Vogogna.

Conflicts of Interest: The authors declare no conflict of interest.

References

1. Remondino, F.; Nocerino, E.; Toschi, I.; Menna, F. A critical review of automated photogrammetric processing of large datasets. *Int. Arch. Photogramm. Remote Sens. Spat. Inf. Sci.* **2017**, *XLII-2/W5*, 591–599. [CrossRef]
2. Hsieh, C.S.; Hsiao, D.H.; Lin, D.Y. Contour Mission Flight Planning of UAV for Photogrammetric in Hillside Areas. *Appl. Sci.* **2023**, *13*, 7666. [CrossRef]
3. Tomaštík, J.; Mokroš, M.; Surový, P.; Grznárová, A.; Merganič, J. UAV RTK/PPK Method—An Optimal Solution for Mapping Inaccessible Forested Areas? *Remote Sens.* **2019**, *11*, 721. [CrossRef]
4. Mildenhall, B.; Srinivasan, P.P.; Tancik, M.; Barron, J.T.; Ramamoorthi, R.; Ng, R. NeRF: Representing Scenes as Neural Radiance Fields for View Synthesis. *Commun. ACM* **2020**, *65*, 96–106. [CrossRef]
5. Hooja, R.; Jain, S. *Conserving Fortified Heritage: The Proceedings of the 1st International Conference on Fortifications and World Heritage, New Delhi, 2015*; Cambridge Scholars Publishing: Cambridge, UK, 2016.
6. Tancik, M.; Weber, E.; Ng, E.; Li, R.; Yi, B.; Kerr, J.; Wang, T.; Kristoffersen, A.; Austin, J.; Salahi, K.; et al. Nerfstudio: A Modular Framework for Neural Radiance Field Development. In Proceedings of the ACM SIGGRAPH 2023 Conference Proceedings—SIGGRAPH'23, Los Angeles, CA, USA, 6–10 August 2023.
7. Fassi, F.; Achille, C.; Fregonese, L. Surveying and modelling the main spire of Milan Cathedral using multiple data sources. *Photogramm. Rec.* **2011**, *26*, 462–487. [CrossRef]
8. Oreni, D.; Brumana, R.; Della Torre, S.; Banfi, F.; Barazzetti, L.; Previtali, M. Survey turned into HBIM: The restoration and the work involved concerning the Basilica di Collemaggio after the earthquake (L'Aquila). *ISPRS Ann. Photogramm. Remote Sens. Spat. Inf. Sci.* **2014**, *II-5*, 267–273. [CrossRef]
9. Reina Ortiz, M.; Weigert, A.; Dhanda, A.; Yang, C.; Smith, K.; Min, A.; Gyi, M.; Su, S.; Fai, S.; Santana Quintero, M. A theoretical framework for multi-scale documentation of decorated surface. *Int. Arch. Photogramm. Remote Sens. Spat. Inf. Sci.* **2019**, *XLII-2/W15*, 973–980. [CrossRef]
10. Cabarcas Granados, M.M.; Hamp, E.; Santana Quintero, M.; Reina Ortiz, M.; Montejo Gaitán, F.; Leguizamón, L.P. Digitizing and documenting heritage for conservation, a case study: Chiribiquete national park archaeological site. *Int. Arch. Photogramm. Remote Sens. Spat. Inf. Sci.* **2023**, *XLVIII-M-2-2023*, 341–347. [CrossRef]
11. Bruno, N.; Roncella, R. A restoration oriented hbim system for cultural heritage documentation: The case study of parma cathedral. *Int. Arch. Photogramm. Remote Sens. Spat. Inf. Sci.* **2018**, *XLII-2*, 171–178. [CrossRef]
12. Barazzetti, L.; Mezzino, D.; Santana Quintero, M. Digital workflow for the conservation of bahrain built heritage: The sheik isa bin ali house. *Int. Arch. Photogramm. Remote Sens. Spat. Inf. Sci.* **2017**, *XLII-2/W5*, 65–70. [CrossRef]
13. Garramone, M.; Jovanovic, D.; Oreni, D.; Barazzetti, L.; Previtali, M.; Roncoroni, F.; Mandelli, A.; Scaioni, M. Basilica di san giacomo in como (italy): Drawings and hbim to manage archeological, conservative and structural activities. *Int. Arch. Photogramm. Remote Sens. Spat. Inf. Sci.* **2023**, *XLVIII-M-2-2023*, 653–660. [CrossRef]
14. Cardaci, A.; Versaci, A.; Azzola, P. The palazzo dell'ateneo in the upper city of bergamo (città alta): New documentation and conservation studies. *Int. Arch. Photogramm. Remote Sens. Spat. Inf. Sci.* **2023**, *XLVIII-M-2-2023*, 365–372. [CrossRef]
15. D'Agostino, P.; Antuono, G.; Elefante, E.; Amore, R. Digital management for the restoration project. The case of the temple of venus in baia. *Int. Arch. Photogramm. Remote Sens. Spat. Inf. Sci.* **2023**, *XLVIII-M-2-2023*, 461–471. [CrossRef]
16. Bonora, V.; Meucci, A.; Conti, A.; Fiorini, L.; Tucci, G. Knowledge representation of built heritage mapping an ad hoc data model in ogc standards: The case study of pitti palace in florence, italy. *Int. Arch. Photogramm. Remote Sens. Spat. Inf. Sci.* **2023**, *XLVIII-M-2-2023*, 281–288. [CrossRef]
17. Koehl, M.; Steiner, V.; Guillemin, S.; Degenève, F.; Zabollone, A.; Bignon, I.; Taufflieb, C.; Tisserand, L.; Hedtmann, L. 3D and hbim models: Digital tools for the diagnostic study of the stair turret of the south-east corner of the main tower of strasbourg cathedral. *Int. Arch. Photogramm. Remote Sens. Spat. Inf. Sci.* **2023**, *XLVIII-M-2-2023*, 871–878. [CrossRef]
18. Prizeman, O.E.C.; Davis, J.; Tam, L. Digitisation of retreating industrial heritage; modelling the decommissioning of the coal washeries of onllwyn. *Int. Arch. Photogramm. Remote Sens. Spat. Inf. Sci.* **2023**, *XLVIII-M-2-2023*, 1251–1260. [CrossRef]
19. Tirinzoni, S. *Restauro del Castello di Mancapane, Progetto Esecutivo*; Relazione: Milano, Italy, 1998.
20. Bascapè, G.C.; Sertoli, R.; Perogalli, C. *Torri e Castelli di Valtellina e Valchiavenna*; Edizioni Banca Piccolo Credito Valtellinese: Sondrio, Italy, 1966.
21. Tirinzoni, S. Castello de Piro al Grumello. Fai Fondo per l'Ambiente Italiano, Indagini e scavo archeologico. Notiziario archeologico 2001–2002 della Soprintendenza per i Beni Archeologici della Lombardia, 135–136.
22. Babbini, M. Indagini di archeologia dell'architettura. La Rocca di Vogogna (VB). *Quad. Della Soprintend. Archeol. Del Piemonte* **2014**, *29*, 61–80
23. Airoldi, A. *Storia di Vogogna. Il borgo, Grossi, Domodossola,* 1992; Volume 1.
24. Luhmann, T. *Close Range Photogrammetry: Principles, Techniques and Applications*; Whittles Pub: Dunbeath, UK, 2006.
25. Hirschmuller, H. Stereo processing by semiglobal matching and mutual information. *IEEE Trans. Pattern Anal. Mach. Intell.* **2007**, *30*, 328–341. [CrossRef]
26. Kraus, K. *Photogrammetry: Geometry from Images and Laser Scans*; Walter de Gruyter: Berlin/Heidelberg, Germany, 2011.
27. Liu, S.; Chen, W.; Li, T.; Li, H. Soft rasterizer: Differentiable rendering for unsupervised single-view mesh reconstruction. *arXiv* **2019**, arXiv:1901.05567.

28. Gao, K.; Gao, Y.; He, H.; Lu, D.; Xu, L.; Li, J. NeRF: Neural Radiance Field in 3D Vision, A Comprehensive Review. *arXiv* **2023**, arXiv:2210.00379.
29. Debbagh, M. Neural Radiance Fields (NeRFs): A Review and Some Recent Developments. *arXiv* **2023**, arXiv:2305.00375.
30. Zhu, F.; Guo, S.; Song, L.; Xu, K.; Hu, J. Deep Review and Analysis of Recent NeRFs. *Apsipa Trans. Signal Inf. Process.* **2023**, *12*, e6. [CrossRef]
31. Croce, V.; Caroti, G.; Luca, L.; Piemonte, A.; Véron, P. Neural radiance fields (nerf): Review and potential applications to digital cultural heritage. *Int. Arch. Photogramm. Remote Sens. Spat. Inf. Sci.* **2023**, *XLVIII-M-2-2023*, 453–460. [CrossRef]
32. Balloni, E.; Gorgoglione, L.; Paolanti, M.; Mancini, A.; Pierdicca, R. Few shot photogrametry: A comparison between nerf and mvs-sfm for the documentation of cultural heritage. *Int. Arch. Photogramm. Remote Sens. Spat. Inf. Sci.* **2023**, *XLVIII-M-2-2023*, 155–162. [CrossRef]
33. Mazzacca, G.; Karami, A.; Rigon, S.; Farella, E.M.; Trybala, P.; Remondino, F. Nerf for heritage 3D reconstruction. *Int. Arch. Photogramm. Remote Sens. Spat. Inf. Sci.* **2023**, *XLVIII-M-2-2023*, 1051–1058. [CrossRef]
34. Murtiyoso, A.; Grussenmeyer, P. Initial assessment on the use of state-of-the-art nerf neural network 3D reconstruction for heritage documentation. *Int. Arch. Photogramm. Remote Sens. Spat. Inf. Sci.* **2023**, *XLVIII-M-2-2023*, 1113–1118. [CrossRef]
35. Condorelli, F.; Rinaudo, F.; Salvadore, F.; Tagliaventi, S. A comparison between 3D reconstruction using nerf neural networks and mvs algorithms on cultural heritage images. *Int. Arch. Photogramm. Remote Sens. Spat. Inf. Sci.* **2021**, *XLIII-B2-2021*, 565–570. [CrossRef]
36. Li, Z.; Müller, T.; Evans, A.; Taylor, R.H.; Unberath, M.; Liu, M.Y.; Lin, C.H. Neuralangelo: High-Fidelity Neural Surface Reconstruction. In Proceedings of the IEEE Conference on Computer Vision and Pattern Recognition (CVPR), Vancouver, BC, Canada, 20–22 June 2023.
37. Turki, H.; Ramanan, D.; Satyanarayanan, M. Mega-NERF: Scalable Construction of Large-Scale NeRFs for Virtual Fly-Throughs. In Proceedings of the IEEE/CVF Conference on Computer Vision and Pattern Recognition (CVPR), New Orleans, LA, USA, 18–24 June 2022; pp. 12922–12931.

drones

MDPI

Article

Multi-Sensor 3D Survey: Aerial and Terrestrial Data Fusion and 3D Modeling Applied to a Complex Historic Architecture at Risk

Marco Roggero [1,*] and Filippo Diara [2]

1 Department of Architecture and Design (DAD), Polytechnic of Turin, Viale P.A. Mattioli 39, 10125 Turin, Italy
2 Historical Studies Department, University of Turin, Via Sant'Ottavio 20, 10124 Turin, Italy; filippo.diara@unito.it
* Correspondence: marco.roggero@polito.it

Abstract: This work is inscribed into a more comprehensive project related to the architectural requalification and restoration of Frinco Castle, one of the most significant fortified medieval structures in the Monferrato area (province of Asti, Italy), that experienced a structural collapse. In particular, this manuscript focuses on data fusion of multi-sensor acquisitions of metric surveys for 3D documenting this structural-risky building. The structural collapse made the entire south front fragile. The metric survey was performed by using terrestrial and aerial sensors to reach every area of the building. Topographically oriented Terrestrial Laser Scans (TLS) data were collected for the exterior and interior of the building, along with the DJI Zenmuse L1 Airborne Laser Scans (ALS) and Zenmuse P1 Photogrammetric Point Cloud (APC). First, the internal alignment in the TLS data set was verified, followed by the intra-technique alignments, choosing TLS as the reference data set. The point clouds from each sensor were analyzed by computing voxel-based point density and roughness, then segmented, aligned, and fused. 3D acquisitions and segmentation processes were fundamental for having a complete and structured dataset of almost every outdoor and indoor area of the castle. The collected metrics data was the starting point for the modeling phase to prepare 2D and 3D outputs fundamental for the restoration process.

Keywords: 3D survey; data fusion; LiDAR; photogrammetry; TLS; UAS

Citation: Roggero, M.; Diara, F. Multi-Sensor 3D Survey: Aerial and Terrestrial Data Fusion and 3D Modeling Applied to a Complex Historic Architecture at Risk. *Drones* **2024**, *8*, 162. https://doi.org/10.3390/drones8040162

Academic Editor: Pablo Rodríguez-Gonzálvez

Received: 20 February 2024
Revised: 6 April 2024
Accepted: 17 April 2024
Published: 19 April 2024

1. Introduction

In complex and risky architectural contexts, multi-sensor data fusion overcomes the limitations of a single sensor and the time constraints for an investigation. It is a must-have solution for outdoor and indoor surveying, as well as for surveying roofs and surroundings. This study discusses the topic of the fusion of point clouds from different sensors and methodologies [1–3], optimally combining LiDAR scans both from terrestrial (4.5 billion points) and Unmanned Aerial Systems (UAS, 200 million points), as well as photogrammetric point clouds (350 million points), to obtain an improved and complete 3D model of a complex historic building at risk.

The analyzed building is Frinco Castle, as shown in Figure 1 (province of Asti, Italy). Despite its imposing medieval defensive structure, it suffered different structural weaknesses and collapsed—combining several buildings throughout the ages contributed to its architectural complexity.

Surveying Frinco Castle required different surveying techniques due to the morphological complexity of the building. The integrated GNSS/total station control network was set up around the castle, and many compound traverses were used to connect the inside survey to the external reference. A complete 3D scan of the castle was acquired by integrating the outside/inside terrestrial laser scanner (TLS) survey with UAV (Unmanned Aerial Vehicle)/UAS (Unmanned Aircraft System) flights.

Figure 1. Frinco Castle from the UAS P1 sensor. Overview of the southwest side and the collapsed area.

The collected data generated from different sensors and methodologies represent the castle's global and punctual metric documentation. However, digital 2D and 3D outputs for the restoration process required a complete and homogeneous overview of point clouds. For this reason, the main goal of this work focuses on metric data refinement, alignment, fusion (from terrestrial and aerial sensors) and metric evaluation of these processes.

The segmentation and classification steps were also carried out in the indoor scans, where different object classes (floor, vault, wall, stairs, and wooden structures) can be selected and performed manually, automatically, or in an algorithm-assisted manual mode. The vaulted rooms related to the collapsed area, also surveyed by UAS, experienced mixed segmentation and alignment.

Finally, the fused metric data constituted the backbone for a parametric modeling design: in fact, initial and partial parametric modeling was performed for wooden beams on the rooftop (especially for the vaulted rooms of the collapsed area); then, NURBS modeling was conducted for having a lightweight model on which mapping stratigraphic features of the masonries and generating detailed photographic renders.

The Case Study: The Medieval Castle of Frinco

The surveyed context is related to the defensive medieval structure in Frinco municipality (AT-Italy). The castle is a fragile building resulting from architectural modifications since the 13th century [4–7]. In fact, architectural extensions and renovations (indoor and outdoor) have created a complex maze that is difficult to decode. The recent stratigraphic analysis [8] confirmed the initial nucleus of tower houses that experienced merging operations during the time. Architectural elements such as bichrome pointed arches, single lancet windows, and the Ghibelline decorations (dovetail battlements) represent important chronological markers related to a period between the 13th and the 15th centuries [8]. Though the medieval period is well-attested in the overall building stratigraphy, the most detectable phase is related to a chronological range between the latter half of the 18th century and the end of the 19th century [8]. The stratigraphic analysis was fundamental for understanding the evolution of the construction site as well as for having a detailed overview of fragile masonry units regarding the outlook for the reconstruction. In fact, a collapse in 2011 endangered the security of the castle and the houses beneath it. A large chunk of the manor collapsed into the populated area in 2014, flooding the churchyard and the public road and encircling the residences. The municipality bought the castle in 2019 and started consolidation and restoration projects in 2020. For the rehabilitation project, the municipality of Frinco commissioned the castle's topographical and architectural 3D surveys in 2021.

2. Metric Survey and Quality Check

2.1. Surveying Equipment

The control network traverse surveying and the ground 3D scans were performed by the Trimble SX10 scanning station [9], with the advantage of handling a unique device on the field instead of operating a laser scanner and a total station. This instrument combines the topographic survey with LiDAR acquisitions, allowing a direct topographic point cloud orientation and avoiding scan registration at the office. It reaches an accuracy of 1 mm + 1.5 ppm with prism and 2 mm + 1.5 ppm di DR mode.

The aerial survey was operated by DJI Matrice 300 RTK equipped with DJI Zenmuse L1 LiDAR and P1 photogrammetric sensors [10]. An exhaustive comparison of terrestrial and aerial point clouds related to Frinco Castle 3D survey can be found in previous work [11]. Also, the reliability and accuracy of the UAS systems [12–14] and scanning stations [15] have been demonstrated by similar analysis.

2.2. Working Planning

Spanning 70 by 40 m and standing 30 m tall from its base to the rooftop, Frinco Castle is a grand structure comprising four primary levels and several intermediate floors. Given the building's intricate design and substantial size, the metric survey was carried out using the Trimble SX10 scanning station. This integrated total station allowed us to benefit from the direct alignment of the scanning stations via traverse adjustment, thereby enabling a seamless connection between the indoor and outdoor point clouds.

A closed traverse was established and examined to serve as a control network composed of eight GNSS points, which were then linked via a high-precision topographic survey (Figure 2). The Control Points (CPs), positioned in redundant numbers around the building area and used as reference points for all subsequent survey work, are stainless steel blocks stamped with the point's name.

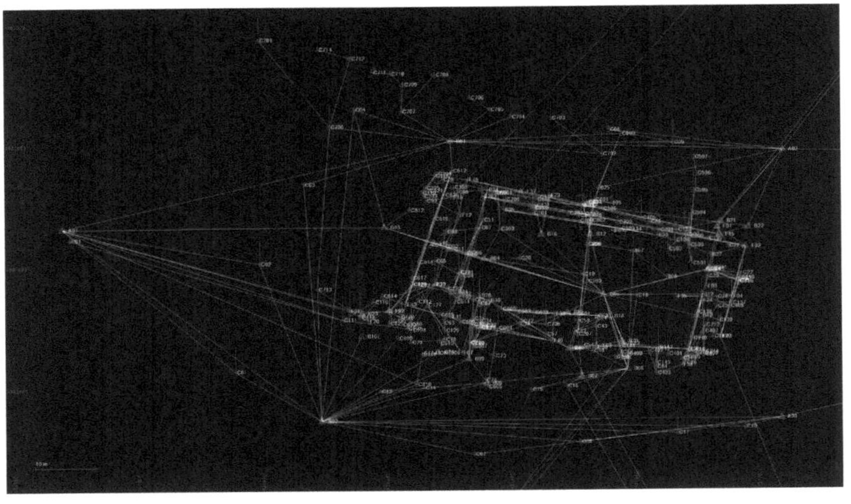

Figure 2. The Frinco Castle control network (Green) and CP error ellipses (Red).

Given the average dimensions of the castle's rooms, the expected scan density in standard mode indoors is one point every 2–6 mm, which is sufficient for the architectural survey's deliverables. The full-dome scan in coarse mode, coupled with the panorama capture, takes approximately 15 min per scan station. Each scan station necessitates about 20–30 min, factoring in the landmarks set up by stainless steel studs, equipment relocation along a traverse, and station setup. Connecting an indoor traverse to the outdoor control network extends the survey duration due to the lengthy path required to transport the equipment from inside to outside and vice versa. Accessing some rooms, especially in the basement and attic, was occasionally challenging. The need to acquire more than 300 scans led to a 32-day survey campaign, meticulously planned using a Gantt Chart. During this campaign, 25 days were allocated explicitly for capturing scans, with an average of 12 scan stations per day. An additional seven days were required for other tasks: the aerial photogrammetric and LiDAR survey, the manual verification and photographic documentation of openings, doors, windows, and floors, and the photographic survey of the facades to produce high-resolution orthophoto deliverables.

2.3. Control Network Adjustment

The user can control SX10 with the Trimble Access™ (TA) field software (version 2023.00) operating on the Trimble Tablet Rugged PC. The tablet can communicate with the SX10 via radio, Bluetooth, or Wi-Fi. During scan acquisition, WIFI communication is mandatory due to the large amount of data to be transmitted, which requires a high bit rate. Optical, leveling, and GNSS data can be adjusted and processed in the office by Trimble Business CenterTM (TBC) software (version 2023.10). All Trimble software is available on the official website [9].

Throughout a 32-day campaign, the survey work was executed, sequentially adjusting the control network and various indoor traverses according to the following general workflow:

1. Data collection and surveying using Trimble Access™ (TA)
2. Exportation of *.job file and subsequent importation into Trimble Business Center™ (TBC)
3. Establishment of control points and execution of network adjustment
4. Optional exportation of scans to Trimble Realworks™ (version 12.0)
5. Repetition of steps 1–4 until the completion of the survey.

To check the adjustment quality, the final adjusted coordinates were extracted and RMS from TBC adjustment reports. Table 1 notices RMSs for each surveyed area: they are minimal outdoors, while much higher values are reported indoors due to often tricky operating conditions, such as confined spaces or nonstable floors, but above all due to the impossibility of closing some traverse.

Table 1. Control points RMSs of the complete control network.

		CP Number	Mean		Max	
			σ_{hor}	σ_{vert}	σ_{hor}	σ_{vert}
Outdoor		15	0.005	0.002	0.009	0.003
Indoor	Basement	110	0.011	0.003	0.095	0.008
	Level 1	52	0.008	0.003	0.022	0.008
	Level 2	45	0.007	0.003	0.022	0.009
	Garret	45	0.015	0.002	0.044	0.004

Two sections of the structure, an underground icebox (12 CP) and a portion of the garret (13 CP) were surveyed by georeferencing the scans via open traverses. Upon excluding these two open traverses from the statistical analysis, the horizontal mean RMS values were found to be less than 10 mm, and the vertical mean RMS values were less than 3 mm. This data can be noticed in Table 2.

Table 2. Control points RMSs of the close traverse control network.

		CP Number	Mean		Max	
			σ_{hor}	σ_{vert}	σ_{hor}	σ_{vert}
Outdoor		15	0.005	0.002	0.009	0.003
Indoor	Basement	98	0.008	0.003	0.022	0.008
	Level 1	52	0.008	0.003	0.022	0.008
	Level 2	45	0.007	0.003	0.022	0.009
	Garret	32	0.010	0.002	0.030	0.004

3. Point Cloud Alignment Methods Outdoor/Indoor

Cloud alignment can be achieved by setting standard targets in the scanned scene, which are typically spheres or reflective adhesive targets used to merge point clouds from adjacent stations. However, when standard targets are impossible or time-consuming, point cloud registration necessitates a cloud-to-cloud alignment that often employs a coarse-to-fine registration strategy. The initial registration parameters for the rigid body transformation of two-point clouds are estimated using a feature-based method. These features can encompass point features, building corners, flat walls or floors, pipes, and other similar site features. Surface feature-based methods are less susceptible to noise, but they demand substantial overlapping areas, with at least three pairs of surfaces needing to be present in the clouds to be registered.

In the fine registration phase, the primary goal is to maximize the overlap of two-point clouds, which is primarily achieved using the Iterative Closest Point method (ICP), RANdom SAmple Consensus method (RANSAC), Normal Distribution Transform method (NDT), or methods that utilize auxiliary data, such as GNSS scan position or standard targets. The most widespread process is the iterative approximation, which primarily refers to the ICP algorithm and a series of enhanced algorithms. However, this method presumes a reasonable estimation of the initial scan location and necessitates a high overlap between the two point clouds.

In cloud-to-cloud alignment, even using standard targets, one cloud is set as a reference, and one is moved to perform a pairwise registration. This causes the movable scan stations to align with the reference scan station, resulting in a station group that can be further refined if necessary. The aligned scan station moves to the reference scan group, and other scans are then aligned and sequentially moved to the reference group. Finally, the overall registration refinement can be performed. The method can be time-consuming and lead to misalignments, due to alignment error propagation, especially in indoor environments and with many scans. The alignment errors can be reduced by using known targets and determining their coordinates by conventional surveying techniques. In any case, aligning the scans without a topographic control network does not seem convenient, not even for small buildings. The control-network adjustment accuracy is undoubtedly related to the scan alignment accuracy. The workflow focused on the following steps and issues:

- Methods for checking the scan alignment;
- Possibility of checking and refining the scan direct orientation by scan-to-scan alignment;
- Accuracy checks for direct scan alignment for the purposes of surveying the architectural context.

Scan Alignment Check

The outcomes of the cloud-to-cloud alignment encompass the residual error and the percentage of overlap, which can be evaluated against predetermined tolerances. However, such a tool is not available in direct topographic scan orientation. The project cloud can be segmented into slices, and the scan alignment can be managed by visually inspecting each slice. The visual inspection suffices for standard applications in conjunction with an accuracy check of the control-network adjustment (Figure 3).

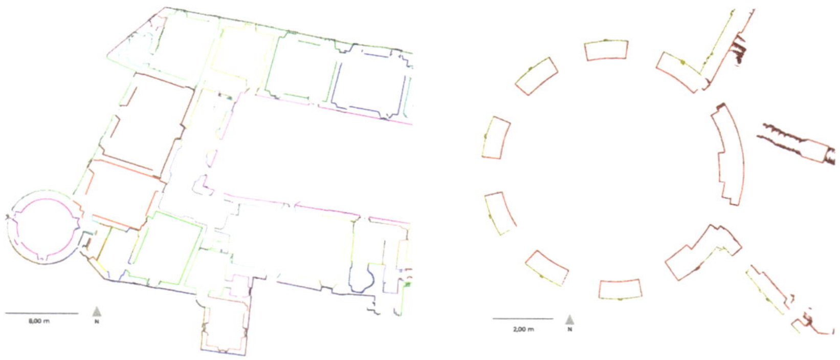

(a) (b)

Figure 3. Visual alignment check on a horizontal slice (**a**). Visual indoor/outdoor point cloud alignment check on a horizontal slice. Detail of the southwest circular tower (**b**). Each scan is in a different color.

Another method for assessing misalignments is the cloud-to-cloud analysis, which is typically applied outdoors when substantial overlap between scans exists. This tool is incorporated in CloudCompare [16]. It has been used with the following parameters:

- Octree level: Auto;
- Maximum distance: 0.02 m (larger distances are attributed to non-overlapping cloud areas);
- Local model: quadric (in order to reduce the noise effect);
- Points (kNN): 6.

The outdoor survey comprised 20 scans, and eight pairs of well-overlapping clouds were analyzed, both prior to and following an ICP-enhanced alignment. Cloud-to-cloud distances were subdivided into classes, computing for each class the statistics:

$$f_i = \frac{n_i}{\max(n_i)}$$

$$F_i = \frac{\sum_{j=1}^{i} n_j}{N}$$

where f_i and F_i are the relative frequency and cumulative frequency, n_i is the number of points inside a class and N is the total number of points. Note that the relative frequency f_i is normalized by the maximum number of points inside a class. The function maximum is always 1 to quickly compare the maximum value position with other point cloud pairs, as in Figure 4.

The worst alignment was detected in the couple C05-A02. The computed statistic f_i presents a maximum around the distance of 2.5 mm, representing the misalignment mean error, a misalignment that, from a practical point of view, is negligible, among other things, cannot be noticed in a visual inspection. Note that:

- The horizontal coordinates of C05 were, however, determined with an accuracy of 4 mm with respect to A02, and the vertical coordinate with an accuracy of 2 mm;
- Station C05 has been oriented on B01 and not on A02, so the two stations are not directly connected;
- A station orientation error can lead to cloud-to-cloud misalignment, even if the coordinates of the instrumental origin are known without errors.

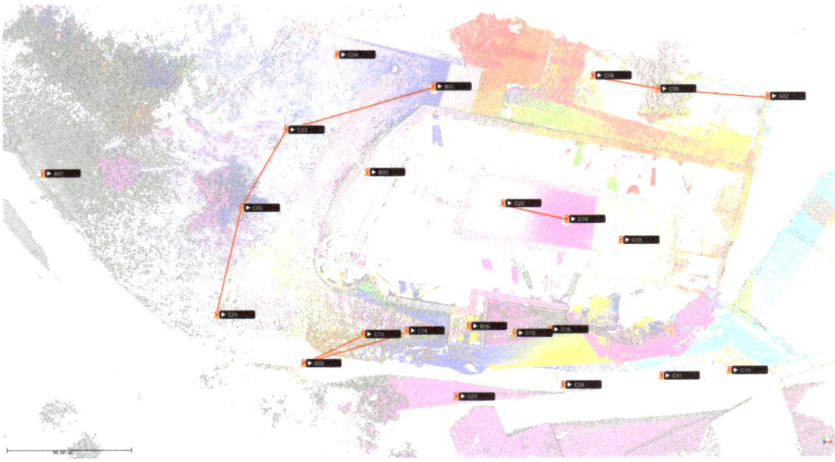

Figure 4. 3D scans outside the castle. The connections in red highlight the eight scan couple alignments verified in this work.

In Figure 5, the statistics f_i and F_i before alignment refinement are represented in red, and after in green.

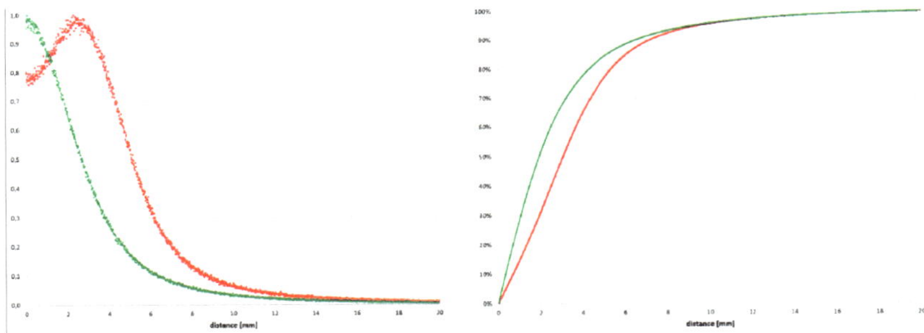

(**a**) (**b**)

Figure 5. (**a**) f_i statistic. (**b**) F_i statistic. Red: Before alignment refinement. Green: After ICP alignment refinement.

The statistic F_i shows that 95% of the cloud-to-cloud distances are under 10 mm. It also shows that before the alignment refinement, only 65% were under 4 mm, and after, it increased to 78%. Plotting the three components X, Y, and Z, of the cloud-to-cloud distance before the alignment, a shift can be observed around 2 mm in Y, while X and Z have zero mean. The shifts estimated by ICP are −1.1, 16.5, and −0.3 mm, according to the shifts observed in the cloud-to-cloud distances (Figure 6).

Similar statistics were computed over the ten examined scan couples before and after the alignment refinement. To understand the impact of the ICP refinement, here the differences in the cloud-to-cloud distance, both in f_i and mostly in F_i: f_i maximum shifts to zero distance, while F_i must necessarily increase on the left side of the curve to have an alignment refinement (Figures 7–9).

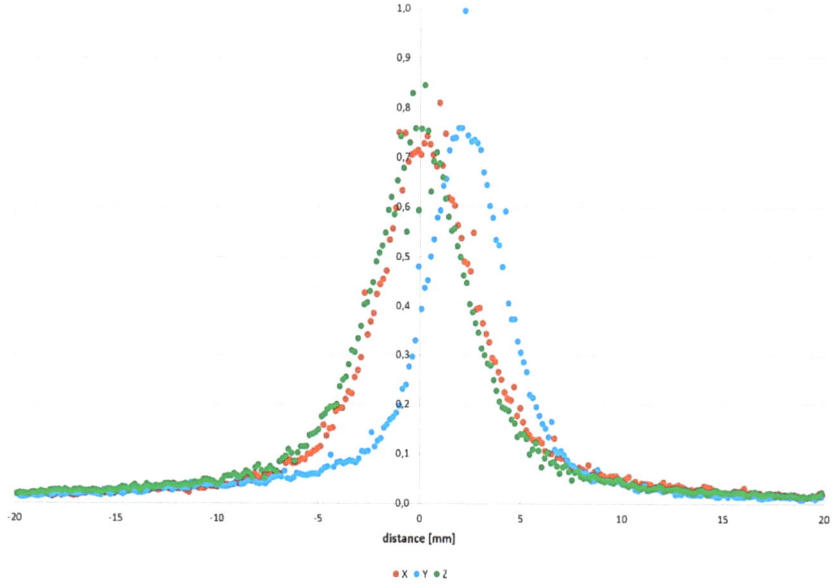

Figure 6. Cloud-to-cloud distance components.

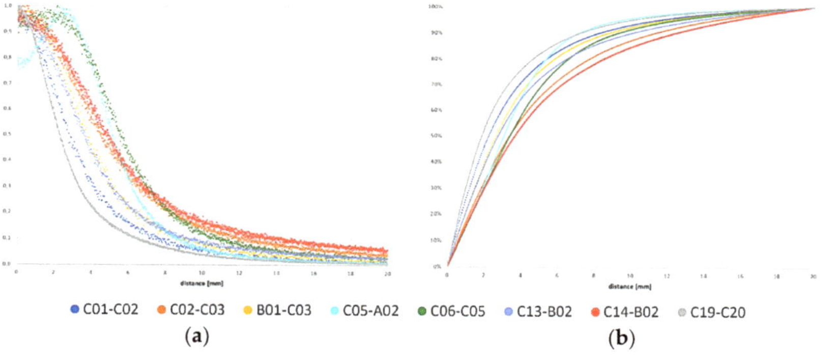

Figure 7. Statistics of the scan couples before the alignment refinement. (**a**) f_i, (**b**) F_i.

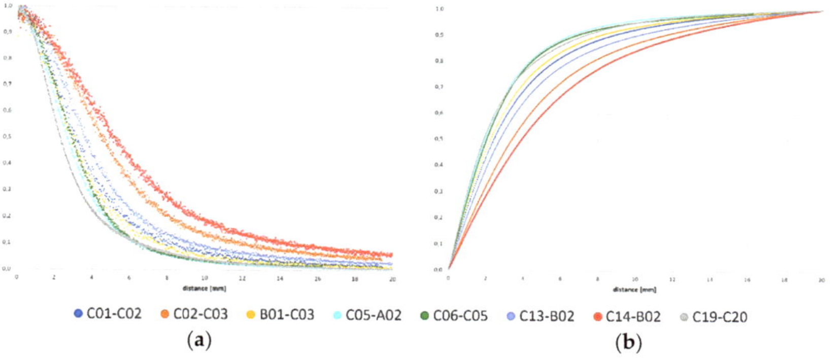

Figure 8. Statistics of the scan couples after the alignment refinement. (**a**) f_i, (**b**) F_i.

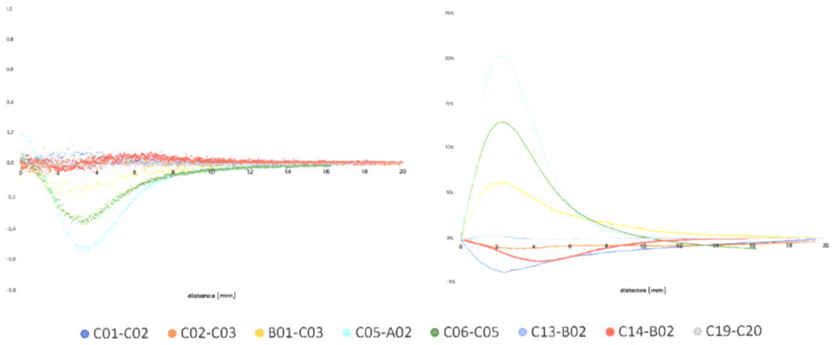

(a) (b)

Figure 9. Differences in the statistics of the scan couples after and before the alignment refinement. (**a**) f_i after ICP minus f_i before and (**b**) F_i after ICP minus F_i before.

Surprisingly, there was no improvement in alignment. This only happened for the couples C05-A02, C06-C05, and B01-C03. There is practically no variation for couples C13-B02 and C19-C20, while a slight worsening was seen for couples C02-C03, C14-B02, and C01-C02. So, cloud-to-cloud alignment can improve scan orientation in case of misalignments. Still, its application is not always convenient when an excellent topographic scan orientation is performed.

Predominantly in indoor surveying scenarios, the registration from one scan to another proves unfeasible due to the exiguous or non-existent overlapping regions. Scan stations are typically situated near the room's center and are interconnected via a topographic station, commonly located in the passageway that links the rooms. The existence of frames, doors, and windows introduces complexities; when left open, they hinder the visibility of wall sections, and when closed, they preclude the complete scanning of neighboring rooms (Figure 10).

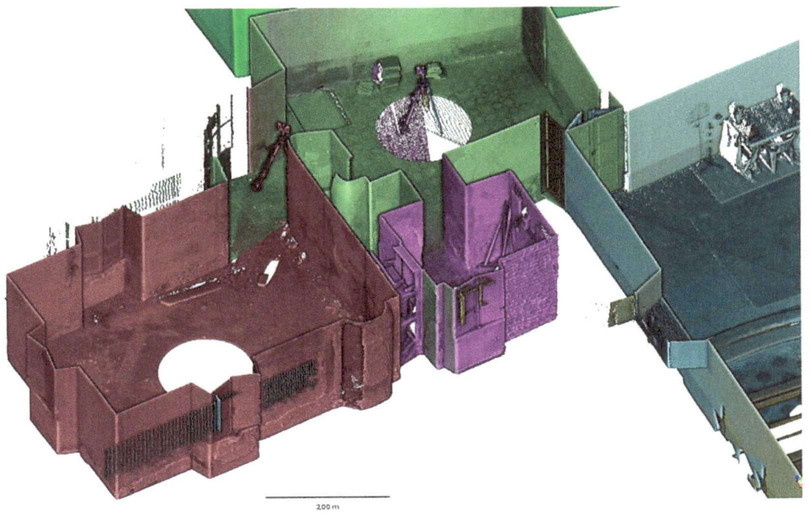

Figure 10. View of a portion of the indoor survey of Frinco Castle. Each scan is in a different color. A minimal overlap between adjacent scans can be observed.

Another common situation is the survey of buildings with multiple overlapping floors. In these instances, the intrados and extrados of the vaults were surveyed to gauge their thickness (Figure 11). The linkage of the scans is exclusively facilitated by topographic measurements, occasionally necessitating extensive traverses to establish a connection with the external control network. Logistically, linking the indoor survey to the external control network can be lengthy, often requiring additional personnel to relocate tripods and prisms outdoors. To streamline this process, the team planned to anticipate the need for outdoor relocation of tripods during the scanning phase, maintaining clear and continuous communication channels between indoors and outdoors.

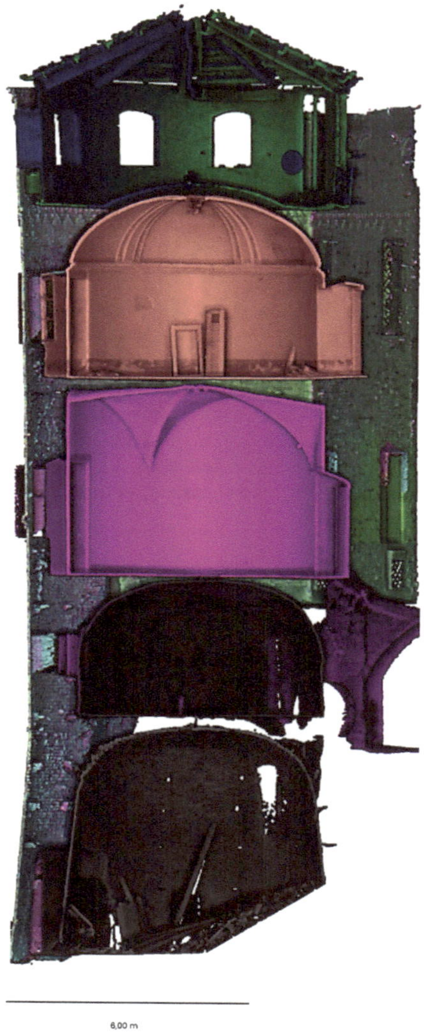

Figure 11. View of a portion of the outdoor and indoor survey of Frinco Castle. Vertical section of the circular tower at the southwest corner of the castle. Each scan is in a different color.

Common features and patches can be used for indoor scan alignment, or the control network adjustment can be checked in other cases. To maximize the possibility of having common entities between indoor and outdoor scans, it is preferable not to set a limit on the scanning distance during the acquisition or when importing the scan data. An estimate of the cloud-to-cloud distance will be helpful for understanding if common areas are present among scans large enough to provide a meaningful statistic. The scanner can acquire external areas through an open window, for example.

Then, the following control procedure, based on the use of shared features between indoor and outdoor scans, was adopted to check the alignment of a single indoor scan (Figure 12):

1. Exclude the indoor portion from the indoor scan. When combining indoor and outdoor scans, alignment and registration become critical. Excluding the indoor part simplifies this process, reducing the number of variables to consider and focusing on the relevant data.
2. Restrict the outdoor point cloud to the indoor bounding box.
3. Approximatively restrict the outdoor point cloud to the overlapping area, if any.
4. Compute the indoor/outdoor cloud-to-cloud distance, taking the indoor one as the reference cloud.
5. The points with a cloud-to-cloud distance larger than 0.05 m are excluded from the outdoor cloud to extract the common patch.
6. Compute the indoor/outdoor cloud-to-cloud distance, taking the outdoor one as the reference cloud.
7. Filter the outdoor cloud by distance scalar field, selecting the range (0.00–0.02 m) to be used for the statistical analysis.
8. Compute the cloud-to-cloud distance distribution and the f_i and F_i statistics.

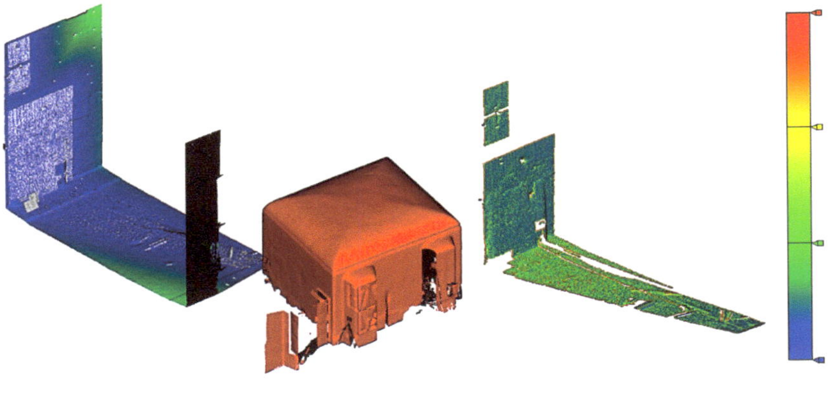

(**a**) (**b**)

Figure 12. Alignment check of an indoor scan, where the outdoor part overlaps with an outdoor scan. (**a**) The indoor part, in Red, is eliminated, and the cloud-to-cloud distance is computed on the overlapping outdoor part (steps 1–2–3) (**b**) Cloud-to-cloud distance of the overlapping part (steps 4–5). The Blue–Green–Yellow–Red color scale represents the distances from the minimum (Blue) to the maximum value (Red).

If the statistic f_i, representing a measure of scan alignment quality, shows a non-zero maximum different from zero, a misalignment to the external point cloud was registered. In this case, it is necessary to assess the severity of the misalignment. Factors to consider include the scale tolerance and evaluating whether the misalignment is within an acceptable scale tolerance. Small discrepancies may be tolerable because minor misalignments may not significantly affect the final deliverables. More significant deviations require correction using algorithms like ICP to perform scan alignment based on point-to-point correspondence iteratively. Note that ICP alignment can be performed if the overlapping parts are sufficiently significant both in the horizontal and the vertical components.

Here is an example of the indoor/outdoor scans alignment check on scans C131, C134, C133, and C135, which share parts of 266, 141, 181, and 166 points in common with the outdoor point cloud: The common parts are not very extensive, but the example is quite demonstrative of the applied control method. As seen in the following graph (Figure 13), misalignments of up to 2 mm can be detected, which are within the scale tolerances of the architectural survey.

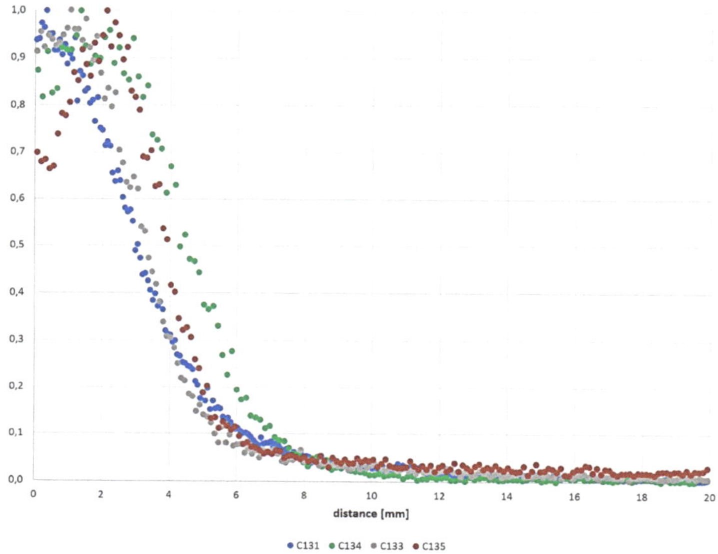

Figure 13. f_i statistic of the indoor/outdoor scans alignment check. Cloud-to-cloud distances are computed with respect to the outdoor point cloud. The series maximum represents the detected 3D shift.

4. Data Fusion

4.1. Terrestrial Scans

Combining different scans of the same object, it is common practice to accept all the data from all the scans to be combined, to select only the points from each scan up to a maximum scanning distance, or to make a manual selection. It is clear, however, that it would be convenient to have a tool capable of selecting the points with the highest scanning quality from the different scans in the overlapping areas.

Before establishing selection criteria grounded in scanning quality, it is crucial to acknowledge that scanning precision is influenced by several factors. These include the scanning distance and the surface's reflectivity, but most notably, the angle of incidence at which the laser beam intersects with the surface of the object being scanned [17,18]. The angle at which the laser beam strikes the surface determines the quantity of reflected light captured by the scanner. Consequently, the quality of the data gathered and its underlying rationale are fundamentally geometric:

- Specular reflection: The reflectivity of a surface can change depending on the angle of incidence of the laser beam. For example, when the laser beam hits the surface at an acute angle, there is a greater probability that the beam will not be reflected toward the scanner but in a different direction, leading to a lack of data from that particular surface area.

- Geometric distortion: Sharp angles of incidence can cause geometric distortion because the laser spot is larger than when the beam hits the surface perpendicularly.

The angle of incidence has been studied primarily for its effect on the intensity value. However, it was exploited to filter out points measured with an angle of incidence more significant than a threshold value. Calculating the laser's incidence angle with respect to the normal to the surface is very simple. For a terrestrial scan with scan origin X_0, Y_0 and Z_0, the local coordinates were used:

$$\vec{p}_i = \begin{pmatrix} x_i \\ y_i \\ z_i \end{pmatrix} = \begin{pmatrix} X_i - X_0 \\ Y_i - Y_0 \\ Z_i - Z_0 \end{pmatrix}$$

To compute for each point the distance from the scan origin:

$$D_i = \sqrt{x_i^2 + y_i^2 + z_i^2}$$

and the components of the scan origin direction:

$$\vec{d}_i = -\vec{p}_i / D_i$$

The unit vector $\vec{n}_i$ normal to the surface is computed in CloudCompare, setting the scan origin (X_0, Y_0, Z_0) as preferred orientation. The incidence angle δ_i, that is, the angle between the vectors $\vec{d}_i$ and $\vec{n}_i$, can be computed as:

$$\delta_i = \cos^{-1}\left(\vec{d}_i \cdot \vec{n}_i\right)$$

The angle δ_i ranges from 0°, when the scan laser beam is normal to the surface, to 90°, when the scan beam is tangent to the surface. As demonstrated in [8], the scan standard error increases exponentially at incidence angles larger than about 60°–70°.

The angle δ_i must be considered by combining two scans. Observe the two scans, A and B, of the same building (Figures 14–18). They have non-overlapping areas that must be retained. It is in the overlapping area where a quality-based selection can be performed. The following steps were designed:

(1) Select and retain the non-overlapping point cloud areas from A and B;
(2) On the overlapping areas:
    (a) Compute the incidence angle δ for the two scan A and B;
    (b) Choose from A or B the points that have incidence angle lower than a fixed threshold (i.e., 70°).
(3) Merge with non-overlapping point cloud areas from A and B.

Figure 14. Point clouds of the original scans: focus on the Scan (**A**,**B**).

Figure 15. Non common parts of Scan (**A**,**B**).

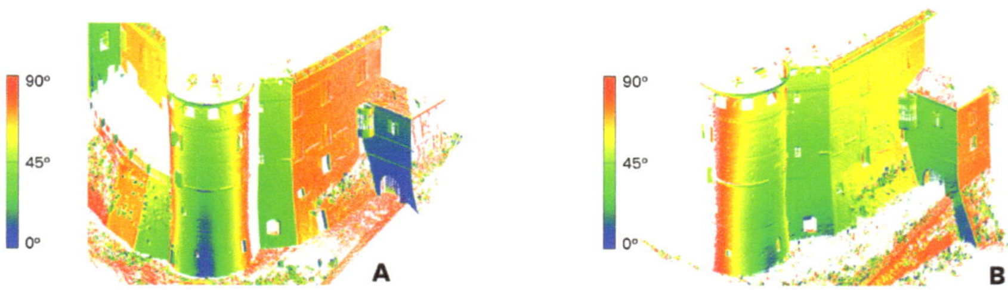

Figure 16. Scan (**A**,**B**): Incidence angle δ (angle between the scan direction and the normal to the object surface).

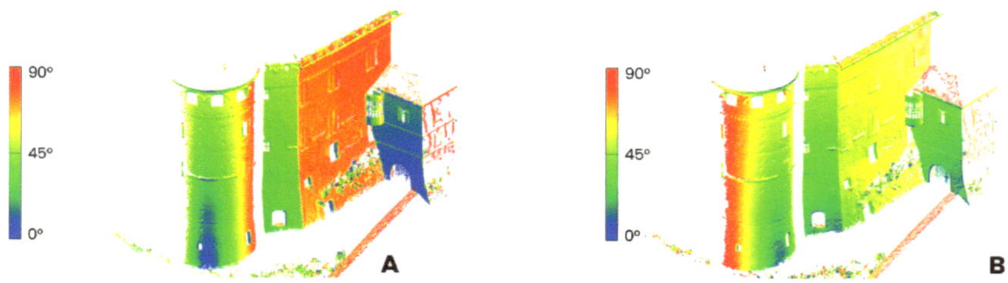

Figure 17. Scan (**A**,**B**): Incidence angle δ on the overlapping area.

Figure 18. Extracted point clouds from Scan (**A**,**B**).

The result is a merged cloud of the A and B scans, cleaned by noisy points in the areas scanned at high incidence angles (Figure 19).

Figure 19. Fused point clouds related to different scans.

4.2. Terrestrial and Airborne Scans

Airborne point clouds can be acquired by photogrammetry or by scanning. At Frinco Castle, the metric survey was carried out both by DJI (Shenzhen, China) L1 LiDAR and P1 camera, and the data comparison has already been discussed by the authors in [6]. The photogrammetric process was operated by using Agisoft Metashape (Figure 20) [19].

Figure 20. The photogrammetric point cloud acquired by the DJI P1 camera.

The merging process of ground and airborne point clouds is simple, assuming that the ground point clouds are always more accurate and less noisy than the airborne ones.

Using the cloud-to-cloud comparison in CloudCompare, it is possible to cancel out the airborne point cloud in the voxels occupied by ground scans (Figure 21). The size of the voxels depends on the scan's sampling distance, and in architectural survey, 1–2 cm is a good value. The procedure is much simpler and faster than the manual segmentation of the airborne point cloud, but alignment errors can lead to undesired point selection (Figure 22). Point cloud alignment can be easily controlled through the cloud-to-cloud distance on the overlapping area. In this way, the combined point cloud was obtained (Figures 23 and 24).

Figure 21. Cloud-to-cloud distances computed by CloudCompare on the airborne/terrestrial scan couple. The grey areas belong to the airborne scan and will be selected for data fusion.

Figure 22. Photogrammetric point cloud of the selected points.

Figure 23. Airborne/terrestrial combined point cloud.

Figure 24. Detail of airborne/terrestrial combined point clouds.

5. 3D Modeling

Point clouds resulting from the multi-sensor data fusion were the structural backbone of the 3D modeling phase. In this regard, reality-based modeling was performed to achieve a detailed polygonal mesh with RGB values, starting from data fused: terrestrial and aerial acquisition, indoor and outdoor. Then, a digital twin (from a graphical and metric point of view) of the castle was obtained. In addition to the polygonal mesh modeling, two different approaches have been operated: parametric modeling of structural elements of roof parts (wooden beams); NURBS modeling of the entire castle.

RGB values are not calibrated. All the terrestrial and aerial photography was acquired on cloudy days to produce uniform and diffuse illumination. The deliberate use of cloudy days ensures that the RGB values in the acquired images are not affected by extreme lighting variations, contributing to better data quality.

5.1. Parametric Design

TLS 3D data was exploited for creating parametric components (structural features) related to the internal parts of the castle, specifically those related to the collapsed area (where the aerial data were also used). The parametric design, using Trimble Real Work software (v.12.0) [4], included the structural parts of the roof, such as wooden beams (Figure 25). This process was carried out through cloud-based adaption: parametric surfaces and geometries were dynamically adapted to 3D point clouds. The parametric modeling process was relatively quick (approximately 2 min for beam) due to point cloud simplification and segmentation before the geometric adaptation: in this way, dynamic cubes were adapted (length–height–thickness values) to segmented point cloud of beams.

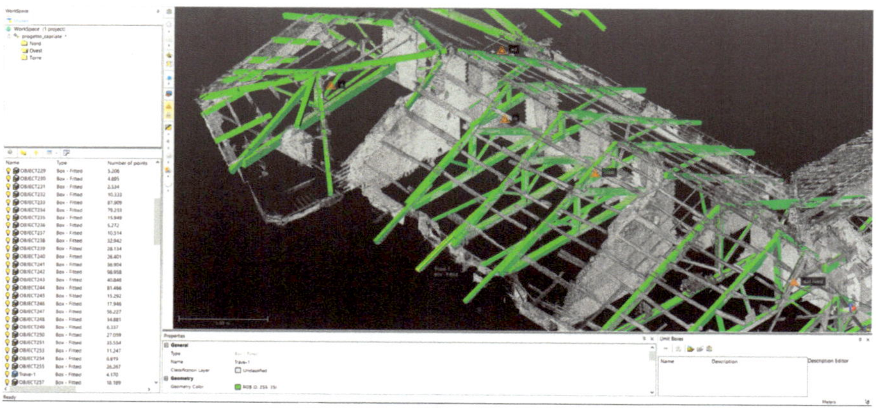

Figure 25. Parametric design related to wooden beams of the rooftop: geometries adapted to point clouds.

This modeling choice has been made especially to have a clear and simplified overview of the structural conformation of the internal rooftop, which is fundamental for creating detailed 2D drawings of indoor situations essential for future reconstruction.

5.2. NURBS Modeling

Metric data related to point clouds and wires from planar sections were the focus of the NURBS (Non-Uniform Rational Basis-Splines) modeling, detailed explained in previous work [15]. A simplified and lightweight model was developed that maps the features of masonries and the architectural stratigraphy of the castle. In fact, the complexity of the castle is given not only by the architectural components but also by the stratigraphic layers of masonries. In fact, the project presented here required extensive free-form modeling that was useful for different purposes.

By using Rhinoceros software (v.7.0) [20], curves and surfaces were interpolated and adapted to wired profiles resulting from point clouds: the complete reconstruction was assigned, combining *extrusions*, *loft*, *sweep*, and *patch* commands.

Firstly, the current situation of the southern front relative to the collapsed area was modeled by reconstructing the collapsed portion (volume) from historical photographs and actual measurements from the integrated survey and then simulating the structural collapse through the use of cutting planes (Figure 26). This modeling phase was fundamental for formulating a new hypothesis of reconstruction (according to local authorities).

Figure 26. NURBS model of Frinco Castle: Simulation of the structural collapse through the use of cutting planes.

The second modeling step concerned the complete free-form reconstruction of the defensive building, including the collapsed area and the internal vaulted rooms. Modeling the entire castle enabled the possibility of a dynamic and lightweight 3D representation on which mapping detailed HD orthophotos and the stratigraphic analysis performed on masonries (Figure 27). Reading stratigraphic units, which are construction site layers (including additions and modifications), was fundamental to formulating a chronological interpretation and then architectural evolution of the castle. The masonry stratigraphy and its interpretation were discussed in a previous work [15]. The multi-sensor approach and data fusion revealed the necessity for the complete 3D documentation of the external and internal covering elements of the castle, especially related to the collapsed area.

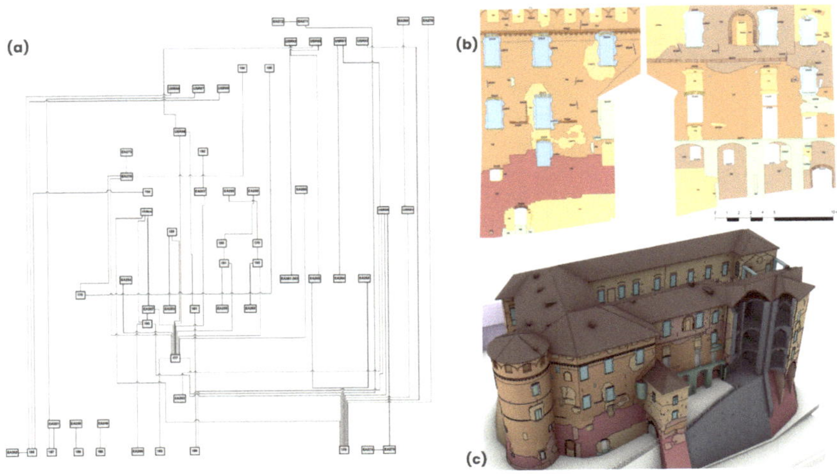

Figure 27. From stratigraphic analysis to 2D drawings and 3D mapping NURBS model. This figure shows a focus on the south front and the collapsed area of the castle: (**a**) stratigraphic diagram (matrix); (**b**) historical interpretation of stratigraphy; (**c**) stratigraphic units mapped into the NURBS model depending on historical periods.

6. Discussion

The suggested approach for verifying scan alignment, applicable to outdoor and indoor scenarios, can be employed for random checks or to identify misalignments that necessitate refinement via ICP. Our case study identified more pronounced misalignments of a few millimeters (2–4 mm), and in this case, ICP refinement is not mandatory.

The Trimble SX10 surveying solution has proven its reliability and precision across all tested conditions for outdoor and indoor architectural surveys. A robust and accurate control network guarantees the direct alignment of scans with an accuracy sufficient for creating architectural drawings and other deliverables typically required in architectural surveys. The primary limitation related to a TLS is the scan rate. However, the system's ability to avoid post-survey scan registration partially offsets the extended scanning duration. The benefit of minimal noise in scan point clouds was evaluated and further reduced by eliminating points scanned at a high incidence angle.

The principal advantage of direct scan referencing by SX10 lies in the indoor scan alignment, mainly when shared features or targets can be challenging to locate or place.

The quality-based selection criteria used to merge different scans can be easily applied by processing in CloudCompare. The scan distance and incidence angle can be computed using Scalar Fields Arithmetic, while to compute the Normal Vector, the standard algorithm implemented in CloudCompare was used, selecting the scan origin as the preferred orientation (previously shifting the scan origin to the coordinate origin). The cloud-to-cloud distance calculation and the command "filter points by value" were exploited to select a subset of the point cloud. The selection of the maximum threshold value for the incidence angle depends on many factors, at first on the availability of redundant data in overlapping scan regions, then on the quality of the data and their level of noise, which are functions of the scanned materials reflectance, of the distance and the scanner quality.

7. Conclusions

This work and manuscript focused on the multi-sensor metric survey of Frinco Castle. A general overview of the techniques used to generate point clouds and the alignment check, segmentation, and data fusion was provided. Outdoor and indoor surveying followed two workflows, similar to other projects [21,22], where only outdoor surveying needed data fusion.

Frinco Castle required careful and ad-hoc survey planning to acquire all the building components. Data integration and fusion were the only possible ways to have a complete metric overview of the castle: multi-sensor surveys, performed with aerial and terrestrial data, demonstrated its reliability for complex architectural contests [23,24].

Integrated and fused point cloud data was the starting point for preparing the architectural parametric and NURBS modeling, designed for achieving different outputs: 2D detailed drawings (floorplans, sections) and NURBS dynamic model.

The 3D survey and NURBS modeling were also designed to help the stratigraphic investigation of external masonries, a needed workflow for building archaeology projects. This information was essential for producing highly detailed orthophotos that were used to map the stratigraphic layers of architectural fronts and, subsequently, to comprehend how the defensive building's architecture evolved [15,25].

Lately, the partial parametric modeling initially planned for having a 2D drawing simplified view of the internal areas (especially the roof) could prove to be fundamental for designing an entire parametric architectural model of the castle for handling metric and semantic data in a dynamic Historic Building Information Modeling (HBIM) environment. The acquired metric data and related digital outputs could be the basis for building up an informative model of the castle. This proposal is also under consideration to have a standard and intelligent platform for the upcoming restoration phase.

The HBIM project of the castle should be designed to preserve metric and semantic data, especially related to the historical interpretation and structural information of this fragile building. Being a 3D database and data-exchange platform, the HBIM model would enable the implementation of the castle's collected historical data and synchronic collaborations for further analyses [26–28]. Professional figures involved in the restoration project could analyze and exchange sensitive data before hands-up operations. This proposal will also be designed through a tailor-made Common Data Environment (CDE), which enables users and collaborators to access, read, edit, and query historical information inside the parametric informative model. This possibility breaks data accessibility barriers and eases forthcoming restoration processes.

Author Contributions: Conceptualization, M.R. and F.D.; methodology, M.R. and F.D.; software, F.D.; validation, M.R. and F.D.; formal analysis, M.R. and F.D.; investigation, M.R. and F.D.; resources, M.R. and F.D.; data curation, M.R. and F.D.; writing—original draft preparation, M.R. and F.D.; writing—review and editing, M.R. and F.D.; visualization, M.R. and F.D.; supervision, M.R. and F.D.; project administration, M.R.; funding acquisition, M.R. All authors have read and agreed to the published version of the manuscript.

Funding: This research was funded by Fondazione Cassa di Risparmio di Asti.

Data Availability Statement: Data are contained within the article.

Acknowledgments: The authors would like to acknowledge the Frinco municipality for granting and supporting topographic and aerial investigations. Special thanks go to Giorgio Viazzo and Chiara Viazzo for contributing and supporting the terrestrial acquisition phase and for the production of charts and elevations.

Conflicts of Interest: The authors declare no conflicts of interest.

References

1. Teppati Losè, L.; Chiabrando, F.; Giulio Tonolo, F.; Lingua, A. UAV photogrammetry and VHR satellite imagery for emergency mapping. The October 2020 flood in Limone Piemonte (Italy). *Int. Arch. Photogramm. Remote Sens. Spat. Inf. Sci.* **2021**, *XLIII-B3-2021*, 727–734. [CrossRef]
2. Shen, W.; Yang, Z.; Yang, C.; Li, X. A LiDAR SLAM-Assisted Fusion Positioning Method for USVs. *Sensors* **2023**, *23*, 1558. [CrossRef] [PubMed]
3. Njimi, H.; Chehata, N.; Revers, F. Fusion of Dense Airborne LiDAR and Multispectral Sentinel-2 and Pleiades Satellite Imagery for Mapping Riparian Forest Species Biodiversity at Tree Level. *Sensors* **2024**, *24*, 1753. [CrossRef] [PubMed]
4. Bordone, R. *Andar per Castelli da Asti Tutt'intorno*; Milvia: Torino, Italy, 1976; ISBN 2020010144721.
5. Sorisio, R. Ricerche Storico-Giuridiche su Frinco. Master's Thesis, University of Turin, Turin, Italy, 1979.
6. Gabiani, N. *Le Torri le Case-Forti ed i Palazzi Nobili Medievali in Asti*; A. Forni: Bologna, Italy, 1978.
7. Malfatto, V. *Asti Antiche e Nobili Casate*; Il Portichetto: Cuneo, Italy, 1982.
8. Diara, F.; Roggero, M. The Frinco Castle: From an Integrated Survey to 3D Modelling and a Stratigraphic Analysis for Helping Knowledge and Reconstruction. *Remote Sens.* **2023**, *15*, 4874. [CrossRef]
9. Trimble Website. Available online: https://www.trimble.com/en (accessed on 16 April 2024).
10. DJI Website. Available online: https://www.dji.com (accessed on 16 April 2024).
11. Diara, F.; Roggero, M. Quality Assessment of DJI Zenmuse L1 and P1 LiDAR and Photogrammetric Systems: Metric and Statistics Analysis with the Integration of Trimble SX10 Data. *Geomatics* **2022**, *2*, 254–281. [CrossRef]
12. Štroner, M.; Urban, R.; Línková, L. A New Method for UAS Lidar Precision Testing Used for the Evaluation of an Affordable DJI Zenmuse L1 Scanner. *Remote Sens.* **2021**, *13*, 4811. [CrossRef]
13. Kersten, T.; Wolf, J.; Lindstaedt, M. Investigations into the accuracy of the UAS system DJI Matrice 300 RTK with the sensors Zenmuse P1 and L1 in the Hamburg test field. *Int. Arch. Photogramm. Remote Sens. Spat. Inf. Sci.* **2022**, *XLIII-B1-2022*, 339–346. [CrossRef]
14. Teppati Losè, L.; Matrone, F.; Chiabrando, F.; Giulio Tonolo, F.; Lingua, A.; Maschio, P. New developments in lidar UAS surveys. Performance analyses and validation of the DJI Zenmuse L1. *Int. Arch. Photogramm. Remote Sens. Spat. Inf. Sci.* **2022**, *XLIII-B1-2022*, 415–422. [CrossRef]
15. Lachat, E.; Landes, T.; Grussenmeyer, P. Investigation of a Combined Surveying and Scanning Device: The Trimble SX10 Scanning Total Station. *Sensors* **2017**, *17*, 730. [CrossRef] [PubMed]
16. CloudCompare Software. Available online: https://www.danielgm.net/cc/ (accessed on 16 April 2024).
17. Soudarissanane, S.; Lindenbergh, R.; Menenti, M.; Teunissen, P. Incidence angle influence on the quality of terrestrial laser scanning points. In Proceedings of the ISPRS Workshop Laserscanning 2009, Paris, France, 1–2 September 2009.
18. Kaasalainen, S.; Jaakkola, A.; Kaasalainen, M.; Krooks, A.; Kukko, A. Analysis of Incidence Angle and Distance Effects on Terrestrial Laser Scanner Intensity: Search for Correction Methods. *Remote Sens.* **2011**, *3*, 2207–2221. [CrossRef]
19. Agisoft Metashape. Available online: https://www.agisoft.com (accessed on 16 April 2024).

20. Rhinoceros. Available online: https://www.rhino3d.com (accessed on 20 February 2024).
21. Ramalho de Oliveira, L.F.; Lassiter, H.A.; Wilkinson, B.; Whitley, T.; Ifju, P.; Logan, S.R.; Peter, G.F.; Vogel, J.G.; Martin, T.A. Moving to Automated Tree Inventory: Comparison of UAS-Derived Lidar and Photogrammetric Data with Manual Ground Estimates. *Remote Sens.* **2021**, *13*, 72. [CrossRef]
22. Zhang, F.; Hassanzadeh, A.; Kikkert, J.; Pethybridge, S.J.; van Aardt, J. Comparison of UAS-Based Structure-from-Motion and LiDAR for Structural Characterization of Short Broadacre Crops. *Remote Sens.* **2021**, *13*, 3975. [CrossRef]
23. Mugnai, F.; Masiero, A. Integrating UAS Photogrammetry and Digital Image Correlation for High-Resolution Monitoring of Large Landslides. *Preprints* **2022**, 2022010248. [CrossRef]
24. Gaspari, F.; Ioli, F.; Barbieri, F.; Belcore, E.; Pinto, L. Integration of UAS-LiDAR and UAS-photogrammetry for infrastructure monitoring and bridge assessment. *Int. Arch. Photogramm. Remote Sens. Spat. Inf. Sci.* **2022**, *XLIII-B2-2022*, 995–1002. [CrossRef]
25. Stanga, C.; Banfi, F.; Roascio, S. Enhancing Building Archaeology: Drawing, UAV Photogrammetry and Scan-to-BIM-to-VR Process of Ancient Roman Ruins. *Drones* **2023**, *7*, 521. [CrossRef]
26. Banfi, F.; Oreni, D.; Bonini, A.J. The Arch of Peace of Milan and its historic memory: From 3D survey and HBIM to mixed reality (VR-AR). In *Connecting. Drawing for Weaving Relationships, Proceedings of the 42th International Conference of Representation Disciplines Teachers, Reggio Calabria, Italy, 16–18 September 2021*; FrancoAngeli: Milan, Italy, 2020; pp. 1660–1677. [CrossRef]
27. Diara, F. HBIM Open Source: A Review. *ISPRS Int. J. Geo-Inf.* **2022**, *11*, 472. [CrossRef]
28. Banfi, F.; Roascio, S.; Mandelli, A.; Stanga, C. Narrating Ancient Roman Heritage through Drawings and Digital Architectural Representation: From Historical Archives, UAV and LIDAR to Virtual-Visual Storytelling and HBIM Projects. *Drones* **2023**, *7*, 51. [CrossRef]

drones

MDPI

Article

Drones and Real-Time Kinematic Base Station Integration for Documenting Inaccessible Ruins: A Case Study Approach

Daniele Treccani *, Andrea Adami and Luigi Fregonese

UNESCO Research Lab, Department of Architecture, Built Environment and Construction Engineering (DABC), Politecnico di Milano, Mantova Campus, Piazza C. d'Arco, 3, 46100 Mantova, Italy; andrea.adami@polimi.it (A.A.); luigi.fregonese@polimi.it (L.F.)

* Correspondence: daniele.treccani@polimi.it

Abstract: Ruins, marked by decay and abandonment, present challenges for digital documentation due to their varied conditions and remote locations. Surveying inaccessible ruins demands innovative approaches for safety and accuracy. Drones with high-resolution cameras enable the detailed aerial inspection and imaging of these inaccessible areas. This study investigated how surveying technologies, particularly Unmanned Aerial Vehicles (UAVs), are used to document inaccessible ruins. Integration with Real-Time Kinematic (RTK) technologies allows direct georeferencing in photogrammetric processing. A case study of the Castle of Terracorpo in Italy was used to demonstrate UAV-only surveying feasibility in inaccessible environments, testing two different scenarios. The first scenario involved the use of a DJI Matrice 300 RTK coupled with the D-RTK2 base station to survey the Castle; both direct and indirect georeferencing were exploited and compared through the photogrammetric process. This first scenario confirmed that this approach can lead to a centimetre-level accuracy (about three times the GSD value for indirect georeferencing and seven times the GSD value for direct georeferencing exploting RTK). The second scenario testing the integration of data from drones at varying resolutions enabled the comprehensive coverage of ruinous structures. In this case, the photogrammetric survey performed with the dji Mavic 3 Cine drone (indirect georeferencing) was integrated with the photogrammetric survey performed with the dji Matrice 300 RTK drone (direct georeferencing). This scenario showed that GCPs extracted from a direct georeferencing photogrammetric survey could be successfully used to georeference and integrate other drone data. However, challenges persist in surveying underground or enclosed spaces that are inaccessible to UAVs. Future research will explore integrating robotic LiDAR survey systems and advanced data fusion techniques to enhance documentation.

Keywords: UAV photogrammetry; RTK; direct georeferencing; data integration; cultural heritage; ruins

Citation: Treccani, D.; Adami, A.; Fregonese, L. Drones and Real-Time Kinematic Base Station Integration for Documenting Inaccessible Ruins: A Case Study Approach. *Drones* **2024**, *8*, 268. https://doi.org/10.3390/drones8060268

Academic Editor: Pablo Rodríguez-Gonzálvez

Received: 29 April 2024
Revised: 13 June 2024
Accepted: 14 June 2024
Published: 17 June 2024

1. Introduction

Ruins can be defined as the remains of man-made architectural structures, even if they can go beyond the merely destroyed architecture [1]. From the architectural point of view, they generally include a wide range of structures, including abandoned, historic, and crumbling buildings, showing obvious signs of decay, such as unstable walls, partially collapsed walls or roofs, with even entire rooms without roofs. Their geographic location can vary; they can be either in mountainous or lowland areas, or in lake or maritime areas. Generally, what makes them "ruins" is the poor state of preservation of the buildings themselves [2], including portions that have collapsed or are in a severe state of degradation and often abandoned (for various reasons).

The digital documentation of complex structures is challenging, and in many cases, the integration of Geomatic techniques, such as Terrestrial Laser Scanners (TLS), Global Navigation Satellite Systems (GNSS), photogrammetry, Unmanned Aerial Vehicles (UAV), Mobile Mapping Systems (MMS), and the implementation of a supporting topographic network comes into play [3–8]. Plus, surveying and documenting ruined structures can be extremely complex, as they are often located in remote or difficult-to-access areas and have severe damage and deterioration. The condition of the ruins can also include hazardous elements such as debris and piles of materials, making survey work not only complicated but also potentially dangerous for human operators. In this context, the use of drones has proven to be extremely useful, enabling detailed, high-resolution images of the ruins to be obtained from different perspectives and angles, even in areas that are difficult to access or dangerous.

To fully understand the crucial role of survey technology integration, the contribution of different components of the system needs to be explored in more detail. First, the use of different types of drones, each with specific capabilities, amplifies the scope and accuracy of the aerial photogrammetric survey. Drones with high-resolution cameras are used for the overall aerial inspection of ruins, while more compact and manoeuvrable drones can penetrate narrow, hard-to-reach spaces, allowing detailed views of interior areas and walls.

Depending on the conditions and characteristics of the ruins, some technological solutions may be more convenient than others. Alternatively, the circumstances and the survey context may constrain the choice of only certain types of instruments. For example, in the case where the ruins are difficult to reach or even not accessible for safety reasons, the choice of using Unmanned Aerial Vehicles (UAVs) seems to be feasible. In such a case, it becomes interesting to understand the most appropriate approaches on how to conduct the survey, which drone to prefer, and which methods to implement to integrate the data resulting from the different instruments. Certainly, the possibility of using a base GNSS receiver coupled with a UAV makes it possible to take georeferenced photographs, which then allows direct georeferencing to be performed within the photogrammetric process [9,10]. In addition, the use of drones coupled with a ground base station, exploiting Real-Time Kinematic (RTK) or Post-Processing Kinematic (PPK) technologies, would result in a greater accuracy of such data [11–13].

Given this context, it becomes interesting to understand what methods and strategies can be put in place for combining data from drones with different resolutions and flight capabilities. In this study, we investigated a specific working condition: the survey of a ruin under the assumption that the environmental and structural conditions do not allow access to it; so we examined the advantages and limitations of using only UAVs for surveying a ruin. We examined the effectiveness of data integration between UAVs with different resolutions by assuming the impossibility of using traditional topographic instruments, and we will discuss the deliverables that could be produced. In particular, the use of a UAV coupled with RTK technology was tested, and its integration with other UAV photogrammetric survey data will be evaluated and discussed.

This study exploited data obtained during the survey campaign of the castle of Terracorpo (province of Caserta, Italy); its geographical position is shown in Figure 1. The UNESCO Research Lab of the Mantua Campus of the Politecnico di Milano was assigned to carry out a project proposal for the conservation and enhancement of the Terracorpo Castle. To support the conservation project, the lab was also involved in carrying out an integrated survey of the building to extract 2D and 3D products typical for this type of project. The work described here was therefore part of a larger project. The case study was completely accessible at the moment of the survey, so it was documented using various survey techniques, including TLS and UAV photogrammetry, to reach all spaces of the building. For research purposes, instead, a subset of the data acquired was selected, and we decided to simulate a situation in which only UAVs were used because of the inaccessibility of spaces.

The rest of this paper is organised as follows. Section 2 presents related articles concerning the survey of complex architectures, including the integration of multi-platform and multi-sensor surveys, and the use of UAVs with RTK and PPK technologies. Section 3 describes the case study, its historical evolution, and the current state of the building; this section also describes the integrated survey conducted on the building. Section 4 is devoted to the description of the two scenarios put in place for our tests and their materials and methods, under the assumption of surveying using only UAVs. Scenario I analyses the differences between direct and indirect georeferencing using a UAV and a ground base to provide centimetric accuracies with RTK technology. Scenario II analyses the case of data integration between two UAVs with different characteristics and accuracies. Section 5 presents the results of the tests conducted under the two scenarios. It also presents some deliverables that can be produced using the presented approach. Section 6 presents a discussion, and Section 7 presents the conclusions.

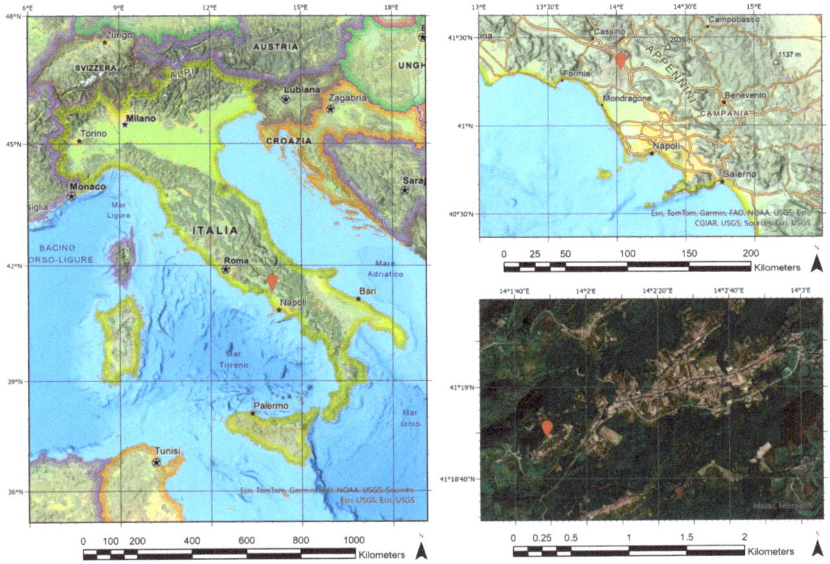

Figure 1. Geographical location of the case study, the Terracorpo Castle (Marzano Appio, Caserta, Italy). The red symbol indicates the position of the framed element at the Italian national, regional, and local scale.

2. Related Works

2.1. Digital Documentation of Ruins

Architectural ruins constitute a part of the existing building stock and can be traced back to both valuable and less valuable buildings. An important step in the management, conservation, and valorisation of these types of buildings is their documentation. The starting point for any intervention on ruins is certainly a historical analysis followed by a survey that allows us to know and correctly determine and interpret their shape, spatial organisation, materials, and degradation [14]. Articles dealing with the documentation of ruined buildings can be found in the scientific literature [14–21], especially when the buildings are of architectural significance.

The authors of [14] discussed the survey stages, which involve the survey campaign on-site, but also a large amount of reading and interpretation skills concerning the architectural spaces surveyed. They focused on fortified architectures in a state of ruin and used a combination of various survey techniques to reach the goal. In the case of reference [15], the authors investigated the ruins of Casa Ippolito, a typical 17th-century Maltese aristocratic country residence. They studied the ruins of the building as the primary source to reconstruct the building's architecture. They studied the materials, the historical evolution of the building, and the phases of the construction. Instead, to fill a gap due to a lack of documental sources, the authors of [16] carried out a survey supplemented by various sources to document the ruins of the castle of Assoro. Focusing instead on burial sites, the authors of [21] developed a 3D digital survey and the consequent high-precision burial model and contour map to enhance the protection of the burial site in Yaoheyuan, Ningxia, one of the top 10 archaeological discoveries in China in 2017.

On a different scale, the authors of [17] investigated an area that includes the ruins of 30 churches and villages in a territory covering Brazil, Argentina, and Paraguay. They then focused on the ruins of one church and carried out an integrated survey in order to produce a geographic database to assist its management. On a similar urban scale, the authors of [18] focused on various buildings in the urban centre of Navelli (L'Aquila, Italy), now in a state of ruin after an earthquake in 2009. The authors undertook a multi-scale and multi-sensor data acquisition–integration task to document these buildings.

Focusing on archaeological ruins, the authors of [19] carried out an integrated survey of the ruins of a Roman aqueduct in order to improve its documentation, but also to enhance its representation through the use of Historic Building Information Modeling (HBIM) and virtual and augmented reality technologies. Also, the authors of [20] focused their attention on HBIM, developing the modelling of a ruined building in a municipality in Umbria (Italy). They evaluated that digital documentation can be an excellent starting point for HBIM, which requires several compromises from a geometric point of view (especially in ruined buildings), but also allows for the better management and analysis of the information content, which leads to virtuous processes for the valorisation of the cultural asset.

What emerges is a particular category of ruins that has been investigated in the literature that concerns the ruins of fortified buildings and castles; this category of ruined buildings often plays an important role in the history of the geographical areas around them. In such cases, the objective is the documentation, also digital, of the building for its future redevelopment. It is also interesting to note that the condition and state of preservation of such objects make them very risky and difficult to access, making the surveys carried out similar to those performed for risk assessment [22–24].

2.2. Integrated Survey of Complex Architectures

Given the environmental complexities and the generally poor state of preservation of ruined buildings, it is often optimal to use a combination of several surveying techniques to fully document them. Various surveying tools and techniques could be used to achieve this goal, such as TLS, UAVs, GNSS receivers, MMS, a total station (TS), and techniques such as close-range photogrammetry. UAV photogrammetry and TLS surveying are prominent among the most widely used technologies. The data from these surveys can then be merged using various methods. Various data integration methods can be found in the literature [18,21,25–32], referring not only to the case of ruins but, in general, to complex buildings.

In general, regardless of the case study analysed, but still concerning integrated surveys of complex historical buildings, we can see that the most commonly used technologies are TLS and close-range photogrammetry (both from the ground and from drones), and through their integration, the creation of a topographic support network is widespread, measured with GNSS receivers [18,25,26] alone or in pairs with TS [16,27–32]. This topographic network is then used to formalise a reference system (local or georeferenced) to which the surveys carried out with the various instruments can be locked, measuring Ground Control Points (GCP) and targets with it. Alternatively, there are also cases where the authors have used different techniques for the integration of data from different sources, such as the use of GCPs identified in the TLS survey for integration with UAV data [16], or the use of point cloud co-registration algorithms such as the Iterative Closest Point (ICP) algorithm [21,32].

In taking a more specific look at the integration techniques employed, several examples of integration between UAV photogrammetry and TLS can be observed. For instance, the authors of [27] surveyed a church using TLS and close-range photogrammetry with images taken using a low-cost UAV; the integration was performed through two methods: measuring the GCP with a TS and retrieving them from a TLS survey. The results showed a slightly higher error when using TLS data. A similar approach was developed by the authors of [28], who developed a survey of a church using TLS, close-range photogrammetry with a DSLR camera, UAV platforms, and a multi-camera system. They used TS to materialise a topographic network to georeference all the data within the same reference system. Due to various reasons (e.g., no complete visibility of all points from TS), not all point clouds were registered using TS-measured GCPs, or using GCPs retrieved from one of the point clouds acquired with other systems. The TLS data were used as a geometric reference for the UAV photogrammetry. Leica register360 cloud-to-cloud was used to assemble all the point clouds.

In shifting the focus to urban areas, integration methods remain similar; for instance, the authors of [25] developed a multi-scale multi-sensor and multi-purpose survey. They used an integration between TLS, UAVs, DSLR cameras, and 360° cameras. GNSS and TLS technologies were used to create a reference network that was used as a standard reference coordinate system. Similarly, the authors of [18] surveyed a damaged area of a village. The survey was carried out using a mobile laser scanner based on SLAM technology and a drone for aerial photogrammetry. The integration of the two technologies was optimised through the placement of GCPs acquired with a GNSS receiver.

Focusing then on a building in ruins, the authors of [30] surveyed a highly degraded historical castle. The purpose of the work was to perform a structural analysis. The survey was performed using a low-cost drone, a DSLR camera, and a smartphone. Data were validated and integrated using TLS. While a GNSS receiver and TS were used to generate a Topographic Reference Network. Similarly, the authors of [26] analysed a castle and its surroundings by performing a multi-scale survey with UAV photogrammetry and a TLS survey. The data were integrated using a GNSS-measured reference topographic network.

2.3. RTK and PPK Technologies for UAVs

From the above, it can be deduced that among the most widely used technologies for surveying ruins and complex cultural heritage, the use of UAVs certainly stands out, often in combination with other technologies (e.g., GNSS, TLS, TS) for georeferencing the data and measuring GCPs. However, there is also the possibility of directly georeferencing the drone images using GNSS stations connected to the UAV system, which then exploits RTK or PPK technologies. In this way, the direct georeferencing of the data can be realised [12,13,33,34], i.e., reducing the need to acquire several GCPs on the ground.

The reliability of a UAV photogrammetric survey in RTK or PPK mode was investigated by the authors of [35], who developed several tests on a forest environment. They developed several tests using both UAVs with RTK and PPK and GCPs. They assessed that within their working conditions, the most accurate results for both horizontal and vertical accuracies were achieved using UAV RTK/PPK technology, even compared with the use of GCPs. Another evaluation was performed by the authors of [36], who tested a DJI Matrice 600 Pro and assessed that the direct georeferenced results of this drone with an onboard GNSS RTK unit can generate a photogrammetric product with decimeter accuracy. Another study, which focused on RTK-based UAV photogrammetry, in terms of interactions between GCP, direct georeferencing, and camera self-calibration, was performed by the authors of [37]. They investigated different case studies with different characteristics (a UAV test field, a moorland, and a building) and used different drones with various flight configurations. They found that the RTK option leads to sufficient results if at least 1 GCP is introduced. Flights without any GCPs lead to a significant height error in the order of up to 30 times the GSD value. Moreover, the authors of [?] aimed at improving the accuracy of UAV-RTK photogrammetric projects without the need to use oblique photographs or GCPs. They used multiple GNSS-fixed base stations correcting the geolocation data of the photographs simultaneously by averaging the differential corrections. The results showed that this methodology considerably reduces the altimetric errors, placing them in limits even below the GSD. Lastly, the authors of [12,39] demonstrated how the integration of a GNSS RTK/PPK module on low-cost commercial UAVs makes the system efficient for the reconstruction of a highly detailed and precise Digital Elevation Model without using GCPs, allowing them to make precision measurements in areas that are difficult to explore and investigate.

Finally, there are practical examples where UAV technologies combined with RTK and PPK technologies have been used to survey case studies of relevant interest, also integrating them with other survey technologies. For example, in reference [40], the authors analysed a small town centre. They integrated 360° videos and UAV image photogrammetry. They used GNSS data acquired by a UAV to perform GNSS-assisted image georeferencing to reduce or even avoid, in specific situations, the need for GCPs. Camera positions were computed either in RTK or PPK mode, with results similar to those using GCP. Another example is related to the survey of an archaeological site, where the authors [41] used aerial photogrammetry (exploiting PPK technology) and TLS data, which were integrated employing the ICP algorithm. Lastly, the authors of [42] surveyed historical defensive structures in mountainous areas. They used traditional photogrammetric solutions and novel methods based on Neural Radiance Fields (NeRFs). They performed direct georeferencing based on PPK and also used traditional GCPs placed on the ground, considering the difficulties in placing and measuring control points in such environments.

2.4. Synthesis and Outlook

Based on what has been presented, it can be seen that for the survey of complex architecture, including ruined areas, the integration of different surveying techniques is the rule, as it allows all areas of the surveyed building to be reached. The integration of such data is mainly realised through the use of topographic networks that measure GCP, either using TS or GNSS receivers. Cases of registration using ICP algorithms have also been presented. Of particular interest is also the possibility of integrated surveys in which the UAV data are coupled with technologies such as RTK or PPK, which allow direct georeferencing in the photogrammetric process.

In this study, we investigated the case of a survey conducted using only UAVs with the use of RTK technologies for direct georeferencing when other surveying technologies cannot be used and in the case where GCPs cannot be detected. Furthermore, in this study, the integration of two acquisitions—carried out with two UAVs of different resolutions, one of which uses an RTK system—without reference to GCPs was tested.

3. Terracorpo Castle: Case Study Description and Integrated Survey

3.1. Description of the Case Study

Terracorpo Castle (see Figure 1) is located in a small settlement that is apparently the original nucleus of the nearby town of Marzano Appio (Caserta, Italy). The earliest certain evidence of the existence of a castle near Marzano is from the Norman–Swabian period (11th and 12th centuries), while the first explicit mention of a castle dates back to 1278. The earliest form, like the layout of the village, was not dissimilar to that of other fortified settlements of the time, that is, with a fortress located in an apical and asymmetrical position with respect to the built-up area, which, together with a place of worship, was enclosed in a domain within descending walls. This assumption rests on the surviving remains of ancient walls on the northern slope of the present fortress and the surviving remains of a fragment of the original enclosure on the southern slope of Terracorpo Hill, where the village is located.

Aerial observations of the village reveal the existence of at least two city walls and the presence of two gateways. The main entrance is located at the western end, identified by a portal with an ogival arch in local Tufo, made in the Gothic style, and can be ascribed to the 14th century. Equally evident are the structural reinforcements, which occurred between 1500 and 1700, reinforcing and strengthening the pre-existing structural elements. The fortress encloses a large inner courtyard with a rectangular plan preceded by a large vestibule. The 1700–1800 appearance of the main layout following the last major transformations is the result of the logical structural evolution affecting all centuries-old structures.

The palace was then fully inhabited, equipped with prisons inside, the hub of all life in the surrounding area. Indeed, the political, commercial, and administrative life of the town took place there.

Although it originated as a fortification and for defensive use, the castle today has the appearance of an austere 18th-century palace that is firmly grafted into the natural scenery of the local area with its long, linear facade of Tufo stone. The castle is currently (see Figure 2) overgrown with weedy vegetation that inhibits its full fruition. It is also in very poor condition due to bombing during World War II, and following collapses due to the 1980s earthquake [43]. As of today, almost the entirety of the main floor no longer exists as the roofs and floors of the main floor have collapsed. Only a few rooms on the piano nobileremain, accessible only through the use of portable ladders to be placed at openings facing the ground floor. There are rooms with vaulted ceilings on the ground floor, corresponding to those on the main floor that still exist. The castle also has a water collection cistern, an underground room, and a room with access from the street level only.

Globally, the castle has five covered rooms (i.e., with roofing) on the ground floor and three covered rooms on the second floor, while eight rooms on the ground floor do not have covers and are at full height. Three of the rooms are on the first floor without roofs. The current conservation condition of the castle allowed a photogrammetric survey with drones, due in part to the absence of roofs in many of the rooms. To complete the survey in the underground rooms and rooms with roofing, a laser scanner survey was also carried out.

Figure 2. Pictures of the Castle of Terracorpo in its actual status.

3.2. Survey of the Case Study

Within this short section, we describe the integrated survey developed in the case study (see Figure 3 for some images of the survey processing). For the tests presented in the next sections, we used a subset of the actual data acquired on-site.

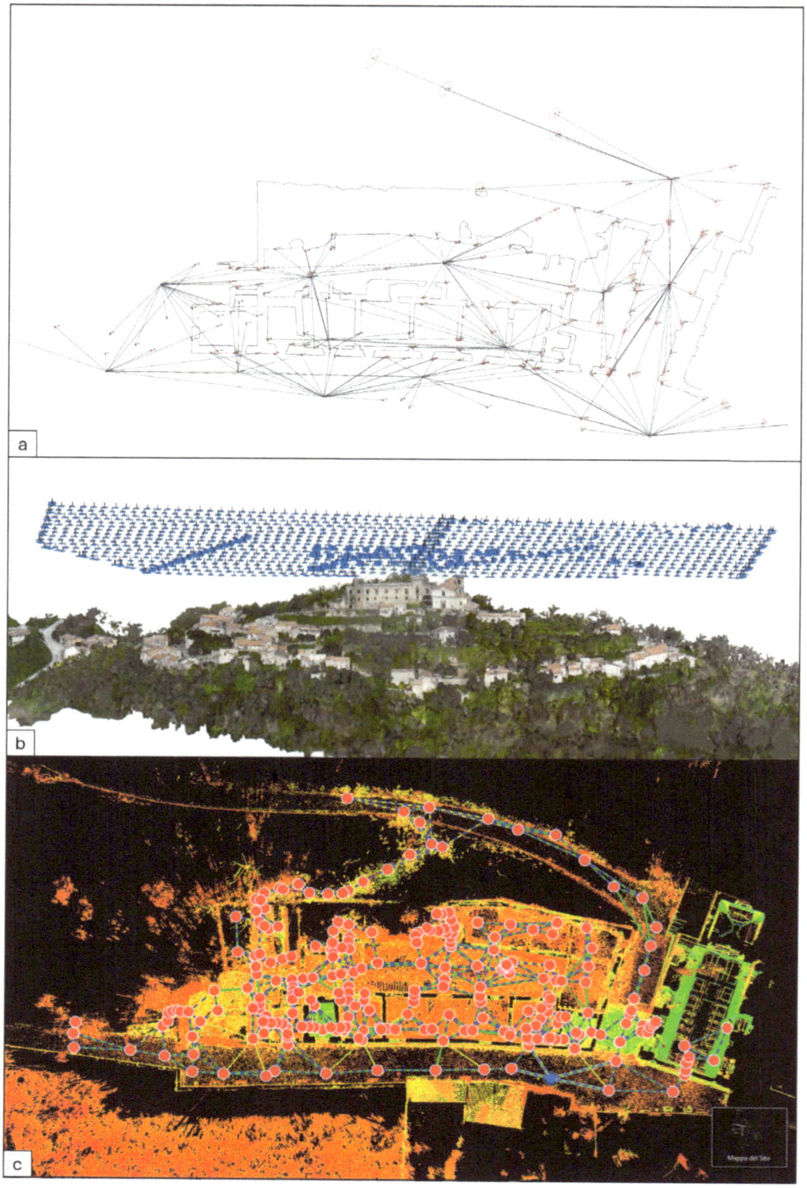

Figure 3. Screenshots of survey processing phases. (**a**) The topographic network developed during the survey to materialise a common reference system for all the data. (**b**) A screenshot of Agisoft Metashape Professional v. 2.0.0 with the reconstructed model of the castle. In blue is the position of the cameras. (**c**) A screenshot of the processing of TLS data with Leica Register 360 Plus (version 2023.0.2) software. Point cloud data are coloured in a false colour scale according to the intensity. The red circles represent the positions of the scanner along the survey execution.

As already mentioned, the purpose of the survey was to support a project for the conservation and enhancement of the castle (see Figure 4 as an example of a deliverable produced). Conducting conservation and restoration work on a historic building generally requires a solid and large amount of information and documentation about the building. In addition to a rich historical documentation of the building's evolution over time, information on the current state of the preservation (materials and state of decay) of the building's surfaces is needed. Therefore, in order to provide good support for the planning of the conservation of the castle, it was necessary to carry out a complete and highly detailed and informative survey of the building and its surroundings. The objective of this survey was, therefore, to provide drawings of plans at various heights and sections at various strategic points of the building, as well as orthophotos for the description of the state of the conservation of the wall surfaces. In order to achieve the appropriate accuracy (the scale of restitution required was 1:50, which admits a drawing error of 1 cm), it was decided to carry out an integrated survey, with laser scanner, photogrammetric techniques (from a drone and from the ground), with the support of a topographic network measured with a total station and GNSS receiver.

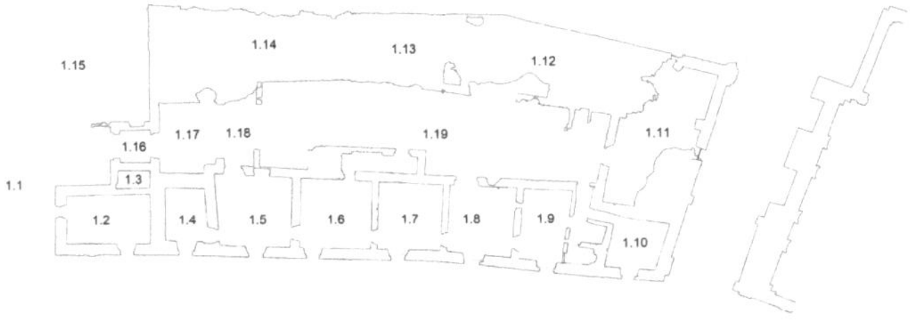

Figure 4. One of the project boards delivered by the UNESCO Research Lab as a result of the survey for the conservation and enhancement project of Terracorpo Castle. The picture shows a board with a plan of the castle.

The survey of Terracorpo Castle was developed by integrating different technologies. A TLS (Leica RTC360), a GNSS receiver (Leica GS-18), a TS (Leica TS30), a Digital Single-Lens Reflex (DSLR) camera (Canon EOS 5DS R), and two UAVs (DJI Matrice 300 RTK equipped with the DJI D-RTK2 base station and DJI Mavic 3 Cine) were used. The choice of the aforementioned surveying techniques and instruments was, first of all, possible because the current condition of the ruined castle left it almost completely accessible. Some rooms on the upper floors are currently not accessible from below (e.g., stairs no longer exist and ceilings closed during previous redevelopment work); therefore, they are only visible with a drone. In addition, due to the presence of underground areas and some rooms with low ceilings on the ground floor, it was also necessary to use TLS technology.

The TS was used for the creation of a topographic network to support the entire survey, while the GNSS receiver was used to measure some GCPs in order to georeference the topographic network in cartographic coordinates. The topographic network was materialised using sixteen points, inside and outside the perimeter of the building, from which the local reference system of the entire complex was generated. The Mean Squared Errors (MSEs) of the coordinates of the adjusted vertices were lower than 5 mm. The measurement operations were carried out with a Leica TS30 total station (declared accuracy of 2 mm + 2 ppm), with redundant measurements such as to allow statistically valid checks.

The calculation of the network was carried out with a least-squares data adjustment program. The topographic network points were also acquired with GNSS instrumentation (exploiting RTK and connected to HxGN SmartNet), to allow the georeferencing of the topographic network on a geographic reference system. The acquired measurements in terms of latitude, longitude, and altitude were then converted into plane geographical coordinates, referring to the WGS84/UTM-33N geographic coordinate system.

The data from the surveying instruments were integrated in various ways, depending also on the areas surveyed:

- The Leica RTC360 TLS was used to survey all areas that could be reached from it, including the underground areas and all accessible rooms of the castle, as well as the exterior fronts. The TLS data were registered based on targets measured using the TS and referred to the topographic network. A total of 256 scans were acquired between the interior, exterior, and underground spaces. It was also possible to detect and register with the rest of the data scans in the spaces of the cistern, accessible from the castle courtyard, and the underground space accessible from the rear of the building. From the latter room, it was possible to acquire information, albeit only a small and incomplete part, of a third underground space under the courtyard.
- The DJI Matrice 300 RTK drone was mainly used for surveying the castle and the urban village built around it. In using an automatic flight plan, 1350 images of the village were acquired in classical nadiral mode from an altitude of 40 m, ensuring a frontal and sidelap overlapping of 80%. A second flight was carried out at an altitude of 30 m from the road in front of the castle; it involved the acquisition of 885 nadiral and oblique images, ensuring a frontal and sidelap overlapping of 80%. Both nadiral and oblique photo flights were carried out using the RTK technology; 16 GCPs were also measured (using both the TS and GNSS receiver) along the area of the survey, to be used within the photogrammetric workflow. The photogrammetric process was performed using Agisoft Metashape Professional v. 2.0.0; images were aligned using "high" resolution on an intel i9 PC, with 256 GB of RAM and an NVIDIA RTX A6000 graphics card.
- The DJI Mavic 3 cine drone was mainly used for surveying the internal fronts of the rooms without ceilings (e.g., the ceilings that had collapsed) and for the external elevations of the castle and the internal elevations of the courtyard. A total of 8300 photographs were taken to describe all the vertical surfaces (interior and exterior) of the castle, ensuring a frontal and sidelap overlapping of 80%. The photogrammetric process was performed by splitting all images into various groups according to the surveyed areas (e.g., one project for each room and one project for each facade) using Agisoft Metashape Professional v. 2.0.0; images were aligned using "high" resolution on an intel i9 PC, with 256 GB of RAM and an NVIDIA RTX A6000 graphics card. The integration of these data with the global reference system was performed by measuring the coordinates of the GCPs with the TS. However, the environmental context of the building did not always allow a sufficient number of GCPs to be measured through the positions of the stations in the topographical network, so the coordinates of some of them were derived from the TLS survey.

4. Materials and Methods

This section and the following ones present two study scenarios (see Table 1), named I and II, developed under the assumption that these ruins were inaccessible and could be surveyed only using UAVs. Various tests were then carried out under these two scenarios. Both scenarios referred only to the use of UAVs, also exploiting the RTK mode for direct georeferencing, without having to refer to GCPs. The tests developed within scenario I had the purpose of assessing the accuracy of a survey performed only with a UAV and RTK technology. The tests of scenario II had the more operational aim of showing the integration of data from two UAVs with different characteristics. The features of the two UAVs used are provided in Table 2, while Figure 5 shows a diagram reporting the development of the tests for the two scenarios; we can note the phases of the photogrammetric process analysed through the various tests in scenario I, as well as the logical development of the test in scenario II. The data exploited for the scenarios were a subset of the data acquired during the survey of Terracorpo Castle.

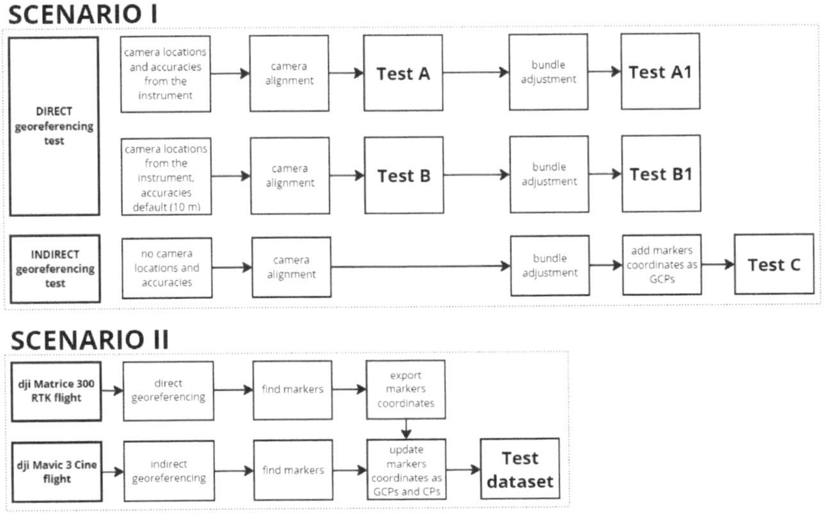

Figure 5. Schematic of the tests carried out in this study. Scenario I analyses the differences between direct and indirect georeferencing, and scenario II analyses the merging of datasets surveyed with two different drones. The steps of the photogrammetric process analysed with the various tests in scenario I and the logical development of the test in scenario II can be seen.

Table 1. Description of the two scenarios exploited in this study, with their purposes and instruments used. The tests designed within the two scenarios were developed under the assumption of using only UAVs to perform a survey of ruins, to assess the reachable reliability and accuracy.

Scenario	Instruments Used	Purpose
I	DJI Matrice 300 RTK (DJI Ltd., Shenzen, China), DJI D-RTK2 station (DJI Ltd., Shenzen, China), GNSS receiver Leica GS18 (Leica Geosystems, part of Hexagon AB, Stockholm, Sweden)	check the accuracy of the photogrammetric workflow using RTK station data to perform direct georeferencing in comparison with indirect georeferencing (checkpoints acquired with Leica GNSS receiver)
II	DJI Matrice 300 RTK (DJI Ltd., Shenzen, China), RTK station DJI D-RTK2 (DJI Ltd., Shenzen, China), DJI Mavic 3 Cine (DJI Ltd., Shenzen, China)	operative workflow followed to integrate data of two drones without other topographic support (one of the two UAV is operating in RTK positioning mode)

Table 2. Features of the UAVs used for the tests performed.

UAV	Camera, Lens	Camera Resolution
DJI Matrice 300 RTK (DJI Ltd., Shenzen, China)	Zenmuse P1 (DJI Ltd., Shenzen, China), 35 mm	45 MP, 8192 × 5460 px
DJI Mavic 3 cine (DJI Ltd., Shenzen, China)	Hasselblad (Hasselblad, Göteborg, Svezia), 12 mm	20 MP, 5280 × 3956 px

4.1. Scenario I: Direct Georeferencing with RTK vs. Indirect Georeferencing with GNSS Surveyed GCPs

To evaluate the effectiveness of RTK technology for a detailed UAV photogrammetric survey, the different stages of the photogrammetric process were analysed to verify the best alternatives according to the final results. The aim was to validate the feasibility of the complete photogrammetric survey of a ruin surveyed solely through the use of drones with RTK systems and to assess the achievable accuracy. In this case, the UAV DJI Matrice 300 RTK employed the RTK technology for accurate measurements of the camera locations and rotations (with accuracies) at the time of image shooting. These data, detected through the RTK approach, has a positioning accuracy, as stated by the manufacturer DJI of 1 cm + 1 ppm (horizontal) and 1.5 cm + 1 ppm (vertical). The advantages of using RTK technology is related to a higher accuracy in the flight plan (especially in following the designed trajectories) and to the possibility of facilitating the photogrammetric process by having the coordinates of the perspective centre of the images.

The image acquisition procedure was fairly straightforward and well defined. In addition to all the necessary pre-flight checks, it is necessary to initialise the ground RTK station on a point of known geographic coordinates (e.g., using a point in the higher-order geodetic networks, or by performing a prior GNSS survey on the same point). After this preliminary step, the connection between the drone and the ground base station must be verified, thus activating the communication between the two systems.

Then, the operator can proceed to the flight, choosing between a flight plan conceived in advance and designed using special flight planning software, or performing a free flight, in which the experienced operator can freely operate the system. However, it is necessary to check the drone–base station distance to avoid signal loss and carefully choose the ground station location (generally in the open and possibly in a high place) to avoid obstacles between the two. At the end of the acquisition phase, the acquired images are obtained; thanks to the RTK technology, every image was acquired with the coordinates of its perspective centre, characterised by a specific planimetric and altimetric accuracy.

The processing phase, in contrast to the acquisition phase, provides several possibilities that could change the final result. The fact that commercial software (in this research, Agisoft Metashape Professional v. 2.0.0 was used) relies on processing algorithms that are not known to the public makes it advisable to carry out some tests to verify the behaviour of the photogrammetric software with images acquired in the manner described above, to identify the procedure that leads to better results.

The coordinates of the perspective centre constitute the approximate coordinates from which the image georeferencing starts. Still, the use of their accuracies, recorded by the RTK and stored within the XML image metadata file, can strongly constrain the calculation. If the RTK link has suffered signal losses or malfunctions, the accuracy value is over-estimated, and thus, the system is over-constrained. In fact, a very high accuracy value allows only minimal "shifts" of the perspective centre in the adjustment phase (within its range). For this reason, at first, we focused on the camera alignment stage. We tried to georeference the same set of nadiral photos using, firstly (Test A), the accuracy values provided by the system and, secondly (test B), a default value (set automatically by the Agisoft Metashape Professional v. 2.0.0 software equal to 10 m—so for test purposes, it was considered almost "unconstrained"). The two solutions were compared by checking the average residuals on image markers locations and on some Check Points (CPs).

Furthermore, the coordinates and accuracy of the image perspective centre coordinates can also affect the bundle adjustment process [44,45]. Again, overestimating the accuracy could constrain the system excessively (and incorrectly). In this case, the second group of tests involved starting from the same set of oriented photos and applying the bundle adjustment process by exploiting, firstly (test A1), the coordinates with accuracy from the RTK base and, secondly (test B1), the coordinates with a generic accuracy (as previously used, 10 m as pre-set by the software). Again, the result was evaluated based on the reprojection error on the image markers and Check Points.

After determining the effect of the RTK accuracy on image georeferencing, the direct and indirect georeferencing processes of the same set of photos were compared. In test C, the images were imported without the coordinates of the perspective centres, and then they were oriented, and later, the photogrammetric model was scaled and georeferenced through bundle adjustment, exploiting the GCPs acquired by the GNSS receiver Leica GS18.

4.2. Scenario II: Data Integration between UAV-RTK and UAV

The purpose of these tests was to evaluate the potential and integration of photogrammetric data from two UAVs with different characteristics. For this test, the data acquired by the DJI Matrice 300 RTK drone (DJI Ltd., Shenzen, China) (connected with the DJI D-RTK2 base for accurate corrections to the measured locations of the images) and the data acquired by the DJI Mavic 3 Cine were used.

The DJI Matrice 300 RTK was used to survey the entire building and surrounding area. For the photogrammetric process, direct georeferencing was used under the same conditions as the best of the tests performed in scenario I, and without the use of GCPs to preserve the assumption of working only with UAVs without ground support. The DJI Mavic 3 Cine was used to survey the castle fronts. In the context of this test, the flight of this UAV was carried out only for the survey of the castle's main exterior elevation.

This test was more operative and intended to evaluate the results achievable under the aforementioned conditions. It is difficult to analyse the behaviours only from a statistical point of view since many variables are running, for example, different image resolutions of the 2 diverse camera systems and different heights of flight (i.e., different distances from objects). For these reasons, this test was intended only to verify the feasibility of the entire survey.

For this test, the integration of the two data was tested. To do so, architectural points and targets were identified on the building facade to be used as GCPs. The coordinates of the GCPs were obtained from the survey data performed with the DJI Matrice 300 (georeferenced using the RTK system and in geographic coordinates) and were then used and entered as constraint coordinates in the photogrammetric process followed for the DJI Mavic 3 Cine. Some of these points were then used as GCPs and others as CPs, and the error values were verified. The two point clouds from the two drones were also imported into CloudCompare (v. 2.13.1) to verify their exact overlap using the C2C method and the M3C2 plug-in [46] that verifies the mutual distances of points on different point clouds.

5. Results

5.1. Scenario I: Direct Georeferencing with RTK vs. Indirect Georeferencing with GNSS Surveyed GCPs

The experimental tests were performed on a subset of images from a nadiral flight conducted by a Matrice 300 RTK equipped with a Zenmuse P1 camera. This subset was made of 560 images, each one with a dimension of 8192 × 5460 pixels. Figure 6 shows the distribution of GCPs and CPs on the surveyed area used within the tests of this scenario. For the purposes of this study, the processing time was neither analysed nor compared.

The first analysis, concerning the influence of the RTK accuracies of the perspective center coordinates within the georeferencing process, demonstrated that accuracy plays a role in the final results, even if the values are not so different. The error is slightly larger in Test case A, compared to B (see Table 3), but in looking at the coordinates and the average error, the use of accuracies in the camera locations leads to a lower error.

In the second case, Tests A1 and B1, we tried to evaluate the contribution of accuracies in the bundle adjustment process. The same sets of pictures, already used in the previous tests, were adjusted with a bundle adjustment process (the command "optimize" in Agisoft Metashape Professional v. 2.0.0, with all parameters selected) in two different cases: in test A1, the bundle adjustment was performed using camera locations and accuracies from an XML file, while in B1, the adjustment used only the default accuracy (set to the default value of 10 m). The comparison of A1 and B1 proved that accuracy affects the results positively, considering an average error from 0.0367 m (in B1) to 0.0320 m (in A1).

Table 3. Accuracy results for the sub-tests developed within scenario I.

Test ID	Georef.	Photogram. Process Stage	Cameras Location Accuracy	Average Pixel Error	Average E-Error	Average N-Error	Average Altitude Error	Average Error
A	direct	camera alignment	imported from XMP metadata	0.426 px	0.048 m	0.023 m	0.026 m	0.0323 m
A1	direct	bundle adjustment	imported from XMP metadata	0.374 px	0.048 m	0.023 m	0.025 m	0.0320 m
B	direct	camera alignment	10 m (default)	0.417 px	0.045 m	0.025 m	0.029 m	0.0330 m
B1	direct	bundle adjustment	10 m (default)	0.375 px	0.045 m	0.027 m	0.038 m	0.0367 m
C	indirect	with GCPs	imported without locations and accuracies	0.388 px	0.018 m	0.013 m	0.021 m	0.0173 m

Figure 6. (**a**) The dataset used for scenario I. In this image, the mesh model is visible on top of the image; it is possible to see the camera locations. (**b**) The GCPs and CPs used throughout the tests performed within scenario I, positioned around the castle area.

The third test compared the previous results (A, B, A1, B1) with the indirect georeferencing process. In test C, which was the case of indirect georeferencing, the accuracy was significantly higher, confirming what has already been shown in the literature. Still, in referring to the coordinates of the Check Points, the average error was about 0.032–0.033 m for RTK (in different conditions) and 0.017 m for the orientation based on the GCPs. Since the Ground Sample Distance was 0.005 mm within the project, we can say that in the indirect georeferencing process, the error was about three times the GSD value, while for the RTK approach (e.g., direct georeferencing), the average quality was about seven times the GSD value.

5.2. Scenario II: Data Integration between UAV-RTK and UAV

In this scenario, the possibility of data integration between photogrammetric models obtained from two different drones was analysed. The first flight (Figure 7a), performed with the DJI Matrice 300 RTK, obtained 885 photographs and covered the entire castle area, and was processed by performing direct georeferencing. The second flight (Figure 7b) was performed with the DJI Mavic 3 Cine, obtaining 8300 photographs of all the reachable vertical surfaces of the castle, of which 1300 photos (used in this scenario) covered the main facade of the castle (the one facing the small road leading to the castle). The photogrammetric process exploited for reconstructing this facade was performed through indirect georeferencing.

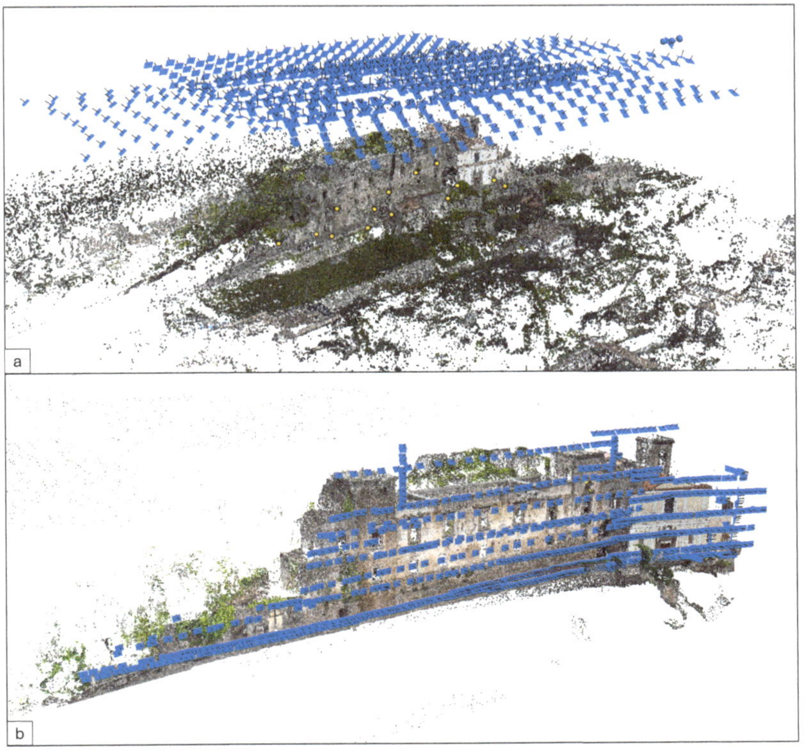

Figure 7. Screenshots of two photogrammetric projects for scenario II. Both figures show the positions of the cameras (in blue) and the tie points identified by the software. (**a**) Survey with DJI Matrice 300 RTK. (**b**) Survey with DJI Mavic 3 Cine.

For the test, the data obtained with the DJI Matrice 300 RTK were processed following direct georeferencing. From this, notable points and markers located on the facade of the building and in common with the flight performed with DJI Mavic 3 Cine were identified. The estimated coordinates of the notable points were then exported to be used as GCPs in the indirect georeferencing of the Mavic 3 Cine data. These estimated coordinates were then assigned an accuracy value of 2 cm, which is compatible with the technique used. From the results of the indirect georeferencing photogrammetric process, it was then possible to identify errors in the CPs, which were 0.0057 m in East, 0.0075 m in North, and 0.0071 m in Elevation. The distribution of errors for each GCP and CP, showed as error ellipses, are reported in Figure 8.

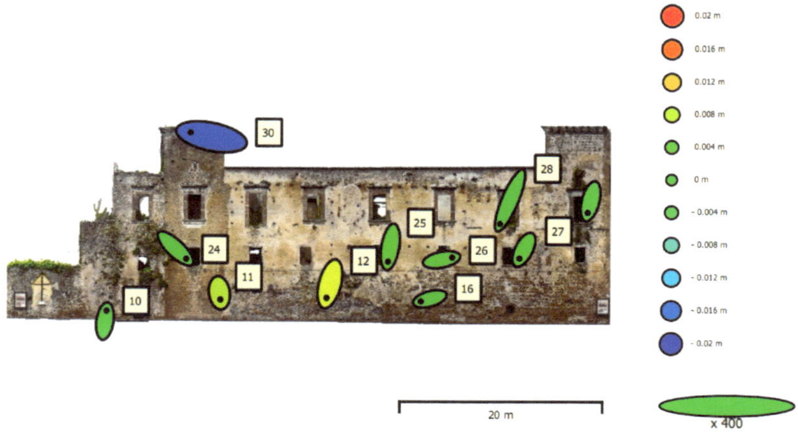

Figure 8. Position residuals represented as error ellipses based on GCP and CP positions for the analysed castle facade. Error ellipses are enlarged by a factor of 400.

In addition, for a further comparison of the two datasets, two algorithms available in the CloudCompare (v. 2.13.1) software were used. We used the M3C2 plug-in (Figure 9), which primarily allows the assessment of significant changes between two point clouds, and we used the Cloud2Cloud (C2C) algorithm (Figure 10), native to CloudCompare (v. 2.13.1), which allows the distance between two point clouds to be calculated. Figure 9 shows the results of the comparison made using the M3C2 method, which clearly (in red) shows the major differences between the two clouds. It can be noted that the most divergent elements are mainly related to some eaves overhanging and some spots horizontally aligned in the middle of the facade, which correspond to holes that were used for the construction of the building and the position of the scaffolding. As the two surveys were carried out from two different positions (one from above and one orthogonal to the facade), it can be observed that in the survey carried out with the Mavic 3 Cine and carried out orthogonally to the facade, these holes were photographed from more angles and better reconstructed. Other discrepancies can be attributed to the presence of weedy vegetation on the facade. Finally, for the rest of the points, the differences were not extensive and were mainly less than 2 centimetres, as it is also visible in the histogram shown in Figure 11 on the left. A similar trend was confirmed through the comparison made with the C2C method (see Figure 10), where a good fit between the two compared point clouds can be seen, with a greater deviation (although contained within 2 cm) in the upper left portion. It can be seen that the distribution of distances also conforms to the distribution of GCPs (see Figure 8) along the facade itself. The GCP to which a greater error corresponds is in fact located at the top left of the facade. Again, however, as can be seen from the distribution histogram of the distances (Figure 11 right), they are within 2 cm.

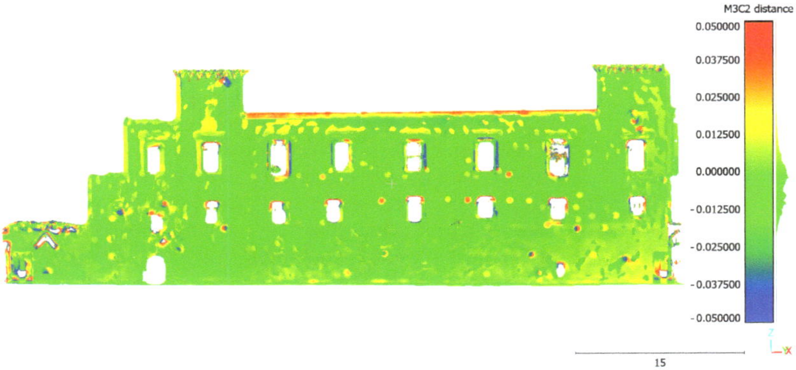

Figure 9. Comparison between the two point clouds of the castle's main facade, obtained through a photogrammetric process and using two drones: DJI Matrice 300 RTK and DJI Mavic 3 Cine. The two photogrammetric point clouds were compared in CloudCompare (v. 2.13.1) using the M3C2 algorithm. The scale bar goes from −0.05 (blue) m to +0.05 m (red).

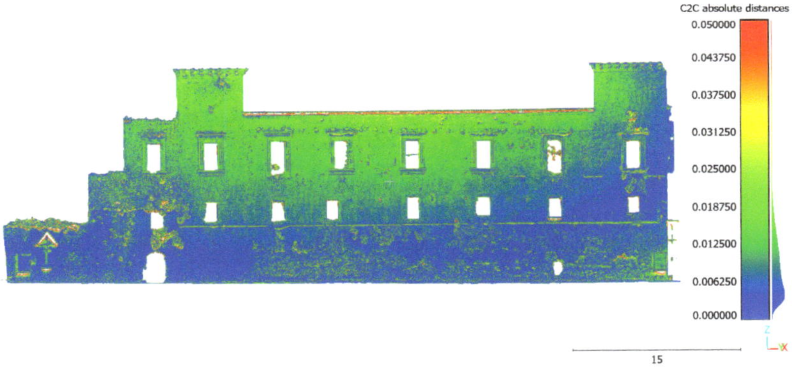

Figure 10. Comparison between the two point clouds of the castle's main facade obtained through a photogrammetric process and using two drones: DJI Matrice 300 RTK and DJI Mavic 3 Cine. The two photogrammetric point clouds were compared in CloudCompare (v. 2.13.1) using the C2C algorithm. The scale bar goes from −0.05 (blue) m to +0.05 m (red).

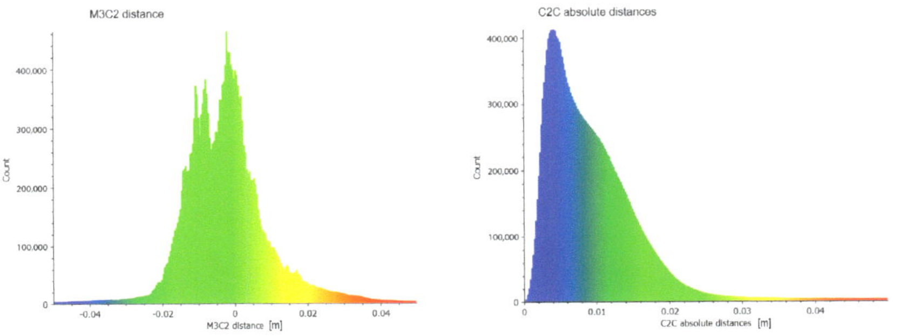

Figure 11. Histograms related to the previous figures (Figures 9 and 10), showing the distribution of distances between points in the two compared point clouds. It can be seen that both histograms, relating to the analyses with the M3C2 algorithm (**left**) and with the C2C algorithm (**right**), show a range tending to be within 2 cm.

5.3. 2D and 3D Deliverables

Continuing with the assumption that the ruins of Terracorpo Castle were only surveyed with UAVs, we describe in this section the products (2D and 3D) that could be obtained. Therefore, we proceed as in scenario II, i.e., assuming that the DJI Matrice 300 RTK drone was used to survey the castle and its surrounding village, and the DJI Mavic 3 Cine drone was used to survey all the interior faces. The subset of data was obtained directly from the survey actually carried out. With these data (RTK georeferenced images and non-georeferenced images), it is possible to develop the photogrammetric process, using direct georeferencing (for the Matrice 300) and indirect georeferencing (for the Mavic 3 Cine) as described in scenario II.

The results that can be obtained are the classic ones of a photogrammetric process, i.e., orthophotos of the area of interest both as a top view (as in a planimetry) or elevation view for facades. In addition, in using the photogrammetric point cloud, it is also possible to create sections of the building, showing the front of the inner rooms in perspective. In using the textured mesh, it could also be possible to realise three-dimensional visualisations (to create renders, walkthrough videos, and Virtual Reality experiences) for the dissemination of the current state of the building or design hypotheses. Figure 12 shows various images of the deliverables.

Figure 12. Examples of deliverables obtained through the photogrammetric process in the case study exploiting UAV data. (**a**) Orthophoto of the castle and its surroundings. (**b**) Ortophoto of the facade of the castle. (**c**) Three-dimensional model (texturised mesh) of the castle.

6. Discussion

The first scenario demonstrated that RTK data allow centimetric accuracies to be obtained while ensuring UAV operator safety. From the tests performed, the values of the accuracies are relevant, especially in the bundle adjustment step of the optimisation. A fundamental condition for the use of RTK data (especially location accuracies) is, in a way, the quality of the data, because overestimated accuracies do not bring real advantages to the final results. On the other hand, if there is a possibility, indirect orientation with ground-based GCPs has shown better effectiveness in terms of achievable accuracy (about three times the GSD value compared to seven times the GSD value of the RTK approach) as also observed in the literature.

As is always the case, it is not possible to define a single, specific scope, as the results obtained, at least in terms of quality and accuracy, must be verified according to the purposes of the survey and the scale of representation.

The test of scenario II demonstrated the possibility of surveying even the most difficult areas (even those accessible by UAV such as narrow rooms) by structuring the survey into two phases: a first UAV survey with RTK to determine, in addition to the general documentation from the top, the reference system to be used, and a second survey that relies on the previous one for the georeferencing stage.

With regard to the products realised from the data generated only with UAVs, several elements can be realised: orthophotos of the building area and its fronts, 2D drawings based on the photogrammetric point cloud, and textured 3D models for dissemination. The realised products suffer from the fact that only the surfaces actually visible from the UAVs are present in the final product. Vegetation with very large canopies can affect the visibility of the instrument and, consequently, the completeness of the final data. The same reasoning applies to all areas that are underground or in closed and/or dark rooms, which are difficult to reach using such instruments.

In general, however, the products that can be realised are compatible with the needs of a preservation project and the digital documentation of a ruined building. Thus, if site conditions do not allow it, a survey performed only with UAVs can still generate documentation of significant use.

Therefore, one advantage of drone-only surveying for areas that are not easily accessible is that it is still possible to carry out a survey with good accuracy (at the centimetre level) despite not being able to inspect the area with other surveying technologies. The execution of the survey requires a good scheduling of the flight plan, as well as the execution of the entire survey, but this still allows for good results while keeping both the survey time in the field and the number of people required limited. The disadvantages are certainly linked to the dependence (at the moment) on the presence of an RTK station positioned on the ground, which must always be in connection with the drone and must therefore be positioned either at a point of known coordinates or must be able to connect to a network of permanent stations to receive corrections. In addition, it should be pointed out that this methodology does not allow for reaching and surveying closed, underground, or generally non-illuminated areas. Finally, it should be noted that integrating a GNSS system to measure a few Ground Control Points (GCPs) greatly enhances accuracy, as demonstrated by various studies cited in the literature and the state-of-the-art reviews.

7. Conclusions

Ruined buildings constitute an important part of built heritage. To ensure the proper management, maintenance, and regeneration of such buildings, it is necessary to carry out adequate documentation of them. The survey of ruins is often characterised by the use of various survey technologies and tools that are integrated using various methods. Among the most widespread techniques, there is the use of TLS and close-range photogrammetry from UAVs or from the ground, integrated using topographic networks surveyed with TS or GNSS receivers.

In this study, we decided to analyse a particular case study, the survey of a ruined building when it is not fully accessible (e.g., for security reasons or due to the environmental conditions surrounding it). In such conditions, the use of UAVs has only been identified as the most suitable survey method. Indeed, such systems make it possible to reach all parts of the ruined building without endangering the survey operator. In addition, if the UAV system is used in combination with RTK technologies, the photogrammetric processing of the survey performed can be carried out using direct georeferencing by exploiting the locations of the images.

The DJI Matrice 300 RTK ground-based D-RTK2 drone and the DJI Mavic 3 Cine drone were used in this study. Two specific scenarios were set up with distinct purposes. The first scenario (I) was aimed at identifying the possible accuracies achievable by a drone used with RTK technology using direct georeferencing following the photogrammetric process. The tests performed showed that this approach allows centimetric accuracies to be achieved. The second scenario (II) tested the integration of two sets of photogrammetric data collected with drones at different resolutions and for different purposes. One drone was used to survey the entire building and its surroundings, georeferencing the data with RTK technology. And a second drone was used to closely survey the building fronts, and the data were georeferenced by deducing GCPs from the data taken with the first drone. This second test showed that the integration of data from drones at different resolutions is possible and optimal for working conditions where survey areas are not easily accessible. These tests also confirmed that underground areas or enclosed rooms are difficult to survey using UAVs alone. In such cases, different surveying technologies must be used, such that they allow the operator to enter these rooms, thus making it necessary to set up security conditions to make the building accessible.

This work demonstrates that the surveying of inaccessible ruins is possible through the use of various UAVs with varying resolutions and technologies in which the use of RTK systems makes it possible to not materialise a topographical network in situ, without the need to use other surveying instruments but inevitably achieving lower accuracies. This research is a first step in analysing survey modes for the digital documentation of areas or buildings that are not easily accessible (i.e., at risk of collapse, in locations that are not easily accessible, or in areas that are dangerous to humans). The use of direct georeferencing with RTK technology has been shown to be effective, but it is important to identify and test methodologies that can achieve higher accuracies, such as the use of PPK or others. In addition, the use of RTK technologies requires the presence of a telephone network to connect to the permanent stations with the GNSS receiver. Such conditions are not always guaranteed. Future research, therefore, may investigate the coupled use of other technologies that, while allowing the user not to enter hazardous (or otherwise unreachable) areas, allow surveys with high accuracy. As an example, integration with robotic LiDAR (Light Detection and Ranging) survey systems could be interesting for the inclusion of underground areas not visible through the drone, without requiring the operator to approach inaccessible areas.

Author Contributions: All authors contributed equally to the paper. All authors have read and agreed to the published version of the manuscript.

Funding: This research received no external funding.

Data Availability Statement: The data presented in this study are available upon reasonable request from the corresponding author.

Acknowledgments: The authors wish to thank the Municipality of Marzano Appio for providing assistance during the survey of Terracorpo Castle.

Conflicts of Interest: The authors declare no conflicts of interest. The funders had no role in the design of the study; in the collection, analyses, or interpretation of data; in the writing of the manuscript; or in the decision to publish the results.

Abbreviations

The following abbreviations are used in this manuscript:

CP	Check Point;
GCP	Ground Control Point;
GNSS	Global Navigation Satellite System;
MMS	Mobile Mapping System;
MSE	Mean Squared Errors;
PPK	Post-Processing Kinematic;
RTK	Real-Time Kinematic;
TLS	Terrestrial Laser Scanner;
TS	Total Station;
UAS	Unmanned Aerial System;
UAV	Unmanned Aerial Vehicle.

References

1. Frydryczak, B. Ruins: Between Past and Present, Between Culture and Nature. *Eidos A J. Philos. Cult.* **2023**, *7*, 9–16. [CrossRef]
2. Sanzaro, D.; Trovato, M.R.; Circo, C. An Interpretive Ruination Model of the Built Heritage in Inner Areas: The Case Study of the Neighbourhood Granfonte in Leonforte. *Heritage* **2023**, *6*, 6965–6992. [CrossRef]
3. Campana, S.; Sordini, M.; Remondino, F. Integration of geomatics techniques for the digital documentation of heritage areas. In Proceedings of the 1st International EARSeL Workshop, CNR, Rome, Italy, 30 September–4 October 2008; pp. 309–312.
4. Conti, A.; Pagliaricci, G.; Bonora, V.; Tucci, G. A comparison between terrestrial laser scanning and hand-held mobile mapping for the documentation of built heritage. *Int. Arch. Photogramm. Remote Sens. Spat. Inf. Sci.* **2024**, *XLVIII-2/W4-2024*, 141–147. [CrossRef]
5. Treccani, D.; Adami, A.; Brunelli, V.; Fregonese, L. Mobile mapping system for historic built heritage and GIS integration: A challenging case study. *Appl. Geomat.* **2024**, *16*, 293–312. [CrossRef]
6. Argyridou, E.; Karaoli, A.; Hadjiathanasiou, M.; Karittevli, E.; Panagi, I.; Mateou, M.; Ioannides, M.; Patias, P.; Efstathiou, K. The first attempt for standardisation in 3D digitisation. The EU study on quality in 3D digitisation of tangible cultural heritage. *Int. Arch. Photogramm. Remote Sens. Spat. Inf. Sci.* **2023**, *XLVIII-M-2-2023*, 103–109. [CrossRef]
7. Chiabrando, F.; Pietra, V.D.; Lingua, A.; Maschio, P.; Noardo, F.; Sammartano, G.; Spanò, A. TLS models generation assisted by UAV survey. *Int. Arch. Photogramm. Remote Sens. Spat. Inf. Sci.* **2016**, *XLI-B5*, 413–420. [CrossRef]
8. Šašak, J.; Gallay, M.; Kaňuk, J.; Hofierka, J.; Minár, J. Combined Use of Terrestrial Laser Scanning and UAV Photogrammetry in Mapping Alpine Terrain. *Remote Sens.* **2019**, *11*, 2154. [CrossRef]
9. Famiglietti, N.A.; Miele, P.; Memmolo, A.; Falco, L.; Castagnozzi, A.; Moschillo, R.; Grasso, C.; Migliazza, R.; Selvaggi, G.; Vicari, A. New Concept of Smart UAS-GCP: A Tool for Precise Positioning in Remote-Sensing Applications. *Drones* **2024**, *8*, 123. [CrossRef]
10. Cirillo, D.; Zappa, M.; Tangari, A.C.; Brozzetti, F.; Ietto, F. Rockfall Analysis from UAV-Based Photogrammetry and 3D Models of a Cliff Area. *Drones* **2024**, *8*, 31. [CrossRef]
11. Stott, E.; Williams, R.D.; Hoey, T.B. Ground Control Point Distribution for Accurate Kilometre-Scale Topographic Mapping Using an RTK-GNSS Unmanned Aerial Vehicle and SfM Photogrammetry. *Drones* **2020**, *4*, 55. [CrossRef]
12. Famiglietti, N.; Cecere, G.; Grasso, C.; Memmolo, A.; Vicari, A. A test on the potential of a low cost unmanned aerial vehicle rtk/ppk solution for precision positioning. *Sensors* **2021**, *21*, 3882. [CrossRef] [PubMed]
13. Cirillo, D.; Cerritelli, F.; Agostini, S.; Bello, S.; Lavecchia, G.; Brozzetti, F. Integrating Post-Processing Kinematic (PPK)–Structure-from-Motion (SfM) with Unmanned Aerial Vehicle (UAV) Photogrammetry and Digital Field Mapping for Structural Geological Analysis. *ISPRS Int. J. Geo-Inf.* **2022**, *11*, 437. [CrossRef]
14. Fallavollita, F.; Ugolini, A. New methodologies for the documentation of fortified architecture in the state of ruins. *Int. Arch. Photogramm. Remote Sens. Spat. Inf. Sci.* **2017**, *XLII-5/W1*, 411–418. [CrossRef]
15. Bianco, L. Architectural ruins: Geoculture of the anatomy of buildings as illustrated by Casa Ippolito, Malta. *Herit. Sci.* **2021**, *9*, 27. [CrossRef] [PubMed]
16. Versaci, A.; Cardaci, A.; Fauzìa, L.R.; Russo, M. The castle of Assoro: the integrated survey for the knowledge and conservation of ruins. *Int. Arch. Photogramm. Remote Sens. Spat. Inf. Sci.* **2019**, *XLII-2/W11*, 1127–1135. [CrossRef]
17. Reiss, M.L.L.; da Rocha, R.S.; Ferraz, R.S.; Cruz, V.C.; Morador, L.Q.; Yamawaki, M.K.; Rodrigues, E.L.S.; Cole, J.O.; Mezzomo, W. Data integration acquired from micro-UAV and terrestrial laser scanner for the 3D mapping of jesuit ruins of Aão miguel das Missões. *ISPRS-Int. Arch. Photogramm. Remote Sens. Spat. Inf. Sci.* **2016**, *XLI-B5*, 315–321. [CrossRef]
18. Marra, A.; Savini, F.; Giallonardo, M.; Fabbrocino, G.; Trizio, I. Multi-sensors acquisition for digital documentation. Application to a damaged area of Navelli village. *DISEGNARECON* **2022**, *15*, 41–415. [CrossRef]
19. Stanga, C.; Banfi, F.; Roascio, S. Enhancing Building Archaeology: Drawing, UAV Photogrammetry and Scan-to-BIM-to-VR Process of Ancient Roman Ruins. *Drones* **2023**, *7*, 521. [CrossRef]

20. Paris, L.; Rossi, M.L.; Cipriani, G. Modeling as a Critical Process of Knowledge: Survey of Buildings in a State of Ruin. *ISPRS Int. J. Geo-Inf.* **2022**, *11*, 172. [CrossRef]
21. Guo, M.; Fu, Z.; Pan, D.; Zhou, Y.; Huang, M.; Guo, K. 3D Digital protection and representation of burial ruins based on LiDAR and UAV survey. *Meas. Control* **2022**, *55*, 555–566. [CrossRef]
22. Versaci, A.; Fauzìa, L.R.; Russo, M.; Cardaci, A. The integrated fast survey for the risk assessment. A proposal for the safeguarding of the medieval castles in central Sicily. *Int. Arch. Photogramm. Remote Sens. Spat. Inf. Sci.* **2020**, *XLIV-M-1-2020*, 893–900. [CrossRef]
23. Roggero, M.; Diara, F. Multi-Sensor 3D Survey: Aerial and Terrestrial Data Fusion and 3D Modeling Applied to a Complex Historic Architecture at Risk. *Drones* **2024**, *8*, 162. [CrossRef]
24. Fino, M.D.; Galantucci, R.A.; Fatiguso, F. Condition Assessment of Heritage Buildings via Photogrammetry: A Scoping Review from the Perspective of Decision Makers. *Heritage* **2023**, *6*, 7031–7067. [CrossRef]
25. Grasso, N.; Spadavecchia, C.; Pietra, V.D.; Belcore, E. LiDAR and SfM-MVS Integrated Approach to Build a Highly Detailed 3D Virtual Model of Urban Areas. In Proceedings of the 9th International Conference on Geographical Information Systems Theory, Applications and Management GISTAM, Prague, Czech Republic, 25–27 April 2023 ; Volume 1, pp. 128–135. [CrossRef]
26. Russo, M.; Panarotto, F.; Flenghi, G.; Russo, V.; Pellegrinelli, A. Ultralight UAV for steep-hill archaeological 3D survey. *DISEGNARECON* **2022**, *15*, 1–17. [CrossRef]
27. Sobura, S.; Bacharz, K.; Granek, G. Analysis of two-option integration of unmanned aerial vehicle and terrestrial laser scanning data for historical architecture inventory. *Geod. Cartogr.* **2023**, *49*, 76–87. [CrossRef]
28. Perfetti, L.; Spettu, F.; Achille, C.; Fassi, F.; Navillod, C.; Cerutti, C. A multi-sensor approach to survey complex architectures supported by multi-camera photogrammetry. *Int. Arch. Photogramm. Remote Sens. Spat. Inf. Sci.* **2023**, *XLVIII-M-2-2023*, 1209–1216. [CrossRef]
29. Naanouh, Y.; Stanislava, V. Digital Documentation and a 3-D Model of Beaufort Castle via RTK GNSS, Terrestrial Laser Scanner and UAS-based Photogrammetry. *E3S Web Conf.* **2021**, *310*, 05002. [CrossRef]
30. Fabris, M.; Granotto, P.F.; Monego, M. Expeditious Low-Cost SfM Photogrammetry and a TLS Survey for the Structural Analysis of Illasi Castle (Italy). *Drones* **2023**, *7*, 101. [CrossRef]
31. Calisi, D.; Botta, S.; Cannata, A. Integrated Surveying, from Laser Scanning to UAV Systems, for Detailed Documentation of Architectural and Archeological Heritage. *Drones* **2023**, *7*, 568. [CrossRef]
32. Mutiarasari, W.M.; Rahman, M.Z.A.; Rahman, A.A. Investigating Data Fusion from Three Different Point Cloud Datasets Using Iterative Closest Point (ICP) Registration. In *Recent Advances in 3D Geoinformation Science, Proceedings of the 18th 3D GeoInfo Conference, Munich, Germany, 12–14 September 2023*; Springer: Cham, Switzerland , 2024; pp. 299–308. [CrossRef]
33. Pargieła, K. Optimising UAV Data Acquisition and Processing for Photogrammetry: A Review. *Geomat. Environ. Eng.* **2023**, *17*, 29–59. [CrossRef]
34. Teppati Losè, L.; Chiabrando, F.; Maschio, P. Direct georeferencing approaches for close-range and UAV photogrammetry in the built heritage domain. *Int. Arch. Photogramm. Remote Sens. Spat. Inf. Sci.* **2023**, *XLVIII-M-2-2023*, 1557–1564. [CrossRef]
35. Tomaštík, J.; Mokroš, M.; Surový, P.; Grznárová, A.; Merganič, J. UAV RTK/PPK Method—An Optimal Solution for Mapping Inaccessible Forested Areas? *Remote Sens.* **2019**, *11*, 721. [CrossRef]
36. Ekaso, D.; Nex, F.; Kerle, N. Accuracy assessment of real-time kinematics (RTK) measurements on unmanned aerial vehicles (UAV) for direct geo-referencing. *Geo-Spat. Inf. Sci.* **2020**, *23*, 165–181. [CrossRef]
37. Przybilla, H.J.; Bäumker, M.; Luhmann, T.; Hastedt, H.; Eilers, M. Interaction between direct georeferencing, control point configuration and camera self-calibration for RTK-based UAV photogrammetry. *Int. Arch. Photogramm. Remote Sens. Spat. Inf. Sci.* **2020**, *XLIII-B1-2020*, 485–492. [CrossRef]
38. Martínez-Carricondo, P.; Agüera-Vega, F.; Carvajal-Ramírez, F. Accuracy assessment of RTK/PPK UAV-photogrammetry projects using differential corrections from multiple GNSS fixed base stations. *Geocarto Int.* **2023**, *38*, 2197507. [CrossRef]
39. Memmolo, A.; Famiglietti, N.A.; Moschillo, R.; Grasso, C.; Vicari, A. UAS-LC-GNSS: Precision Surveying with a Low-Cost GNSS System for Commercial Drones. *Rend. Online Della Soc. Geol. Ital.* **2023**, *60*, 134–139. [CrossRef]
40. Previtali, M.; Barazzetti, L.; Roncoroni, F. GNSS assisted photogrammetric reconstruction from combined 360° videos and UAV mages. *Int. Arch. Photogramm. Remote Sens. Spat. Inf. Sci.* **2024**, *XLVIII-2/W4-2024*, 365–372. [CrossRef]
41. Kafataris, G.; Skarlatos, D.; Vlachos, M. Fusion of direct georeferenced aerial drone with terrestrial laser scanner data the case of the roman baths of Amathus, Cyprus. *Isprs Ann. Photogramm. Remote Sens. Spat. Inf. Sci.* **2023**, *X-1/W1-2023*, 1057–1064. [CrossRef]
42. Barazzetti, L.; Previtali, M.; Cantini, L.; Oteri, A.M. Digital Recording of Historical Defensive Structures in Mountainous Areas Using Drones: Considerations and Comparisons. *Drones* **2023**, *7*, 512. [CrossRef]
43. Bello, S.; de Nardis, R.; Scarpa, R.; Brozzetti, F.; Cirillo, D.; Ferrarini, F.; di Lieto, B.; Arrowsmith, R.J.; Lavecchia, G. Fault Pattern and Seismotectonic Style of the Campania—Lucania 1980 Earthquake (Mw 6.9, Southern Italy): New Multidisciplinary Constraints. *Front. Earth Sci.* **2021**, *8*, 608063. [CrossRef]
44. Hu, Z.; Tao, P.; Wang, L.; Zhang, S.; Ji, S. Toward a unified theoretical framework for photogrammetry. *Geo-Spat. Inf. Sci.* **2020**, *23*, 75–86. [CrossRef]

45. Triggs, B.; McLauchlan, P.F.; Hartley, R.I.; Fitzgibbon, A.W. Bundle Adjustment—A Modern Synthesis. In *Vision Algorithms: Theory and Practice, Proceedings of the International Workshop on Vision Algorithms Corfu, Corfu, Greece, 21–22 September 1999*; Triggs, B., Zisserman, A., Szeliski, R., Eds.; Springer: Berlin/Heidelberg, Germany, 2000; pp. 298–372.
46. Lague, D.; Brodu, N.; Leroux, J. Accurate 3D comparison of complex topography with terrestrial laser scanner: Application to the Rangitikei canyon (N-Z). *ISPRS J. Photogramm. Remote Sens.* **2013**, *82*, 10–26. [CrossRef]

MDPI AG
Grosspeteranlage 5
4052 Basel
Switzerland
Tel.: +41 61 683 77 34

Drones Editorial Office
E-mail: drones@mdpi.com
www.mdpi.com/journal/drones

www.ingramcontent.com/pod-product-compliance
Lightning Source LLC
Chambersburg PA
CBHW041411210926
51850CB00013B/3240
9783725827633